T0146180

# Opossums

# Opossums

*An Adaptive Radiation of New World Marsupials*

Robert S. Voss and Sharon A. Jansa

Johns Hopkins University Press
*Baltimore*

© 2021 Johns Hopkins University Press
All rights reserved. Published 2021
Printed in the United States of America on acid-free paper
9   8   7   6   5   4   3   2   1

Johns Hopkins University Press
2715 North Charles Street
Baltimore, Maryland 21218-4363
www.press.jhu.edu

Library of Congress Cataloging-in-Publication Data

Names: Voss, Robert S., author. | Jansa, Sharon A., author.
Title: Opossums : an adaptive radiation of new world marsupials /
    Robert S. Voss and Sharon A. Jansa.
Description: Baltimore : Johns Hopkins University Press, 2021. |
    Includes bibliographical references and index.
Identifiers: LCCN 2020013111 | ISBN 9781421439785 (hardcover) |
    ISBN 9781421439792 (ebook)
Subjects: LCSH: Opossums.
Classification: LCC QL737.M34 V67 2021 | DDC 599.2/76—dc23
LC record available at https://lccn.loc.gov/2020013111

A catalog record for this book is available from the British Library.

*Special discounts are available for bulk purchases of this book. For more
information, please contact Special Sales at specialsales@jh.edu.*

Johns Hopkins University Press uses environmentally friendly book
materials, including recycled text paper that is composed of at least
30 percent post-consumer waste, whenever possible.

*For Louise Emmons*

# Contents

# Acknowledgments

We are very fortunate in having colleagues who were willing to read early drafts of our chapters and suggest improvements. In particular, we are grateful to Robin Beck for reading Chapters 1, 2, and 3; to Vera Weisbecker for reading Chapter 5; to Christine Cooper and John Harder for reading Chapter 6; to Harry Greene for reading Chapters 9 and 11; to Lance Durden, Agustín Jiménez, and Jessica Light for reading Chapter 10; and to Gregory Adler for reading Chapters 8, 12, and 13. Any remaining errors of fact, interpretation, or emphasis are, of course, our own.

Patricia J. Wynne drew all the line art for this volume with her customary skill, attention to detail, and inimitable style. Photographic material was made available by friends, colleagues, and freelance photographers whose generosity we gratefully acknowledge, whether or not we finally used their images for this volume: Víctor Acosta-Chaves, Antoine Baglan, Eduardo Beltrocco, Gerardo Ceballos, Rodrigo Conte, Mónica Díaz, Louise Emmons, Antoine Fouquet, Elí García-Padilla, Mario Gómez-Martínez, Roland Kays, Jorge La Grotteria, José Martínez, José Ochoa, Ricardo Ojeda, Pedro Peloso, Darío Podestá, Fiona Reid, Rogério Rossi, Thiago Semedo, Thiago da Silveira, Pablo Teta, and Christian Ziegler. Rights to reproduce the image in Figure 11.1 were purchased from Superstock, Inc.

Patricia Wynne's drawings of opossum manual osteology (Fig. 5.10) are based on micro-CT scans expertly processed by Abigail Curtis for a grant proposal that was never funded. We thank Kelly Kroft for photographing the microfilariae (Fig. 10.1) in a blood smear that was originally prepared by Kevin Olival, and we thank Agustín Jiménez for identifying adult worms from the body cavity of the same specimen. We are grateful to Craig Chesek for photographing the opossum skin in Fig. 4.10. Jan Ryser very kindly sent us a copy of his 1990 doctoral dissertation on *Didelphis*.

Our decades-long research on opossums would have been far less productive without many substantial contributions from our students, including especially Juan Díaz-Nieto, Tom Giarla, Eliecer Gutiérrez, and Silvia Pavan. Lorissa Fujishin and Carmen Martin helped generate much of the sequence data on which our phylogenetic research was based. We are also grateful to colleagues who collaborated in our work or supported it in other tangible ways over the same period, especially Keith Barker, Robin Beck, Ana Carmignotto, François Catzeflis, Louise Emmons, Dave Fleck, David Flores, Al Gardner, Darrin Lunde, Ted Macrini, Jim Patton, Rogério Rossi, Nancy Simmons, and Sergio Solari. Lastly, we thank Tiffany Gasbarrini at Johns Hopkins University Press for her candid and insightful advice and subsequent editorial support.

# Opossums

# Introduction

> I am sensible how *tedious* I have been, in the *Description* of this *one* Animal,
> and from a *single* Observation too. . . . But it being an *Animal*, so very
> remarkable; and one too, *sui Generis*, or a distinct *Species* from all others,
> I was the more inclin'd to be as particular as I could, though not so much as
> I could have wished.
>
> Tyson (1698: 159)

The scientific study of marsupials began with Edward Tyson's dissection of a female opossum in 1698. Transported alive from Virginia to London, but "of late languishing and falling from its Meat," it had just died in the menagerie of the Royal Society. Alert to this rare opportunity to examine an almost-fabulous North American species (Parrish, 1997), Tyson was determined to make a thorough job of it. Wrapping up a lengthy report that was published in the *Philosophical Transactions* of the Society, Tyson concluded that the opossum was not the chimerical monster described by earlier authors—"with a snowte lyke a foxe, a tayle lyke a marmasette, eares lyke a batte, hands lyke a man, and feete lyke an ape" (Eden, 1555)—but an entirely new kind of animal with organs admirably suited to its own way of life. At one stroke, this pioneering treatise removed marsupials from the realm of myth to the annals of zoology. Despite its quaint diction, archaic punctuation, credulous repetition of travelers' tales, and occasional flights of poetry, Tyson's work was thoroughly modern in conception, his anatomical descriptions clear, his illustrations accurate and useful. It was a good beginning.

Opossums (family Didelphidae, order Didelphimorphia) are New World marsupials. For more than a century after their discovery by Vicente Yáñez Pinzón, the Spanish explorer who landed on the coast of Brazil in 1500 and captured a female with pouch young, they were the only marsupials widely known to Europeans.

Subsequently eclipsed in the public imagination by their more charismatic Australian relatives (kangaroos, koalas, etc.), opossums remained for many years a somewhat obscure group, of interest primarily to taxonomists and students of mammalian reproduction.

All of this changed in the early 1980s with the successful domestication of the short-tailed opossum (*Monodelphis domestica*) and its subsequent popularity as a model organism for laboratory research (VandeBerg, 1989; VandeBerg and Robinson, 1997). In just a few years, knowledge of many fundamental aspects of didelphid biology—especially development, physiology, and behavior—increased dramatically, with important contributions from dozens of research groups. More recently, the publication of "the opossum" genome has produced a wealth of genetic data from the same species (Belov et al., 2007; Gentles et al., 2007; Mikkelson et al., 2007; Samollow, 2008). Welcome as it is, however, much of this new information is difficult to interpret evolutionarily in the absence of relevant information about didelphid phylogeny, comparative biology, and natural history. Therefore, one purpose of this book is to provide essential context for future didelphid research by summarizing current knowledge of evolutionary relationships, phenotypic variation, and ecology. Such information is now widely scattered in the literature, and much of it has yet to be synthesized into any coherent account of opossum diversity.

Another motivation is to correct a strong ecogeographic bias in the marsupial literature. Due to the research productivity of several generations of Australian zoologists, much of the published information about marsupial phenotypes and natural history concerns antipodean species living in dry-temperate or -subtropical habitats. However, marsupials first evolved in South America, where most of their living descendants inhabit humid tropical environments. As quintessentially rainforest marsupials, opossums merit rich phenotypic and natural-history documentation equivalent to that long available for their arid-adapted Australian cousins.

Additionally, opossums are of interest to primatologists and paleontologists as living models of extinct taxa. Some arboreal opossums are thought to resemble early primates in key morphological and ecological characteristics, so knowledge of didelphid natural history might shed light on the adaptations of remote human ancestors. Opossums are also the most diverse group of living mammals to retain a primitive (tribosphenic) dentition, so knowledge of didelphid feeding behavior and diets can provide insights about the trophic adaptations of early mammals. Primatologists and paleontologists with such research interests have previously had no authoritative reference to consult on relevant aspects of opossum biology, and some have focused their attention on the North American species (*Didelphis virginiana*), a very atypical opossum.

Lastly, it has become commonplace in textbooks and other semipopular media to cast marsupials in the role of victims in a global struggle for dominance with placental mammals. In this hackneyed scenario (Gould, 1980), marsupials became extinct in the northern hemisphere due to their inability to compete with placentals, and they only held out in South America as long as that continent—an island throughout much of the Tertiary—was separated from North America by a marine barrier. Once the Panamanian landbridge was formed, South American marsupial diversity declined catastrophically as carnivores and other placental groups rushed in to occupy the niches that marsupials once filled. Only in Australia, still an island continent, did marsupials continue to flourish until the Recent arrival of humans.

Although certain aspects of this narrative remain plausible, the current composition of South American mammal faunas suggests a more nuanced interpretation of marsupial-placental interactions. South American rainforests are, indeed, numerically dominated by placental mammals, but these habitats also support diverse marsupial faunas, including a dozen or more sympatric opossum species at well-sampled Amazonian localities. Clearly, didelphids are still thriving in the midst of placental competitors. Three centuries after Tyson's pioneering monograph, it is time to try and understand how opossums have achieved this unique success.

Few mammalian higher taxa have been so consistently neglected by scientific publishers as Didelphimorphia. In fact, this is the first book-length treatment of opossums, although Hunsaker's (1977) volume broadly overlapped our topic. Useful in its day, Hunsaker's book has long been obsolete. Opossum research has increased impressively over the last several decades, not only with laboratory colonies of *Monodelphis domestica*, but also on wild populations of numerous other species and in the fields of taxonomy and phylogenetic inference. Remarkably, over half the references cited in this volume have been published since the year 2000, impressive testimony to the now very active pan-hemispheric community of opossum researchers.

To keep this book within reasonable page limits, we assume that the reader is familiar with basic concepts in evolutionary biology, ecology, and systematics that would be tedious to explain in context. Fortunately, in this age of electronic media, explanations of unfamiliar terminology can be accessed in a few keystrokes, so it now seems unnecessary to define such essential vocabulary as sexual selection, trophic niches, or monophyly. In effect, this book is written for researchers and for students embarking on careers in research.

A note on our references: faced with an almost overwhelming embarrassment of riches in the scientific literature, we had to be selective. Whenever possible, we cite review articles rather than long lists of relevant studies, we cite only the most trenchant reports (usually flagged with "e.g.") among many alternatives, and we prefer to cite articles readily available from widely disseminated online sources (such as JSTOR) rather than those sequestered behind corporate paywalls. In other words, despite its length, our reference list is not exhaustive, and we apologize in advance for any blameworthy omissions.

# I

# PHYLOGENETIC CONTEXT AND
# HISTORICAL BIOGEOGRAPHY

# 1

# Metatheria and Marsupialia

Marsupials and placentals are sister taxa that include all living members of Theria, the diverse (>6000 species) and familiar group of mammals that give birth to live young (Fig. 1.1). Egg-laying monotremes (Prototheria) are another major branch of mammals, but monotremes now consist of just a few living taxa (echidnas and the platypus). Although marsupials and placentals are both viviparous, these clades differ in several important reproductive traits. Among others, gestation (intrauterine development) is short in marsupials, which are born at an early stage of development and complete their ontogeny externally, attached to teats that may or may not be contained in a protective pouch; therefore, most of the total energy transfer between a marsupial mother and her offspring is accomplished via lactation rather than placentation. By contrast, placentals typically have a much longer gestation and are born at a later stage of development, so placentation accounts for a correspondingly larger fraction of total energy transfer between mother and offspring (Chapter 6). Unfortunately, reproductive physiology and the soft tissues that support it are not preserved in the fossil record, so it is not known when the distinctive reproductive differences between marsupials and placentals evolved. Lacking such knowledge, many systematists restrict the names Marsupialia and Placentalia to the "crown groups" of their respective clades. We follow this convention and explain its use to distinguish marsupials (sensu stricto) from their extinct relatives below.

Metatheria is the appropriate name for therian mammals that are more closely related to marsupials than to placentals (Rougier et al., 1998; Kielan-Jaworowska et al., 2004; Williamson et al., 2014); therefore, all marsupials are metatherians, but not all metatherians are marsupials. In this book we use Marsupialia and "marsupials" as equivalent terms for the metatherian crown group: the clade that contains living metatherians, their most recent common ancestor, and all of its descendants. Note that genealogy—not extinction—is the key concept here, because

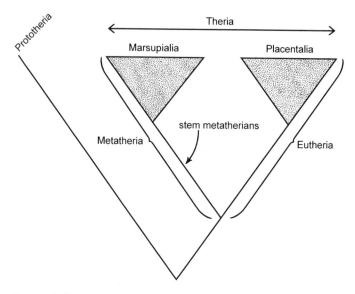

*Fig. 1.1.* Phylogenetic relationships among higher taxa of mammals. Therian crown groups are shaded. See text for explanations of taxon names.

many marsupials (descendants of the last common ancestor of living metatherians) are extinct. Extinct metatherians that do not belong to the crown group are here referred to as stem metatherians. Thus, †*Diprotodon optatum*, an extinct rhinoceros-sized (ca. 2800 kg; Wroe et al., 2004) Pleistocene wombat relative from Australia, is a marsupial because it is descended from the most recent common ancestor of living metatherians, but †*Mayulestes ferox*, a South American Paleocene taxon, is a stem metatherian because it is not a member of the crown clade (Rougier et al., 1998; Forasiepi, 2009; Oliveira and Goin, 2015; Beck, 2017).*

The earliest known metatherian fossil was formerly thought to be †*Sinodelphys* from the Early Cretaceous of China (Luo et al., 2003), but a recent reanalysis of this and other contemporaneous Chinese fossils suggests that they are all eutherians; apparently, the oldest known metatherians are from somewhat younger (Albian, ca. 110 million years old [Ma]) deposits in western North America (Bi et al., 2018). By the Late Cretaceous, however, metatherians were widespread in Laurasia, as indicated by numerous fossil taxa from Asia, North America, and Europe (Kielan-Jaworowska et al., 2004; Williamson et al., 2014). Although several

---

* Here, and subsequently, we use a printer's dagger (†) to indicate extinct taxa.

*Table 1.1*   Classification and Geographic Distribution of Recent Marsupials

| Order & family | Common name(s) | Species[a] | Distribution |
|---|---|---|---|
| **DIDELPHIMORPHIA** | | | |
| Didelphidae | opossums | 116 | New World |
| **PAUCITUBERCULATA** | | | |
| Caenolestidae | shrew-opossums | 7 | New World |
| **MICROBIOTHERIA**[b] | | | |
| Microbiotheriidae | monito del monte | 1 | New World |
| **DASYUROMORPHIA**[b] | | | |
| Dasyuridae | quolls, dunnarts, etc. | 77 | Australasia |
| Myrmecobiidae | numbat | 1 | Australasia |
| Thylacinidae | thylacine | 1 | Australasia |
| **NOTORYCTEMORPHIA**[b] | | | |
| Notoryctidae | marsupial moles | 2 | Australasia |
| **PERAMELEMORPHIA**[b] | | | |
| Chaeropodidae | pig-footed bandicoot | 1 | Australasia |
| Peramelidae | bandicoots | 20 | Australasia |
| Thylacomyidae | bilby | 1 | Australasia |
| **DIPROTODONTIA**[b] | | | |
| Acrobatidae | feather-tailed possums | 2 | Australasia |
| Burramyidae | pygmy possums | 5 | Australasia |
| Hypsiprymnodontidae | rat-kangaroo | 1 | Australasia |
| Macropodidae | kangaroos, wallabies, etc. | 66 | Australasia |
| Petauridae | gliders | 12 | Australasia |
| Phalangeridae | possums & cuscuses | 30 | Australasia |
| Potoroidae | potoroos, etc. | 11 | Australasia |
| Pseudocheiridae | ring-tailed possums, etc. | 17 | Australasia |
| Tarsipedidae | honey possum | 1 | Australasia |
| Vombatidae | wombats | 3 | Australasia |

[a] Numbers of living species after Wilson and Reeder (2005), but including additional taxa from subsequent publications cited in this book and by Eldridge et al. (2019: Supplementary Data SD1).
[b] Member of the supraordinal clade Australidelphia.

metatherian taxa from the Late Cretaceous and Paleogene of North America—including †*Alphadon*, †*Herpetotherium*, †*Pediomys*, and †*Peradectes*—were once considered marsupials (and some were even classified as didelphids; e.g., by Simpson, 1945), most phylogenetic analyses now suggest that these were stem taxa rather than members of the crown clade.[1] The Laurasian fauna appears to have been much reduced by the end-Cretaceous (KPg) mass extinction event, and by the middle Miocene, metatherians had become extinct throughout the northern hemisphere (Eldridge et al., 2019).

Living marsupials are currently classified in seven orders, of which three are restricted to the New World and four to Australasia (Table 1.1). Opossums (Didelphidae) are placed in their own order, Didelphimorphia, which is by far the most diverse of the three extant New World clades. The other two New World orders—Microbiotheria and Paucituberculata—were moderately diverse in the South American Tertiary (Chapter 2) but now include just a handful of Recent taxa. Microbiotheria contains only one living species, *Dromiciops gliroides*, which inhabits the wet-temperate forests of Chile and southern Argentina (Hershkovitz, 1999; Patterson and Rogers, 2008; Fontúrbel et al., 2012; Suárez-Villota et al., 2018). Living paucituberculatans, sometimes known as "shrew-opossums," include seven currently recognized species in three genera that occur in wet-temperate Patagonian forests and in climatically similar montane vegetation of the tropical Andes (Patterson, 2008; Ojala-Barbour et al., 2013).

Molecular phylogenetic studies consistently recover Didelphimorphia in one or the other of two positions near the root node of Marsupialia. In particular, recent analyses based on nuclear exon sequences (Meredith et al., 2008, 2009; Duchêne et al., 2018) and other analyses based on retrotransposon insertions (Nilsson et al., 2010; Gallus et al., 2015) place Didelphimorphia as the sister group to all other marsupials (Fig. 1.2A), whereas analyses of whole mitochondrial genomes (Mitchell et al., 2014) and amino-acid alignments (Meredith et al., 2011) recover Paucituberculata as the sister group to other marsupials (Fig. 1.2B). In the former scenario, Paucituberculata is the sister group to Australidelphia, a supraordinal clade that includes Microbiotheria and the Australasian radiation, whereas in the latter scenario Didelphimorphia and Australidelphia are sister taxa. In both topologies, the first three interordinal cladogenetic events gave rise to lineages that are exclusively or primarily restricted to South America, so it seems probable that much of early marsupial evolution occurred on that continent.

To avoid confusion with other taxonomic conventions, we use Didelphidae and "didelphids" as equivalent terms for the didelphimorphian crown clade: the most recent ancestor of living opossums and all of its descendants (Voss and Jansa, 2009). Other authors have referred living opossums to as many as four families, but no multi-family opossum classification has gained wide currency, and all proposed multi-family schemes include paraphyletic taxa (e.g., "Marmosidae" sensu Hershkovitz, 1992). Also, to avoid confusion with other conventions, we reserve Didelphimorphia and "didelphimorphians" for marsupials that are more closely related to didelphids than to other Recent clades currently ranked as Linnaean orders. Alternative concepts of Didelphimorphia that include superficially

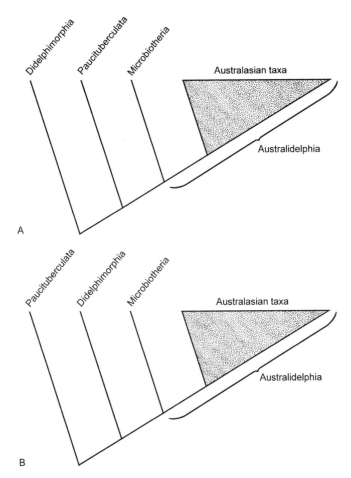

*Fig. 1.2.* Alternative hypotheses of phylogenetic relationships among New World marsupial orders and Australasian taxa (Dasyuromorphia + Diprotodontia + Notoryctemorphia + Peramelemorphia): **A**, results from analyses of nuclear exon sequences and retrotransposon insertions; **B**, relationships based on analyses of whole mitochondrial genomes and amino-acid alignments (see text). Australidelphia is the supraordinal clade that includes microbiotherians and Australasian marsupials.

opossum-like Mesozoic and Paleogene fossil taxa (e.g., Goin, 2003; Case et al., 2005; Gelfo et al., 2009) are not monophyletic.

Molecular estimates of key dates in didelphimorphian evolution are important because most metatherian fossils consist of fragmentary dentitions that are difficult to place phylogenetically, and because neither didelphimorphians nor didelphids are unambiguously diagnosable by dental characters (Voss and Jansa, 2009). Unfortunately, molecular estimates of relevant dates vary widely due to differences in calibration points and analytic methods. Whereas some node-based relaxed-clock methods suggest that didelphimorphians split away from other marsupials sometime in the Late Cretaceous (e.g., 71–86 million years ago [Ma]; Meredith et al., 2008; Mitchell et al., 2014; Duchêne et al., 2018), other node-based estimates place the origin of Didelphimorphia in the early Paleocene (ca. 64 Ma; Jansa et al., 2014), as does a recent "tip-and-node" total-evidence estimate (Maga and Beck, 2017). Similar uncertainty applies to molecular dating of the didelphimorphian crown clade, which is usually construed as the split between the Recent subfamilies Caluromyinae and Didelphinae. Node-based relaxed-clock methods place this event anywhere from the late Eocene (ca. 35–38 Ma; Meredith et al., 2008) to the late Oligocene (ca. 26 Ma; Jansa et al., 2014), and a late Oligocene date for the didelphimorphian crown clade is also recovered by total-evidence dating, with or without taking the fossilization process and sampling bias into account (Maga and Beck, 2017; Beck and Taglioretti, 2019).

Taken together, these estimates imply a long interval of stem-didelphimorphian evolution that preceded the first appearance of modern (crown-clade) opossums. Whether this interval extended from the Late Cretaceous to the late Eocene (as suggested by the dates in Meredith et al., 2008) or from the early Paleocene to the late Oligocene (Jansa et al., 2014; Beck and Taglioretti, 2019), the probable locus of stem-didelphimorphian evolution, together with most of the subsequent Neogene diversification of didelphids, was South America. A South American origin for Didelphimorphia is suggested both by the absence of unequivocally marsupial fossils in the North American record and by ancestral-area reconstructions based on molecular phylogenies (e.g., Voss and Jansa, 2009). Therefore, the only plausible source of tangible clues about the biogeographic context of early opossum evolution is the geological record of South America.

NOTE

1. Current differences of opinion about the biogeographic origin of Marsupialia are based on the contested relationships of several North American fossil taxa. Among recent publications, Case et al. (2005) suggested that several Late Cretaceous fossils from the western interior of North America are early representatives of "typically South American" marsupial groups. In particular, they claimed that †*Nortedelphys*, a genus they referred to the family †Herpetotheriidae, represents the "Cohort Ameridelphia" (a group within which they also included didelphids), and that †*Ectocentrocristus* and several other extinct genera belong to the "Cohort Australidelphia" (which includes Australasian marsupials).

These inferences are problematic, however, because Case et al. (2005) did not perform any phylogenetic analyses, and subsequent research has not supported their conclusions. Williamson et al.'s (2012, 2014) analyses, for example, suggest that whatever †*Nortedelphys* might be, it is not a herpetotheriid, whereas †*Ectocentrocristus* probably *is* a herpetotheriid and not a close relative of any australidelphian marsupial. Additionally, no published phylogenetic study supports the monophyly of Case et al.'s "cohorts." Instead, their "Ameridelphia" seems to be a hodgepodge of stem and crown-clade taxa, whereas their concept of Australidelphia includes taxa of genuinely ambiguous relationships (e.g., †Polydolopidae) and others that clearly belong elsewhere (e.g., †Argyrolagidae, which appears to be closely related to "ameridelphian" paucituberculatans; Sánchez-Villagra, 2001; Beck, 2017).

Herpetotheriids, which occur from the Late Cretaceous to the early Miocene of North America, are strikingly similar to marsupials, especially in possessing a well-developed alisphenoid tympanic process (Gabbert, 1998), but the cochlea of †*Herpetotherium* (the only herpetotheriid in which this organ is well preserved) has fewer cochlear turns than the reconstructed morphology of the marsupial common ancestor (Horovitz et al., 2008). Although some phylogenetic reconstructions (e.g., Maga and Beck, 2017) have recovered †*Herpetotherium* within the crown clade, analyses of the most extensive dataset yet assembled for New World metatherians (Muizon et al., 2018) suggest that †Herpetotheriidae is the sister group of Marsupialia.

Peradectids include numerous referred taxa from the Cretaceous and Paleogene of Asia, North America, Europe, and Africa, but the monophyly of this group remains unsupported by phylogenetic analyses of available character data (Williamson et al., 2012, 2014). Although most peradectids are only known from dental fragments, the postcranial skeletons of several ambiguously identified (allegedly peradectid) specimens from the Eocene of Germany were described by Kurz (2007), and a well-preserved cranium of †*Mimoperadectes houdei* from the North American Eocene was described by Horovitz et al. (2009). A phylogenetic analysis based on this new material recovered †Peradectidae as the sister taxon of Didelphidae (i.e., within Marsupialia; Horovitz et al., 2009), but the analyzed character data included at least one substantial error: †*Mimoperadectes houdei* was described (and scored for phylogenetic analysis) as having an alisphenoid tympanic process. We carefully examined the holotype of †*M. houdei* (USNM 482355, the only peradectid specimen in which the ear region is preserved) and observed no trace of an alisphenoid tympanic process. Most subsequent phylogenetic analyses of peradectid relationships (e.g., Maga and Beck, 2017; Muizon et al., 2018) have recovered peradectids as stem metatherians, not marsupials.

In effect, there seems to be no compelling evidence that any Cretaceous or Tertiary North American fossil taxon belongs to the crown clade Marsupialia. However, support for phylogenetically crucial nodes in most published phylogenies is weak, so a North American origin for Marsupialia is hard to rule out. An additional problem for phylogenetic inference about fossil metatherians is the predominance of dental characters in most analyzed datasets, with a consequently high probability of misleading results due to functional and developmental character correlations. Clearly erroneous relationships among Recent marsupials recovered by parsimony analyses of craniodental datasets (e.g., by Wilson et al., 2016; Carneiro, 2018) almost certainly result from this problem. Methods exist for dealing with homoplasious character correlations (Dávalos et al., 2014), but they have yet to be applied to metatherian phylogenetic inference.

# 2

# South America, the Island Continent

By the Late Cretaceous, South America was well on its way to becoming the island continent it would remain throughout much of the Cenozoic. Already separated from North America by an open seaway in the Late Jurassic, South America had also drifted away from Africa as the supercontinent of Gondwana began its piecemeal breakup, but landbridge connections between South America, Antarctica, and Australia persisted (Hay et al., 1999; Veevers, 2004; Seton et al., 2012). Although physical geology tells us this much (and may eventually tell us more) about Late Cretaceous geography, additional information can be inferred from fossil faunas.

Not so long ago, the Late Cretaceous mammalian fauna of South America was virtually unknown (Patterson and Pascual, 1972; Simpson, 1980). Subsequent discoveries, however, have replaced ignorance and conjecture with a substantial body of tangible facts. In particular, the rich Late Cretaceous assemblage from Los Alamitos in northern Patagonia contains numerous archaic taxa—mostly "dryolestoids" and gondwanatheres—but not a single therian mammal (Bonaparte, 1990; Goin et al., 2012). Although the geochronology of the Los Alamitos Formation is not well constrained, it is thought to be about 75 Ma (Flynn et al., 2012). Because contemporaneous North American faunas were dominated by therians (Cifelli et al., 2004; Williamson et al., 2014), the Los Alamitos assemblage provides compelling evidence that the mammalian faunas of North and South America were effectively isolated at this time.

By contrast, metatherians are abundantly represented in the exceptionally well-preserved fossil fauna from Tiupampa, Bolivia, which was originally thought to be Late Cretaceous (Marshall and Muizon, 1988), but is now regarded as Paleocene (Marshall et al., 1997; Gelfo et al., 2009). Although the Tiupampa fauna (like the fauna from Los Alamitos) lacks direct geochronological constraints (Flynn and Swisher, 1995), stratigraphic correlations suggest that it is about 65 Ma (early Paleocene; Muizon et al., 2018). Therefore, the most that can be said at present is that marsupial ancestors probably entered South America sometime in the latest Cretaceous (Maastrichian) or earliest Paleocene, perhaps simultaneously with archaic "ungulates" and other taxa of undoubted North American origin (e.g., pantodonts; Muizon et al., 2015), which also make their first appearance in the South American record at about the same time.

Most paleontologists (e.g., Goin et al., 2012; Wilf et al., 2013) agree that these arrivals are evidence for some transient connection between the two American continents, presumably a short-lived Central American archipelago (Redwood, 2019), but from the late Paleocene onward there is no evidence that any mammalian faunal exchange occurred between North and South America for several tens of millions of years. The breakup of the Gondwanan supercontinent was not yet complete by the end of the Mesozoic, however, because South America remained connected to Antarctica. A narrow landbridge extending from Patagonia to the Antarctic Peninsula, the so-called Weddellian Isthmus, seems to have persisted as a corridor for faunal exchange throughout most of the Paleocene and perhaps into the early Eocene (Scher and Martin, 2006; Reguero et al., 2014). Sometime during this interval, one or more South American marsupial lineages must have entered Antarctica and crossed the continent to populate Australia, but that exciting narrative, which is still hypothetical in many important respects (Woodburne and Case, 1996; Beck et al., 2008; Beck, 2012; Black et al., 2012), lies outside the storyline that leads to modern opossums.

Although definitive dates for the last dry-land connection between South America and Antarctica (late Paleocene or early Eocene) and for the final uplift of the Panamanian isthmus (sometime in the Pliocene) are still uncertain, there is no doubt that South America was truly an island continent throughout much of the Tertiary, cut off from other landmasses to the north and south by open seaways (Wilf et al., 2013). During this long episode, on the order of 40–50 million years, the fossil record suggests that only four immigrant mammalian taxa reached South America. Of these, the earliest were ancestors of living platyrrhine primates and caviomorph rodents, both of which arrived sometime in the Eocene, almost certainly by overwater dispersal, and probably from Africa (Antoine et al., 2011; Bond et al., 2015). The other two, procyonid carnivorans (which first appear in the late Miocene record) and cricetid rodents (in the early Pliocene), heralded the mass invasion of the continent by North American taxa across the Panamanian landbridge at the end of the Tertiary (Chapter 3).

Canonical accounts of metatherian evolution in insular South America emphasized the morphological diversity said to have evolved from "opossum-like" immigrant ancestors (Simpson, 1971, 1980; Patterson and Pascual, 1972). However, not all contemporaneous researchers were impressed by the phylogenetic evidence for this scenario (McKenna, 1980). Such skepticism, which turns out to have been well founded, was prompted by one of the more bizarre phenotypes assigned to this alleged adaptive radiation: †*Necrolestes*, a burrowing, mole-like taxon. In fact, †*Necrolestes* exhibits no convincingly diagnostic metatherian trait, and recent phylogenetic analyses suggest that it was a late-surviving non-therian "dryolestoid" (Rougier et al., 2012; O'Meara and Thompson, 2014).

Even disembarrassed of †*Necrolestes*, however, the fossil metatherian fauna of South America was impressively diverse (Table 2.1). Five South American metatherian lineages are currently ranked as Linnaean orders (Didelphimorphia, Microbiotheria, Paucituberculata, †Polydolopimorphia, and †Sparassodonta), and others might well be recognized as such. The latter include a number of early (mostly Paleocene and Eocene) groups of archaic aspect that remain to be convincingly associated with any conventionally recognized higher taxon. †Caroloameghiniidae, for example, has alternatively been classified in Didelphimorphia or †Peradectia, but not on the basis of any explicitly phylogenetic analysis of character data.

Some fossil microbiotherians and paucituberculatans were perhaps not unlike modern representatives of those groups, and most are thought to have been rather small (ca. 20–500 g) insectivorous, frugivorous, or insectivorous-frugivorous species (Abello et al., 2012, 2018). Microbiotherians (living and extinct) are all rather similar morphologically, but fossil paucituberculatans were phenotypically diverse, especially in the Miocene, when several families are known to have occurred sympatrically at some Patagonian localities. Argyrolagids, among the most unusual of all South American metatherians, were small bipedal-hopping forms that may have been herbivorous or granivorous inhabitants of semiarid habitats (Simpson, 1970; Sánchez-Villagra and Kay, 1997). These odd little creatures were once thought to be polydolopimorphians, but new morphological evidence and recent phylogenetic analyses (Sánchez-Villagra, 2001; Beck, 2017) suggest that they were paucituberculatans.

†Polydolopimorphia is a somewhat problematic taxon that formerly contained argyrolagids (now believed to have been paucituberculatans). Most polydolopimorphians have more or less bunodont (blunt-cusped) molars, but they otherwise defy characterization because of the fragmentary preservation of many included taxa. Only recently has sufficient morphological character data become available to assess the relationships of one included species, †*Epidolops ameghinoi*,

Table 2.1  Known Geological Occurrence of South American Metatherian Lineages

| Taxa[a] | Paleocene | Eocene | Oligocene | Miocene | Pliocene | Pleistocene | Recent |
|---|---|---|---|---|---|---|---|
| Metatheria *incertae sedis*[b] | | | | | | | |
| †Caroloameghiniidae | X | X | X | | | | |
| †Derorhynchidae | X | ? | | | | | |
| †Pucadelphyidae[c] | X | ? | | | | | |
| †Protodidelphidae | X | ? | | | | | |
| †Sternbergiidae | X | [X] | X | | | | |
| Didelphimorphia | | | | | | | |
| Didelphidae[d] | | | | X | X | X | X |
| Microbiotheria | | | | | | | |
| Microbiotheriidae | ? | X | X | X | [X] | [X] | X |
| Paucituberculata | | | | | | | |
| †*Bardalestes*, etc.[e] | X | X | X | | | | |
| †Abderitidae | | | X | X | | | |
| †Argyrolagidae | | | X | X | X | | |
| Caenolestidae | | | | X | X | [X] | X |
| †Palaeothentidae | | | | X | | | |
| †Pichipilidae | | X | [X] | X | | | |
| †Polydolopimorphia | | | | | | | |
| †Bonapartheriidae | | X | | | | | |
| †Gashterniidae | X | | | | | | |
| †Polydolopidae | X | X | X | X | | | |
| †Prepidolopidae | | X | | | | | |
| †Rosendolopidae | | | | | | | |
| †Sparassodonta[f] | | | | | | | |
| †Mayulestidae[g] | X | X | | | | | |
| †Hondadelphidae | | | | X | | | |
| †"Hathliacynidae" | | | X | X | X | | |
| †"Borhyaenidae" | | | X | X | | | |
| †Proborhyaenidae | X | X | X | | | | |
| †Thylacosmilidae | | | | X | X | | |
| TOTALS | 10 | 11 | 12 | 12 | 6 | 3 | 3 |

[a] Only taxa currently ranked as families are listed (except as noted), but we omit several dental taxa that remain so poorly known that little can be meaningfully said about their phylogenetic status, and some taxa of dubious South American occurrence (e.g., †Glasbiidae, †Herpetotheriidae, and †Peradectidae) are also omitted. Geological occurrence is based on McKenna and Bell (1997) with additional information from Goin (2006) and other references cited below. Key: X = definitely known from fossils, [X] = unknown from fossils but necessarily extant, ? = fossils possibly of this age.
[b] Archaic lineages unassigned to order or of dubious ordinal assignment.
[c] Sensu Muizon et al. (2018), excluding †*Mayulestes* (a sparassodont).
[d] Including †Sparassocynidae (after Beck and Taglioretti, 2019).
[e] Three plesiomorphic paucituberculatan genera unclassified to family (Goin et al., 2009a; Abello, 2013).
[f] Classification of sparassodonts follows Forasiepi (2009) as modified by Muizon et al. (2018); non-monophyletic taxa are bracketed by quotation marks.
[g] After Muizon et al. (2018); includes †*Patene*.

which appears to have been a metatherian stem taxon rather than a crown-clade marsupial (Beck, 2017).

Sparassodonts were carnivorous or hypercarnivorous (exclusively flesh-eating) metatherians, some of which were impressively large and may have been convergent in appearance, behavior, and diet with placental carnivorans. "Hathliacynid" sparassodonts (not, apparently, monophyletic; Muizon et al., 2018) are often described as dog-like; some borhyaenoids have been compared with mustelids, bears, or hyenas; and thylacosmilids are strikingly—if superficially—similar to saber-toothed cats. Whether or not such comparisons are really meaningful, functional analyses of sparassodont morphology suggests that this clade was adaptively diverse, especially in the Miocene (Argot, 2004; Prevosti et al., 2012; but see Croft et al., 2018). Numerous sparassodonts are represented by well-preserved fossils, and recent analyses have clarified many previously obscure aspects of their evolutionary history. Among other phylogenetic inferences based on current datasets, it now seems clear that sparassodonts were stem metatherians rather than marsupials (Forasiepi, 2009; Engelman and Croft, 2014; Forasiepi et al., 2015; Muizon et al., 2018).

Only a few well-sampled South American paleofaunas are available to illustrate taxonomic patterns of metatherian community composition as these extant and extinct lineages waxed and waned in diversity over time (Table 2.2). The earliest of these is the Paleocene fauna from Tiupampa, Bolivia. This assemblage was originally thought to include didelphimorphians (Marshall and Muizon, 1988; Muizon, 1998), but subsequent research suggests that the two best-preserved Tiupampan "didelphimorphians" (†*Andinodelphys cochabambensis* and †*Pucadelphys andinus*) are stem metatherians rather than close opossum relatives (Engelman and Croft, 2014; Forasiepi et al., 2015; Maga and Beck, 2017), and the latest phylogenetic analyses place these taxa near the base of the sparassodont radiation (Muizon et al., 2018). This important fauna also includes two confirmed sparassodonts (†*Mayulestes ferox* and †*Allqokirus australis*), an alleged microbiotherian (†*Khasia cordillerensis*), one polydolopimorphian (†*Roberthoffstetteria nationalgeographica*), and several species of archaic aspect and uncertain higher-taxonomic affinities. In summary, with one rather dubious exception (†*Khasia*; see Carneiro et al., 2018), most of the Tiupampan metatherians seem to be stem taxa.

By contrast, crown-clade marsupials (paucituberculatans and undoubted microbiotherians) make up almost half of the early Oligocene fauna from La Cancha, Argentina, although the dominant clade in terms of number of species was †Polydolopimorphia (Goin et al., 2010). This remarkable assemblage—regrettably

*Table 2.2*    Taxonomic Composition of Five Metatherian Paleofaunas from South America

| Taxa | Tiupampa[a] (ca. 65 Ma) | La Cancha[b] (ca. 33 Ma) | Santa Cruz[c] (ca. 17 Ma) | La Venta[d] (ca. 13 Ma) | Monte Hermoso[e] (ca. 5 Ma) |
|---|---|---|---|---|---|
| Metatheria *incertae sedis* | 8 | 2 | 0 | 0 | 0 |
| Didelphimorphia | 0 | 0 | 0 | 4 | 8 |
| Microbiotheria | 1? | 3 | 1 | 1 | 0 |
| Paucituberculata[f] | 0 | 5 | 3 | 2 | 2 |
| †Polydolopimorphia[g] | 1 | 7 | 0 | 0 | 0 |
| †Sparassodonta | 2 | 1 | 8 | 5 | 2 |
| TOTAL SPECIES | 12 | 18 | 12 | 12 | 12 |

[a] Early Paleocene of Cochabamba department, Bolivia (ca. 18° S). Fauna from Santa Lucía Formation at Tiupampa, Tiupampan SALMA (South American Land Mammal Age) (Marshall and Muizon, 1988; Muizon, 1994; Muizon and Cifelli, 2001; Muizon et al., 2018).
[b] Early Oligocene of Chubut province, Argentina (ca. 46° S). Fauna from La Cancha unit of Sarmiento Formation, Tinguirirican SALMA (Goin et al., 2010).
[c] Early Miocene of Santa Cruz province, Argentina (ca. 51° S). Fauna from FL 1–7 of Santa Cruz Formation, Santacrucian SALMA (Kay et al., 2012).
[d] Middle Miocene of Huila department, Colombia (ca. 3° S). Fauna from Honda Group sediments, Laventan Stage/Age (Goin, 1997).
[e] Early Pliocene of Buenos Aires province, Argentina (ca. 39° S). Fauna from Monte Hermoso Formation, Montehermosan SALMA (Tomassini et al., 2013).
[f] Includes †Argyrolagidae.
[g] Excludes †Argyrolagidae.

consisting mostly of isolated teeth—includes the last known representatives of two archaic lineages, †Caroloameghiniidae and †Sternbergiidae, each of which is represented by a single species. Sparassodonts, which dominate many younger (late Oligocene to mid-Miocene) faunas, are represented at La Cancha by just one species of "hathliacynid."

The early Miocene fauna from the Santa Cruz Formation contains 25 metatherian species—4 microbiotherians, 10 paucituberculatans, and 11 sparassodonts (Abello et al., 2012; Prevosti et al., 2012)—but it is unlikely that all of these occurred synchronously. A more realistic picture of Santacrucian diversity is provided by the fauna from Fossiliferous Levels 1–7, which appear to be very close in age (Kay et al., 2012). This assemblage contains 12 metatherian species, of which fully two-thirds are sparassodonts. However, other stem metatherians are notably absent, and although marsupials are present, didelphimorphians are not.

The middle Miocene La Venta fauna from Colombia is one of the earliest known assemblages that contains unambiguously identifiable didelphimorphians, and it is the earliest in which marsupials (didelphimorphians, microbiotherians, and paucituberculatans) comprise the majority of the metatherian community. Argyrolagids are absent, possibly because the local paleoclimate (Kay and Madden, 1997) was too humid for open-country bipedal hoppers. A noteworthy aspect of the La Venta mammal fauna (which includes >70 species, not including bats) is the complete absence of any immigrant North American taxa. Despite molecular data suggesting that intercontinental biotic exchange was well underway in plants and other organisms by the middle Miocene (Chapter 3), the La Venta mammal fauna consists exclusively of endemic groups that had been present in South America since the Eocene: metatherians, xenarthrans, archaic "ungulates" (astrapotheres, litopterns, notoungulates), primates, and caviomorph rodents.

The early Pliocene Monte Hermoso Formation is the source of the earliest well-sampled fauna in which didelphimorphians are the dominant metatherian group. Montehermosan didelphimorphians include eight species of didelphids (including one "sparassocynid").[1] Sparassodonts, nearing the end of their long geological tenure, are represented at Monte Hermoso by just one "hathliacynid" and one thylacosmilid. Both of the Montehermosan paucituberculatans are argyrolagids. Caenolestids and microbiotherians are conspicuously absent at Monte Hermoso, both groups probably having retreated by this time to the cold-temperate refugia that both occupy today. Significantly, the Montehermosan fauna also includes continental newcomers—procyonid carnivorans and cricetid rodents—whose arrival signaled an end to the prolonged biogeographic isolation of South America's native mammals.

NOTE

1. "Sparassocynids," which were once commonly assumed to be the sister group of didelphids (e.g., by Forasiepi et al., 2009), are now believed to belong within the didelphid radiation (Beck and Taglioretti, 2019). They appear to have been small hypercarnivorous creatures, probably adapted to open habitats, and they differed most conspicuously from other opossums in the anatomy of their ear region.

# 3

# The Great American Biotic Interchange and Its Aftermath

Closure of the ancient seaways that separated North and South America throughout much of the Tertiary was a prolonged process. Recently reported geological evidence suggests the emergence of a Central American island arc in the late Oligocene, with progressive shallowing of the intervening straits by tectonic uplift throughout the Miocene; however, narrow marine connections between the Pacific Ocean and the Caribbean Sea seem to have persisted until the late Pliocene (Coates and Stallard, 2013; Jackson and O'Dea, 2013). Dispersal rates of terrestrial organisms between North and South America (estimated from molecular phylogenies) appear to show significant increases at about 20 Ma and again at about 6 Ma, presumably in response to the emergence of insular stepping stones and gap-narrowing, and it is now clear that intercontinental dispersal proceeded on different schedules for different groups (Bacon et al., 2015). However, there is no compelling evidence for wholesale immigration of North American mammals to South America—or vice versa—prior to emergence of a definitive landbridge at about 2.8 Ma (Woodburne, 2010; O'Dea et al., 2016).

As discussed in the preceding chapter, nonvolant North American mammals are unlikely to have arrived in South America before about 11 Ma, because none are present in the well-sampled fauna from La Venta, Colombia (Kay et al., 1997b). Additionally, it seems unlikely that many had arrived before about 5 Ma, when only procyonid carnivorans and cricetid rodents are recorded from the even better-sampled fauna of Monte Hermoso, Argentina (Tomassini et al., 2013). Although proboscideans (gomphotheres) and tayassuids (peccaries) have been collected from Amazonian sediments that are said to be at least 9.5 Ma (Campbell et al., 2000; Frailey and Campbell, 2012), these claims are controversial (Mothé and Avilla, 2015), and it seems improbable that such eurytopic large mammals (living proboscideans and tayassuids occur in a very wide range of environments, including savannas and deserts) should have lingered in Amazonia for millions of years before their first appearance in Argentinian sediments. Although the relevant geological record is ecogeographically biased—productive fossil localities in South America are concentrated at temperate latitudes (Carrillo et al., 2015)—and much may still be learned from tropical paleontology (e.g., Antoine et al., 2016; Bloch et al., 2016), the evidence at hand still supports essential aspects of the long-accepted timeline for mammalian participation in the Great American Biotic Interchange (GABI).

Mammalian faunal exchange following landbridge formation extended over about 2 million years and involved numerous clades (Webb, 1976, 1985; Marshall et al., 1982). Immigrants to North America from the south included opossums, sloths, armadillos (including glyptodonts), and caviomorph rodents, whereas South America was invaded by North American carnivorans (mustelids, canids, felids), proboscideans, perissodactyls (tapirs, horses), and artiodactyls (peccaries, camels, deer). These exchanges substantially increased the higher-taxonomic diversity of faunas on both continents, allegedly resulting in ecological disequilibria, but the biotic phenomena responsible for subsequent extinctions have been widely debated.

It is reasonable to suppose that intercontinental dispersals may have had a significant impact on South American metatherians because several North American immigrant clades might have preyed upon or competed with native taxa. Among early-arriving (pre-landbridge) immigrants that might have had such adverse effects were procyonid carnivorans and crotaline viperids (pitvipers). Procyonids first appear in the South American fossil record at about 7 Ma (Reguero and Candela, 2011), but the earliest known South American procyonids belong to already-endemic genera (e.g., †*Cyonasua*) that may have descended from even earlier immigrants. Most Recent procyonids seldom eat opossums, but *Procyon cancrivorus*, the largest living raccoon (ca. 3–7 kg; Emmons, 1997), is known to do so (Gatti et al., 2006), and Tertiary South American procyonids were much larger than *P. cancrivorus*. Species of †*Cyonasua*, for example, are estimated to have weighed between 13 and 28 kg (Tarquini et al., 2018). Additionally, dental morphology suggests that South American fossil procyonids were more carnivorous than Recent taxa (Soibelzon, 2011), so they might well have preyed on opossums and other small metatherians, perhaps including juvenile and subadult sparassodonts.

The earliest known fossil of a South American pitviper is from Huayaquerian sediments (probably >6.8 Ma; Flynn and Swisher, 1995) in central Argentina (Albino and Montalvo, 2006), but molecular dating based on phylogenies of Recent taxa suggest that bothropoid pitvipers began to radiate in South America between 10 to 23 Ma (Wüster et al., 2002; Alencar et al., 2016). Although some living opossums are known to have evolved impressive resistance to pitviper venom (Jansa and Voss, 2011; Voss and Jansa, 2012), Miocene metatherians were presumably not venom resistant and might have been easy prey for ophidian predators endowed with novel sensory abilities and biochemical weaponry (Chapter 14).

Other immigrant clades that could have had an adverse impact on South American metatherians as predators or competitors (estimated arrival times after Woodburne, 2010) include cricetid rodents (6 Ma); canids and mustelids (2.6 Ma); and cervids, felids, and tapirids (1.8 Ma). Living canids, felids, and mustelids are known to eat Recent opossums (Chapter 11), and it is reasonable to assume that their extinct ancestors also preyed on metatherians. Cricetid rodents, deer, and tapirs might seem like unlikely opossum competitors, but living members of all three clades exhibit substantial trophic-niche overlap with Recent opossums (Chapter 12). The arrival times of many other immigrant placental taxa—especially those that inhabit rainforest environments—are unknown, but it seems probable that the South American metatherian fauna experienced monotonically increasing predatory and competitive pressures throughout the episode of faunal exchange that followed landbridge completion. Whether or not such interactions were directly responsible for any extinctions, the fact remains that, of a formerly diverse continental fauna of endemic metatherian lineages, only didelphids, paucituberculatans, and microbiotherians are known to have survived to the end of the Pliocene.

During the Pleistocene (an interval of almost 2.6 million years), the South American mammal fauna continued to receive new immigrant groups and acquired a distinctively new continental character. Besides the admixture of immigrant and endemic clades, one of the salient features of the Pleistocene fauna was the extraordinary diversity and ecological dominance of very large mammals. According to the latest tally (Cione et al., 2009), over 30 species of late Pleistocene South American mammals weighed >1000 kg, including ground sloths, glyptodonts, gomphotheres, a camel, and a few surviving native ungulates (litopterns and notoungulates). Equally impressive was the roster of large-but-not-gigantic (44–1000 kg) mammals, which included species of bears, horses, tapirs, deer, peccaries, and caviid rodents. Of this astonishing late Pleistocene megafauna, totaling over 80 species, nearly all became extinct within a few thousand years, approximately coincident with, or following shortly after, the arrival of humans in South America (Cione et al., 2009).

Species in several other groups not conventionally recognized as megafauna also went extinct at the end of the Pleistocene, including several monkeys (Cartelle and Hartwig, 1996; Halenar and Rosenberger, 2013), several canids (Prevosti et al., 2009), several caviomorph rodents (Kerber et al., 2016; Mayer et al., 2016; Gomes et al., 2019), and a vampire bat (which probably fed on megafauna; Trajano and de Vivo, 1991). Of these, all were the largest in their respective lineages, and all but the vampire bat probably weighed in excess of 10 kg. By contrast, opossums may have emerged unscathed from the end-Pleistocene extinction event, presumably because none was large.[1] The largest known fossil didelphid appears to have been †*Thylophorops lorenzinii*, a hefty (ca 8.6 kg; Zimicz, 2014) Pliocene species, but no late Pleistocene opossum is known to have been larger than the living North American *Didelphis virginiana*. Insofar as can be determined from fossilized fragments, all of the didelphid species recorded from the late Pleistocene are alive today, with the problematic exception of †*Sairadelphys tocantinensis*.[2]

South America's indigenous (pre-GABI) mammals are sometimes alleged to have been competitively inferior to late-arriving immigrant taxa from North America (e.g., by Webb, 1976; MacFadden, 2006) or to have been more vulnerable to predation (Faurby and Svenning, 2016). However, although faunal interchange and subsequent survival may have favored North American taxa in savanna habitats, the reverse was true in forested environments. Today, descendants of North American immigrant groups (carnivorans, perissodactyls, artiodactyls, squirrels, and cricetid rodents) account for fewer than half the species of nonflying mammals in well-sampled Amazonian faunas (Table 3.1).[3] In these diverse communities, didelphids are the second-most speciose group (after rodents), with a dozen or more sympatric species recorded at several localities (Voss et al., 2019).

Not only did opossums survive contact with a host of invading placental mammals before and after landbridge formation, but they successfully invaded North America as well. As in South America, productive fossil deposits in North America are concentrated at temperate latitudes, where the first opossum (*Didelphis*) is not known to have appeared until about 0.8 Ma (Woodburne, 2010). However, at least 19 didelphid species in nine genera now occur in tropical North America (Central America and southern Mexico; Appendix 1), and one species (*Didelphis virginiana*) has extended its geographic range in recent decades all the way to Canada, apparently having become a human commensal in the process (Kanda et al., 2009; Wright et al., 2012).

The stubborn survival of didelphids in placental-dominated faunas throughout most of the Western Hemisphere, together with the extant diversity of didelphids in modern Amazonian rainforests, effectively refute the widespread stereotype that marsupials are poor competitors or hapless prey. In the following

*Table 3.1*   Nonvolant Rainforest Mammals of Paracou, French Guiana[a]

| DIDELPHIMORPHIA | PRIMATES | RODENTIA |
|---|---|---|
| *Caluromys philander* | *Alouatta seniculus* | *Guerlinguetus aestuans** |
| *Chironectes minimus* | *Ateles paniscus* | *Sciurillus pusillus** |
| *Didelphis imperfecta* | *Cebus apella* | *Euryoryzomys macconnelli** |
| *Didelphis marsupialis* | *Pithecia pithecia* | *Hylaeamys megacephalus** |
| *Gracilinanus emiliae* | *Saguinus midas* | *Hylaeamys yunganus** |
| *Hyladelphys kalinowskii* | *Saimiri sciureus* | *Neacomys dubosti** |
| *Marmosa demerarae* | CARNIVORA | *Neacomys paracou** |
| *Marmosa lepida* | *Speothos venaticus** | *Nectomys rattus** |
| *Marmosa murina* | *Herpailurus yagouaroundi** | *Neusticomys oyapocki** |
| *Marmosops parvidens* | *Leopardus pardalis** | *Oecomys auyantepui** |
| *Marmosops pinheiroi* | *Leopardus wiedii** | *Oecomys bicolor** |
| *Metachirus nudicaudatus* | *Panthera onca** | *Oecomys rex** |
| *Monodelphis touan* | *Puma concolor** | *Oecomys rutilus** |
| *Philander opossum* | *Eira barbara** | *Rhipidomys nitela** |
| CINGULATA | *Galictis vittata** | *Coendou melanurus* |
| *Cabassous unicinctus* | *Nasua nasua** | *Coendou prehensilis* |
| *Dasypus kappleri* | *Potos flavus** | *Dasyprocta leporina* |
| *Dasypus novemcinctus* | PERISSODACTYLA | *Myoprocta acouchy* |
| *Priodontes maximus* | *Tapirus terrestris** | *Cuniculus paca* |
| PILOSA | ARTIODACTYLA | *Echimys chrysurus* |
| *Bradypus tridactylus* | *Mazama americana** | *Makalata didelphoides* |
| *Choloepus didactylus* | *Mazama nemorivaga** | *Mesomys hispidus* |
| *Cyclopes didactylus* | *Pecari tajacu** | *Proechimys cuvieri* |
| *Myrmecophaga tridactyla* | *Tayassu pecari** | *Proechimys guyannensis* |
| *Tamandua tetradactyla* | | |

[a] A Recent fauna from northeastern Amazonia. Species list from Voss et al. (2001), Adler et al. (2012), and previously unpublished observations (F. Catzeflis, personal commun.).
* Species marked with an asterisk are descended from North American immigrant (GABI) ancestors.

chapter, we document the taxonomic diversity of living didelphids and sketch the outlines of the ecological niches that they currently occupy in communities of Recent mammals.

NOTES

1. Statistical analyses of Quaternary faunal data suggest that size is the only significant correlate of late Pleistocene mammalian extinctions in the Western Hemisphere (Lessa and Fariña, 1996; Lessa et al., 1997), a result consistent with the inference that the end-Pleistocene extinctions were primarily anthropogenic (Cione et al., 2009; Sandom et al., 2014).

2. This strange taxon (described by Oliveira et al., 2011) exhibits no compelling didelphid synapomorphies, has several dental traits not seen in any living opossum (e.g., absence of an anterior

cingulum, extreme reduction of M4), and has not been included in any phylogenetic analysis sufficiently inclusive to test its alleged didelphid affinities. We are not, in fact, convinced that it is even a marsupial.

3. The participation of bats in the GABI is not well documented in the fossil record, so counting species descended from indigenous versus immigrant lineages in modern chiropteran faunas is problematic. The available evidence from fossils and from phylogenetic reconstructions of New World bat endemism (reviewed by López-Aguirre et al., 2018) suggest complex historical-biogeographic scenarios rather than a clear distinction between North and South American lineages.

# II

# OPOSSUM CLASSIFICATION
# AND DIVERSITY

# 4

# Taxonomic Accounts

The classification of Recent didelphids has undergone many changes over the last two decades, largely as the result of phylogenetic analyses of genetic and phenotypic datasets. Although some nodes in opossum phylogeny remain problematic, all of the monophyletic groups currently ranked as subfamilies, tribes, and genera are robustly supported by a wide range of evidence, including nuclear exon sequences, whole mitochondrial genomes, karyotypes, and morphology (Voss and Jansa, 2009; Mitchell et al., 2014; Amador and Giannini, 2016).

The following accounts provide thumbnail sketches of each genus and higher taxon currently regarded as valid, including key aspects of geographic distribution, habitats, behavior, and diet. Inevitably, much is known about widespread and commonly encountered taxa, whose accounts are correspondingly detailed, whereas less is known about others, whose accounts are unavoidably brief. Unfortunately, the absence of definite natural history information for some taxa has been supplemented with conjecture and imagination in the semipopular literature, and it is useful to identify such fictions here.

## Subfamily Caluromyinae

Caluromyines include two genera, *Caluromys* (with three currently recognized species) and *Caluromysiops* (with one). The monophyly of this subfamily is strongly supported by phylogenetic analyses of both morphological and molecular datasets (e.g., Voss and Jansa, 2009). Three of the four living species are South American endemics, but one (*Caluromys derbianus*) extends its range northward into Central America and southern Mexico. These are medium-size (ca. 200–500 g) opossums with woolly fur, long tails, and large eyes; all are thought to be highly arboreal and predominantly frugivorous. To some primatologists, caluromyines closely resemble cheirogaleid lemurs, and it has been suggested that they are appropriate living models for testing adaptive hypotheses about early primate evolution (Rasmussen, 1990).

## *Caluromys*

Species of *Caluromys* (Fig. 4.1) occur in rainforest and semideciduous forest from southern Mexico to northeastern Argentina. The South American species occur in lowland forests, but Central American *C. derbianus* is said to occur from sea level to about 2500 m (Bucher and Hoffman, 1980). Trapping studies (e.g., Malcolm, 1991; Patton et al., 2000; Lambert et al., 2005), spool-and-line tracking (Miles et al., 1981), nest location (Delciellos et al., 2006; Lira et al., 2018), and direct observations of free-ranging individuals (e.g., Charles-Dominique et al., 1981; Rasmussen, 1990) provide compelling evidence that these opossums are primarily active in the canopy and subcanopy and seldom descend to ground level. Species of *Caluromys* are agile climbers—with strongly prehensile appendages and strikingly primate-like gaits (Schmitt and Lemelin, 2002; Lemelin and Schmitt, 2007; Youlatos, 2008, 2010; Dalloz et al., 2012)—and foraging individuals make extensive use of small-diameter supports, especially fine terminal branches (Rasmussen, 1990).

*Fig. 4.1. Caluromys derbianus* in lowland dry forest near Diriamba, Carazo, Nicaragua (photo by José Martínez)

The diet of *Caluromys* is thought to consist primarily of fruit, but arthropods and small vertebrates are also eaten occasionally (Charles-Dominique et al., 1981; Atramentowicz, 1982; Rasmussen, 1990; Casella and Cáceres, 2006; Lessa and da Costa, 2010).[1] In the most thorough field study of any species of *Caluromys* to date, approximately 75% of the diet of *C. philander* in a French Guianan rainforest was estimated to consist of fruits, which included those of at least 23 species of trees, lianas, and hemiepiphytes (Charles-Dominique et al., 1981; Atramentowicz, 1982, 1988). Additionally, all three species of *Caluromys* are known to visit flowers to feed on nectar, which may be an important carbohydrate resource in the dry season, when fruit is scarce (Chapter 9). *Caluromys philander* has also been observed to eat gum, a possibly important source of complex carbohydrates, for which its enlarged caecum might be a digestive adaptation (Charles-Dominique et al., 1981). Most of the arthropods eaten by *Caluromys* are insects (Charles-Dominique et al., 1981; Lessa and da Costa, 2010), some of which are caught by patient visual and olfactory searches along branches, but flying insects are sometimes snatched in midair (Rasmussen, 1990). Among other reports of vertebrate predation, Casella and Cáceres (2006) found the remains of birds and mammals in dissected digestive tracts of *C. lanatus* from southeastern Brazil, and this species is also known to enter live traps to eat cricetid rodents (Camargo et al., 2017). We have observed *C. philander* to be attracted to bats caught in mistnets, and it seems plausible that bats often fall prey to these opossums while both are foraging in the canopies of fruiting or flowering trees.

## *Caluromysiops*

The only currently recognized species of *Caluromysiops*, *C. irrupta* (Fig. 4.2), is known from fewer than a dozen museum specimens and from several unvouchered observations, all of which were obtained at widely scattered Amazonian rainforest localities in Peru and Brazil (Santori et al., 2016). *Caluromysiops* is morphologically similar to *Caluromys*, so it is probably ecologically similar as well, but field observations of behavior are only available from a single Peruvian site where L.H. Emmons and her colleagues watched free-ranging individuals in mature floodplain forest. At this locality *C. irrupta* was repeatedly observed in the canopy at night (Fig. 4.2), where it visited flowers of the tree *Quararibea cordata* (Bombacaceae) to consume nectar (Janson et al., 1981; Emmons, 2008). Captive individuals eat fruit, and some are reported to also eat crickets, newly hatched chicks, and mice (Collins, 1973).

*Fig. 4.2. Caluromysiops irrupta* in the subcanopy of lowland rainforest at Cocha Cashu, Madre de Dios, Peru (photo by Louise H. Emmons)

## Subfamily Glironiinae
### *Glironia*

The subfamily Glironiinae contains only *Glironia venusta*, the so-called bushy-tailed opossum (Fig. 4.3). Rarely encountered over a wide geographic range, this species seems to occur in both primary forest and secondary vegetation throughout much of lowland Amazonia (Díaz and Willig, 2004; Calzada et al., 2008; Ardente et al., 2013), but observations and collected specimens have also been reported from dry forests (e.g., in eastern Bolivia; Tarifa and Anderson, 1997) and in lower-montane Andean forests (Arguero et al., 2017). This small ($<$200 g)[2] opossum is either the sister taxon of other Recent didelphids, the sister taxon of Hyladelphinae + Didelphinae, or the sister taxon of Caluromyinae (Voss and Jansa, 2009; Mitchell et al., 2014). Whichever position it occupies, *G. venusta* represents a very long branch in didelphid phylogeny (Chapter 14) and exhibits unusual morphological traits not seen or seldom seen in other opossums—including enormous orbital fossae; large, strongly recurved, and laterally compressed claws; and a weak dentition—all of which hint at distinctive behaviors. Unfortunately, most published observations of free-ranging individuals are based on fleeting glimpses;

Fig. 4.3. *Glironia venusta* in lowland rainforest near Paranaíta,
Mato Grosso, Brazil (photo by Thiago da Silveira)

stomach contents of collected specimens have never been examined, and rare opportunities to observe the behavior of living captives have been unproductive.
Therefore, little is known about the natural history of this species.

Free-ranging individuals have almost always been observed in trees (at reported
heights of 2–15 m above the ground), where they seem equally agile climbing on
large-diameter trunks, slender vertical stems, or vines; there is one observation of
an individual emerging from a tree hole at dusk and ascending to the canopy, and
other individuals first observed in subcanopy vegetation subsequently retreated
into the canopy when disturbed (Emmons, 1997; Calzada et al., 2008; Silveira et al.,
2014). The rapid, nervous, squirrel-like movements of *Glironia venusta*, which

include leaping from one aboveground support to another and rapid movements on large-diameter vertical trunks, are quite unlike the slower, more deliberate climbing behavior of most other opossums (Emmons, 1997; Calzada et al., 2008). Dietary information about *G. venusta* is almost entirely lacking. Emmons (1997) reported one individual that seemed to be hunting insects but at one point spent many seconds licking something from a branch, and Silveira et al. (2014) observed a group of three juveniles licking up fluid that seeped from cracks in a tree trunk. Despite such evidence of exudate feeding, *Glironia* exhibits no dental adaptations for bark-gouging, so it seems unlikely that exudates are a major food resource.

## Subfamily Hyladelphinae
### *Hyladelphys*

The subfamily Hyladelphinae contains only *Hyladelphys kalinowskii*, a tiny (ca. 10–20 g) species that was first described in 1992 and is still known from just a few specimens (Fig. 4.4). All known collecting localities and sight records are from the Amazonian lowlands of eastern Peru, northern Brazil, and the Guianas (Astúa, 2006). Phylogenetic analyses of molecular data suggest that *Hyladelphys* is the sister taxon of the subfamily Didelphinae (Chapter 14), and several of its morphological characteristics are consistent with its intermediate position between caluromyines and didelphines (Jansa and Voss, 2005).

The scant available natural history information about *Hyladelphys kalinowskii* comes from haphazard encounters and a few individuals trapped in the course of faunal inventories. To date, all reported nocturnal observations of this species have been of individuals perched on stems, leaves, or lianas within a few meters of the ground, and all diurnal encounters have been of individuals asleep in leaf nests attached to understory vegetation (Voss et al., 2001; P. Peloso, personal commun.; Catzeflis, 2018), so it seems reasonable to infer that *H. kalinowskii* is arboreal but probably not a canopy-dweller. In French Guiana, this species has been found in both well-drained and swampy primary forest, in secondary vegetation, and in native houses surrounded by slash-and-burn garden plots (Catzeflis, 2018).

The nests occupied by *Hyladelphys kalinowskii* are of exceptional interest because the leaves from which they are built are cemented together and anchored to nearby vegetation by a mysterious white substance of unknown origin (Catzeflis, 2018). A single individual maintained alive for a few days at a field station ate orthopterans and fruit (Catzeflis, 2018), but nothing else is known about the diet of this species.

*Fig. 4.4. Hyladelphys kalinowskii* in lowland rainforest near Oiapoque, Amapá, Brazil (*above*; photo by Pedro Peloso) and in a leaf nest at La Trinité Natural Reserve, French Guiana (*below*; photo by Antoine Fouquet)

## Subfamily Didelphinae

The monophyly of Didelphinae, a clade that includes about 95% of living opossums, is strongly supported by phylogenetic analyses of molecular sequences, but it is not consistently recovered by analyses of morphological data (Voss and Jansa, 2009). Several robustly supported clades within Didelphinae are currently ranked as tribes in the Linnaean hierarchy: Marmosini (*Marmosa, Monodelphis,*

*Tlacuatzin*), Metachirini (*Metachirus*), Didelphini (*Chironectes, Didelphis, Lutreolina, Philander*), and Thylamyini (*Chacodelphys, Cryptonanus, Gracilinanus, Lestodelphys, Marmosops, Thylamys*). Although Didelphini was long recognized as a clade by virtue of a distinctive karyotype and large body size (e.g., by Reig et al., 1987), the contents of Marmosini and Thylamyini were not effectively sorted out until the advent of modern molecular systematics.

## Tribe Marmosini

The term "marmosines" was formerly used synonymously with "murine opossums" or "mouse opossums" to refer to the species that Tate (1933) included in the genus *Marmosa* (sensu lato). In this book, however, we use "marmosines" to refer to members of the monophyletic group that includes *Marmosa* (sensu stricto), *Monodelphis*, and *Tlacuatzin*, but we will seldom have cause to do so because this tribe has few diagnostic morphological traits, and none that are unique to it. In effect, this is a genealogical grouping of adaptively divergent forms for which only molecular support is currently available.

### *Marmosa*

Throughout most of the last century, the genus *Marmosa* (sensu Tate, 1933) contained a phylogenetically heterogeneous collection of small, pouchless, long-tailed, black-masked didelphids, including many species that are now referred to other genera (Rossi et al., 2010). As defined by Voss and Jansa (2009), *Marmosa* now includes *Micoureus* (formerly recognized as a valid genus; Gardner, 2008) as a subgenus. Other subgenera (besides the nominotypical subgenus) include *Eomarmosa*, *Exulomarmosa*, and *Stegomarmosa* (Voss et al., 2014). The 18 currently recognized species of *Marmosa* collectively range from northeastern Mexico (Tamaulipas) throughout Central America and tropical South America to northern Argentina. Most species occur in lowland rainforest (Fig. 4.5), but a few inhabit gallery forests in savanna landscapes (Smith and Owen, 2015), desert thornscrub (Thielen et al., 1997a), and montane forest (Nadkarni and Wheelwright, 2000).

Trapping results (e.g., O'Connell, 1979; Malcolm, 1991; Leite et al., 1996; Lambert et al., 2005), spool-and-line tracking (Prevedello et al., 2009), observations of free-ranging animals in the wild (Enders, 1935; Charles-Dominique et al., 1981), and laboratory studies of locomotion (e.g., Delciellos and Vieira, 2006, 2009b) all suggest that species of *Marmosa* are agile climbers that often use small-diameter arboreal substrates. Animals released in capture-mark-recapture studies almost always choose arboreal rather than terrestrial escape routes (O'Connell, 1979;

*Fig. 4.5. Marmosa zeledoni* eating a cicada in understory vegetation at La Selva, Heredia, Costa Rica (photo by Víctor Acosta-Chaves)

Miles et al., 1981; Prevedello et al., 2009), and nests are usually in abandoned bird nests, palm crowns, tree cavities, and other arboreal refugia (Enders, 1935; Miles et al., 1981; Timm and LaVal, 2000; Moraes and Chiarello, 2005b; Delciellos et al., 2006). Apparently, species of the subgenus *Micoureus* are more frequently active in the canopy and subcanopy than sympatric species of the subgenus *Marmosa*, which are more often observed or captured in understory vegetation (Charles-Dominique et al., 1981; Lambert et al., 2005). Some species (e.g., *M. isthmica, M. murina, M. macrotarsus*) seem to prefer dense secondary growth and other kinds of disturbed habitats to climax vegetation (Enders, 1935; Guillemin et al., 2001; Voss et al., 2019). In fact, members of the subgenus *Micoureus* seem to be the only species of *Marmosa* commonly captured in tall primary forest (Malcolm, 1991; Patton et al., 2000; Voss et al., 2001).

Published dietary analyses based on feces recovered from live-trapped animals (e.g., Thielen et al., 1997b; Cáceres et al., 2002; Pinheiro et al., 2002; Pires et al., 2012) invariably report that arthropods (mostly insects) are more frequently eaten than fruit (as identified by ingested seeds). However, these results might be biased (underestimating the importance of fruit) if only the pulp of large-seeded fruits is consumed, and direct observations of feeding behavior by free-ranging individuals suggests that this may be the case (Charles-Dominique et al., 1981). Nevertheless, the feeding behavior of both free-ranging individuals and captives (which cannot be maintained on a fruit-only diet) clearly indicates that large insects are actively stalked and avidly consumed (Enders, 1935; Acosta-Chaves et al., 2018), and some species live in habitats where fruit is only available seasonally, apparently subsisting exclusively on insects when fruit is scarce (Thielen et al., 1997b). Small vertebrates (e.g., birds and amphibians) are known to be eaten occasionally (Charles-Dominique et al., 1981; Cáceres et al., 2002; Casella and Cáceres, 2006; Lessa and da Costa, 2010; Acosta-Chaves et al., 2018), as are certain fruit-like flowers (Sperr et al., 2008) and nectar (Janson and Emmons, 1990).

## *Monodelphis*

Species of *Monodelphis* (Fig. 4.6) are small (<150 g) shrew-like opossums with proportionately smaller eyes and ears than other didelphids; the limbs and tail are short, and the tail lacks external modifications for prehension. Also unlike other opossums—which are almost invariably drab, with unpatterned substrate-matching fur in various shades of dull brown or gray—several species of *Monodelphis* are brightly colored, and other species have chipmunk-like dorsal stripes (see color photographs in Pavan et al., 2014; Pavan and Voss, 2016). This genus includes 24 species in five subgenera (Appendix 1) that collectively range from eastern Panama to northern Argentina. Most species occur in lowland rainforest, but a few (such as *M. gardneri*) occur in montane forest (Solari et al., 2012), some (e.g., *M. domestica*) occur in dry forests (Streilein, 1982a), and at least one (*M. dimidiata*) occurs in temperate grasslands (Pine et al., 1985; Baladrón et al., 2012).

*Fig. 4.6. Monodelphis americana* near Brasilia, Distrito Federal, Brazil (*above*; photo by Rodrigo Conte); *M. touan* from French Guiana (*below*; photo by Antoine Baglan)

Many trapping studies (e.g., O'Connell, 1979; Malcolm, 1991; Patton et al., 2000; Vieira and Monteiro-Filho, 2003; Lambert et al., 2005), one radio-tracking study (Charles-Dominique et al., 1981), and numerous haphazard observations reported by authors (e.g., Goeldi, 1894; Davis, 1947; Voss and Emmons, 1996) convincingly document the terrestrial habits of *Monodelphis*, rainforest species of which forage in the litter layer, inside hollow logs, under fallen trees, and in other dark ground-level refugia sheltered and carpeted by decaying vegetable debris. However, some species are said to occupy nests that may be a meter or more above ground level (Davis, 1947; González and Claramunt, 2000). Although captive individuals are reportedly able to climb (Streilein, 1982a; González and Claramunt, 2000), *M. domestica* is slow and unsteady when walking on small-diameter substrates (Pridmore, 1992; Lemelin et al., 2003; Lammers and Biknevicius, 2004), and it has obvious difficulty maintaining its balance on slender branches that other opossums navigate with ease (Lemelin and Schmitt, 2007). Most unusually (for didelphids), several species of *Monodelphis* are known to be diurnally active—including *M. americana, M. dimidiata, M. emiliae, M. kunsi, M. scalops,* and *M. touan* (see Davis, 1947; Nitikman and Mares, 1987; Charles-Dominique et al., 1981; Pine et al., 1985; Vieira and Paise, 2011; Pheasey et al., 2018; Voss et al., 2019)—whereas only one species (*M. domestica*) is definitely known to be nocturnal (Streilein, 1982a; Seelke et al., 2014).

Consistent with their shrew-like appearance, species of *Monodelphis* are restless, active predators that attack a wide range of invertebrate and vertebrate prey. Examination of digestive tracts of wild-caught animals consistently suggests a primarily insectivorous diet (Busch and Kravetz, 1991; Casella and Cáceres, 2006; Castilheiro and Filho, 2013), and captives readily consume a wide variety of terrestrial arthropods (González and Claramunt, 2000), including scorpions (Streilein, 1982a; Pheasey et al., 2018). However, digestive tracts of *M. dimidiata* frequently also contain rodent remains (Busch and Kravetz, 1991; Casella and Cáceres, 2006), feces of *M. domestica* not infrequently contain squamate bones or scales (Carvalho et al., 2019), and captives of several species of *Monodelphis* have been observed to kill and eat frogs, squamates, and mice (Goeldi, 1894; Streilein, 1982a; González and Claramunt, 2000; Pheasey et al., 2018). Although seeds are sometimes found in stomachs of wild-caught individuals, fruit does not appear to be an important part of natural diets based on the studies cited above.

## Tlacuatzin

A recent revision of *Tlacuatzin*, an endemic Mexican genus, recognized five species (Arcangeli et al., 2018), four of which were formerly treated as subspecies or synonyms of *T. canescens* (Fig. 4.7). These are small opossums (about 40–60 g; Zarza et al., 2003) that externally resemble *Marmosa*—pouchless, with long prehensile tails and blackish facial masks—but they have 22 diploid chromosomes (*Marmosa* has $2n = 14$), and the two genera differ in several technical details of craniodental morphology (Voss and Jansa, 2003). The genus is widely distributed

*Fig. 4.7. Tlacuatzin sinaloae* in dry forest near the Chamela-Cuixmala biological station, Jalisco, Mexico (photo by Elí García-Padilla)

in tropical and subtropical dry forests in western Mexico, the Yucatan Peninsula, and the Tres Marias Islands.

Trapping studies suggest that *Tlacuatzin* is predominantly arboreal, although it is sometimes also captured on the ground (Ceballos, 1990; Kennedy et al., 2013), and some of the habitats in which it is found are so sparsely vegetated that terrestrial locomotion must be frequent. Mating occurs in trees, however, with both partners suspended upside down by their tails (Valtierra-Azotla and García, 1998), a copulatory posture that strikingly resembles that described for captive *Marmosa* (Barnes and Barthold, 1969). Most nests (made of dry leaves and lined with grasses or plant fibers) are in tree cavities or shrubs, but abandoned bird nests are also used as diurnal refugia (Ceballos, 1990). Species of *Tlacuatzin* seem to be primarily insectivorous—feces of trapped individuals contain only insect remains—although small vertebrates (e.g., lizards) and perhaps fruit are also eaten (Ceballos, 1990;

Zarza et al., 2003). At least one species, *T. balsasensis*, is also known to visit cactus flowers to consume nectar (Ibarra-Cerdeña et al., 2007), which is perhaps a seasonally important source of calories and/or water.

## Tribe Metachirini
### *Metachirus*

The tribe Metachirini contains a single genus, *Metachirus*, that occurs in lowland rainforests from southern Mexico to northern Argentina (Fig. 4.8). Only one species, *M. nudicaudatus*, was recognized as valid for many years, but Voss et al. (2019) summarized molecular and morphological evidence for recognizing a second species, *M. myosuros*. These are large (ca. 260–480 g), long-limbed opossums with tails that lack external modifications for prehension. They superficially resemble species of *Philander* (members of the tribe Didelphini) by having a conspicuous pale spot above each eye, but *Metachirus* and *Philander* (both commonly known as "four-eyed" opossums) are dissimilar in many other external and craniodental traits and are not closely related to one another (Voss and Jansa, 2009).

Whereas most rainforest opossums are arboreal or semiarboreal, species of *Metachirus* are almost exclusively terrestrial. This behavior is convincingly documented by spool-and-line tracking of individual animals (Miles et al., 1981; Cunha and Vieira, 2002) and by numerous trapping studies (e.g., Malcolm, 1991; Guillemin et al., 2001; Lambert et al., 2005), all of which suggest that species of *Metachirus* seldom leave the ground. Unusually for rainforest opossums, which typically nest in trees, nests of *Metachirus* are constructed on the ground and carefully concealed in the litter (Miles et al., 1981; Loretto et al., 2005).[3] Equally remarkable is the cursorial gait of *Metachirus*, which flees by running rapidly away on the forest floor, whereas many other sympatric opossums climb rather than run to escape danger (Miles et al., 1981; personal obs.). Several aspects of the postcranial skeleton and musculature of *Metachirus* are plausibly interpreted as cursorial adaptations (Grand, 1983; Argot 2001, 2002, 2003), and *M. myosuros* was found to be the most inept climber of seven rainforest didelphids tested in the laboratory for acrobatic performance (Delciellos and Vieira, 2009a).

Information about the diet of *Metachirus* is almost entirely based on analyses of feces, which, unfortunately, are subject to several known sources of bias (Chapter 9). Nevertheless, three studies with reasonably large sample sizes suggest that *M. myosuros* (*M. "nudicaudatus"* of authors) is predominantly insectivorous but also eats fruit and, occasionally, small vertebrates (Santori et al., 1995; Cáceres,

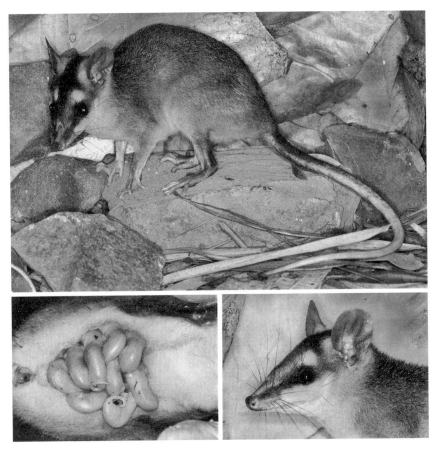

*Fig. 4.8. Metachirus myosuros* captured in lower-montane moist forest near Peñas Blancas Massif Natural Reserve, Jinotega, Nicaragua (*above*). Details of nursing young (*bottom left*) and cranial vibrissae (*bottom right*) are from the same individual. (Photos by Fiona Reid)

2004; Lessa and da Costa, 2010). Because species of *Metachirus* do not climb, most of the fruit they consume probably consists of items picked and discarded by canopy frugivores. Interestingly, *M. myosuros* seems to eat social insects—ants and termites—in larger numbers than other sympatric opossums (Mathews, 1977; Santori et al., 1995; Cáceres, 2004; Lessa and da Costa, 2010), although it exhibits no apparent morphological adaptations for myrmecophagy. Unlike some large opossums (members of the tribe Didelphini, see below), *Metachirus* is not resistant to snake venom (Perales et al., 1994), and it is not known to eat snakes.

## Tribe Didelphini

Members of the tribe Didelphini are large opossums, with average adult weights of 300 g or more, and the largest species (*Didelphis virginiana*) can weigh as much as 5000 g. In most other external traits, however, the four living genera—*Chironectes, Lutreolina, Philander*, and *Didelphis*—are strikingly dissimilar, and each occupies a distinct trophic niche in the Neotropical ecosystems they inhabit. Despite this adaptive diversity, tribal monophyly is supported by an impressive number of unique morphological traits (Voss and Jansa, 2009).

### Chironectes

*Chironectes* includes just the water opossum or yapok (*C. minimus*), which occurs in tropical and subtropical forests from southern Mexico to Uruguay (Fig. 4.9). The world's only semiaquatic marsupial, *C. minimus* is boldly marked dorsally with black and silver bars and exhibits numerous morphological adaptations to its unusual lifestyle. Among others, the fur is water repellent, the hind feet are fully webbed, and the marsupium can be closed to prevent pouch young from drowning (Enders, 1937; Augustiny, 1942). The hands are very large, with greatly elongated digits that end in fleshy pads and vestigial claws; a fleshy tubercle supported internally by an enlarged pisiform bone on the lateral aspect of the wrist may serve as a rudimentary sixth digit; and the plantar epithelium exhibits several unusual features that include unique sensory organs (Chapter 5). The tail is not laterally flattened or otherwise modified for aquatic propulsion; although lacking external modifications for prehension, it can be used for grasping and transporting nest materials (Oliver, 1976).

Water opossums swim with alternate strokes of their webbed hind feet; the forefeet and tail are not used for rectilinear swimming (Mondolfi and Padilla, 1957; Fish, 1993). Although *Chironectes* is sometimes said not to climb, individuals have been reliably reported to do so on occasion, rapidly ascending slender streamside trees to escape human intruders (F.A. Reid, pers. commun.). Water opossums forage in streams, so home ranges are effectively linear and dendritic, tracing the course of local drainage patterns (Galliez et al., 2009). Diurnal refugia are usually in shallow cavities among stones and tree roots near the water's edge (Zetek, 1930; Galliez et al., 2009; Palmeirim et al., 2014), but water opossums have also been seen to enter streamside burrows (presumably those made by other mammals, since the nearly clawless forefeet are conspicuously unsuited for digging). *Chironectes* uses

*Fig. 4.9. Chironectes minimus* on a stream bank in lowland moist forest, Bartola Wildlife Refuge, Río San Juan, Nicaragua (photo by Fiona Reid)

its hands, which are sensitive tactile organs (Chapter 5), to search for food in shallow water, a behavior similar to that observed in raccoons and clawless otters (Mondolfi and Padilla, 1957; Davis, 1966; Oliver, 1976). Anecdotal observations of stomach contents and prey remains found in dens suggest that the natural diet of *C. minimus* consists largely of freshwater crustaceans and fish (Zetek, 1930; Mondolfi and Padilla, 1957). Captive individuals eat shrimp, fish, terrestrial vertebrates (e.g., mice), and raw meat, but not fruit (Zetek, 1930; Davis, 1966; Rosenthal, 1975; Oliver, 1976).[4]

### Didelphis

Species of *Didelphis* (Fig. 4.10), sometimes known as "common opossums," are the largest living New World marsupials (ca. 600–5000 g). They occur from southern Canada to central Argentina (approximately from 49° N to 39° S) and inhabit a wide range of environments, including temperate woodlands and grasslands, tropical rainforests, tropical dry forests, and savannas (Gardner, 1982; Cerqueira, 1985). Although six species are currently recognized, only the Virginia opossum (*D. virginiana*), black-eared opossums (*D. auritus* and *D. marsupialis*), and white-eared opossums (*D. albiventris, D. imperfecta,* and *D. pernigra*) seem to be phenotypically

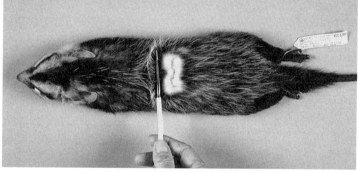

*Fig. 4.10. Didelphis albiventris* in dry forest near Santo Tomé, Santa Fe, Argentina (*above*; photo by Eduardo Beltrocco); a museum specimen of the same species showing white underfur (*below*; photo by Craig Chesek)

and ecologically distinct. Species of *Didelphis* are usually allopatric, but the geographic ranges of *D. virginiana* and *D. marsupialis* overlap in Mexico and northern Central America, where they are sometimes sympatric but not, apparently, syntopic (Gardner, 1982). Similarly, the ranges of black-eared and white-eared opossums sometimes overlap in South America, where sympatry has been documented at several localities (e.g., Cerqueira, 1985; Catzeflis et al., 1997). In South America, black-eared opossums are predominantly lowland rainforest species, whereas white-eared opossums typically inhabit temperate grasslands, savannas, dry forests, montane forests, and alpine habitats (Handley, 1976; Cerqueira, 1985; Mares et al., 1989; Durant, 2002).

Species of *Didelphis* are scansorial opossums that spend most of their time on the ground, but they are also competent climbers. Predominantly terrestrial locomotion is a practical necessity for *D. virginiana* and white-eared opossums, both of which normally live in habitats with broken or discontinuous woodland canopies, and which sometimes live in almost treeless landscapes. Although black-eared opossums live in tall closed-canopy forests, trapping studies (e.g., Malcolm, 1991; Lambert et al., 2005), spool-and-line-tracking (Miles et al., 1981), and direct observations (Charles-Dominique et al., 1981; Janson et al., 1981; Rasmussen, 1990; Tschapka and von Helversen, 1999) suggest that they are primarily active on the ground or in understory vegetation, climbing trees only to rest, groom, eat fruit, or visit nectar-producing flowers. In particular, spool-and-line tracking suggests that *D. marsupialis* does not travel through the canopy; instead, it ascends and descends by the same or adjacent trees (Miles et al., 1981). Both black-eared and white-eared opossums released in the course of capture-mark-recapture studies normally choose terrestrial escape routes, even when trees are nearby (Streilein, 1982a; Fleck and Harder, 1995). However, whereas *D. virginiana* and white-eared opossums normally use terrestrial daytime refugia (burrows, hollow logs, crevices among rocks; Hamilton, 1958; Streilein, 1982a), nests of *D. marsupialis* and *D. auritus* are often in trees (Miles et al., 1981; Delciellos et al., 2006).

Although usually described as omnivores, common opossums are perhaps more accurately described as generalized predators that also eat fruit when it is available. *Didelphis virginiana* and white-eared opossums, in particular, live in seasonal habitats where fruit is scarce or unavailable for much of the year. Natural history studies of *D. virginiana* (reviewed by Gardner, 1982) convincingly establish that the bulk of the natural diet of this temperate species consists of small vertebrates (e.g., rabbits, rodents, shrews, moles, ground-nesting birds, reptiles, and amphibians), but invertebrates (especially insects and earthworms) also comprise a substantial fraction, and fruit is often eaten in the summer and fall. In anthropogenic landscapes, *D. virginiana* also eats a wide range of agricultural products and other items that do not occur in undisturbed natural habitats. The only dietary study of a Neotropical common opossum based on a large sample of stomach contents (Cordero and Nicholas, 1987) found that birds, mammals, insects, and fruit (in that order) were the most important dietary items for *D. marsupialis* in anthropogenic landscapes of northern Venezuela, but fecal analyses suggest that *D. auritus* might eat more fruit in the Atlantic Forest than congeneric species do in other biomes (Santori et al.,

1995; Cáceres and Monteiro-Filho, 2001; Ceotto et al., 2009). All dietary studies of common opossums, however, suggest that they frequently eat small vertebrates. Several species (*D. albiventris, D. aurita,* and *D. virginiana*) are known to eat venomous snakes and to be resistant to snake venom (Voss and Jansa, 2012). Other unusual items in reported diets of common opossums are toads (Hamilton, 1958; Garrett and Boyer, 1993; Laurance and Laurance, 2007) and the terrestrial (eft) stage of newts (Hart et al., 2019), so it seems probable that *Didelphis* is also resistant to the potent cardiotoxic and neurotoxic defenses of amphibians (Chapter 6).

## Lutreolina

The genus *Lutreolina* includes just two species. The more widespread taxon is *L. crassicaudata* (Fig. 4.11), which has a disjunct geographic range in humid tropical and subtropical grasslands north and south of Amazonia. In the northern part of its distribution, *L. crassicaudata* is found in the Llanos and other savanna landscapes of Colombia, Venezuela, and Guyana. South of the Amazon, *L. crassicaudata* inhabits the Cerrado, Pantanal, Pampas, and contiguous grass- and sedge-dominated habitats of Paraguay, Uruguay, eastern Bolivia, southern Brazil, and northern Argentina. However, at least one insular population is known to occur in secondary forest (Cáceres et al., 2002). A second species, *L. massoia,* occurs in montane forests of the Andes in northwestern Argentina and southeastern Bolivia (Martínez-Lanfranco et al., 2014).

Species of *Lutreolina* are strikingly weasel- or mink-like in external appearance, with small eyes and ears, an elongate body, and short limbs; the tail lacks external modifications for prehension. Because *L. crassicaudata* usually occurs in treeless habitats, its normal locomotion must be predominantly terrestrial. Although this opossum is alleged to swim and dive well (e.g., by Marshall, 1978a), it has no morphological adaptations for aquatic locomotion, its fur is not water repellent, and captive individuals do not swim voluntarily (Davis, 1966). When compelled to do so, captive *Lutreolina* swim competently, but they are no faster in the water than other terrestrial opossums, and their locomotor performance does not suggest any aquatic specialization (Santori et al., 2005). Alleged nesting habits (Marshall, 1978a) appear to be unsupported by original observations in the scientific literature and, together with the supposed aquatic skills of this species, may be romantic fabrications.[5]

*Lutreolina crassicaudata* is alert, quick, and predatory in its movements (personal obs.; Davis, 1966), and its short-faced skull (Astúa de Moraes et al., 2000), carnassialized dentition (Voss and Jansa, 2003), and massively developed temporalis

*Fig. 4.11. Lutreolina crassicaudata* at the Reserva Ecológica Costanera Sur, Buenos Aires, Argentina (photo by Jorge La Grotteria)

musculature (Delupi et al., 1997) all suggest carnivorous adaptations. Captive individuals eat a wide range of animals—mice, frogs, fish, earthworms, and shrimp—but not fruit (Davis, 1966). Olrog (1979) conjectured that the local abundance of *L. massoia* (misidentified as *L. crassicaudata* in his study) was correlated with mouse population density, and he noted that these opossums can be trapped by using mice as bait. In the single dietary study based on animals trapped in the typical (natural grassland) habitat of *L. crassicaudata*, most analyzed feces contained vertebrate remains, including the fur of 10 species of small mammals (Monteiro-Filho and Dias, 1990). Unfortunately, other information about the diet of wild *L. crassicaudata* have been obtained from atypical situations—including a forested marine island (where most identified prey consisted of crabs and beetles; Cáceres et al., 2002), more or less urbanized landscapes (where insects, carrion, and garbage were eaten; Facure and Ramos, 2011), and a naturalized landfill adjacent to a major city (where seeds, presumably from fruit, and other plant remains were identified in feces; Muschetto et al., 2011). At most, these studies of ecologically marginal populations suggest that *L. crassicaudata* can subsist on whatever is locally available, but they seem unlikely to indicate the typical diet of this still poorly studied opossum. *Lutreolina crassicaudata* is known to be resistant to snake venom (Perales et al., 1994) and to prey on venomous snakes (Rodrigues, 2005).

### *Philander*

*Philander* is a genus of eight currently recognized species (Fig. 4.12), commonly known as "gray four-eyed opossums" or "pouched four-eyed opossums," that collectively range from southern Mexico to northern Argentina; species of *Philander* occur primarily in lowland rainforest, but they also inhabit gallery forests of the Chaco and Cerrado (Redford and Fonseca, 1986; Emmons et al., 2006a; Huck et al., 2017a; Voss et al., 2018). Most species of *Philander* are allopatric, such that only a single species is locally present, but several species have overlapping geographic distributions in western Amazonia, where as many as three can occur sympatrically (Voss et al., 2018). There is some suggestion from Amazonian trapping studies that sympatric species of *Philander* occur in different habitats, either segregated by vegetational succession (i.e., in secondary growth versus primary forest; Hice and Velazco, 2012) or by flooding regime (Patton et al., 2000).

Although these are large opossums by didelphid standards, species of *Philander* are substantially smaller than species of *Didelphis*, from which they also differ externally by lacking long guard hairs and white underfur, and in having a prominent white spot above each eye—the "four-eyed" phenotype that they share with *Metachirus* (Chapter 5). However, species of *Philander* have grayish or blackish dorsal fur, and adult females have a well-developed pouch, hence the common names that distinguish this genus from brownish-furred and pouchless species of *Metachirus*.

Trapping studies (Malcolm, 1991; Patton et al., 2000; Lambert et al., 2005), spool-and-line tracking (Cunha and Vieira, 2002), and direct observation of radio-collared individuals in the wild (Charles-Dominique et al., 1981) consistently indicate that species of *Philander* are primarily terrestrial opossums that often climb in understory vegetation but which never ascend to the canopy of primary forest. Corroborating evidence includes laboratory tests of climbing ability, which suggest that *Philander* is more agile on aboveground substrates than strictly terrestrial *Metachirus* but less so than fully arboreal taxa like *Marmosa* and *Caluromys* (Delciellos and Vieira, 2009a). Additionally, individuals released in capture-mark-recapture studies run away on the ground more frequently than they

*Fig. 4.12. Philander opossum* from French Guiana (*above*; photo by Antoine Baglan); *P. melanurus* attacking a coral snake in riparian forest near Matiguás, Matagalpa, Nicaragua (*below*; photo by Mario Gómez-Martínez)

choose arboreal escape routes (Fonseca and Kierulff, 1989; Fleck and Harder, 1995). Nests are at ground level (e.g., in cavities among tree buttresses) or in hollow trees, tangled lianas, open tree forks, and palm crowns from 8 to 10 m above the ground; some reported nests are spherical structures made of leaves, possibly by the animal itself (Miles et al., 1981; Lira et al., 2018). Many fieldworkers have remarked a preference for moist microhabitats, noting frequent captures or sightings of *Philander* along the edges of streams, rivers, or swamps (Davis, 1947; Hall and Dalquest, 1963; Handley, 1976; Fonseca and Kierulff, 1989; Julien-Laferrière, 1991).

In French Guiana, *Philander opossum* was observed to forage almost exclusively at ground level, searching for prey, fallen fruit, and other edible items on the ground or among the branches and trunks of fallen trees (Charles-Dominique et al., 1981). About 80% of examined stomach contents in this exemplary study consisted of animal prey, especially earthworms in the wet season, when large worms (up to 80 cm long) are abundant in the leaf litter after heavy rains.[6] Other prey recovered from stomachs in French Guiana by the same investigators included insects and arachnids, but in Central America *P. opossum* has been observed to prey on frogs (Kluge, 1981; Ryan et al., 1981; Tuttle et al., 1981) and venomous snakes (Gómez-Martínez et al., 2008). Fruits eaten by *P. opossum* in French Guiana are those most commonly found on the ground under fruiting trees, and this species is also known to feed on the nectar of root-parasitic terrestrial herbs (Charles-Dominique et al., 1981). Fecal analyses of Brazilian *P. quica* (*P. "frenata"* of authors) likewise indicate a diet that includes vertebrates, invertebrates, and fruit (Santori et al., 1997; Cáceres, 2004; Ceotto et al., 2009; Macedo et al., 2010). Frequency data from these studies suggest that *P. quica* eats vertebrates more frequently than do other sympatric opossums of similar size (e.g., *Metachirus myosuros* and *Didelphis aurita*). Identified vertebrate prey of *P. quica* includes small mammals, birds, lizards, and venomous snakes (Santori et al., 1997; Macedo et al., 2010), and this species is known to be resistant to snake venom (Perales et al., 1994).

## Tribe Thylamyini

Thylamyines include five genera of small mouse-like ("murine") opossums, all of which are endemic to South America. Many thylamyines superficially resemble *Marmosa*, with which they were formerly confused by taxonomists (e.g., Tate, 1933), but molecular phylogenetic analyses (Chapter 14) consistently recover marmosines and thylamyines as separate clades, and the same analyses provide strong support for a sister-group relationship of thylamyines with *Metachirus* and didelphines (*Chironectes*, *Lutreolina*, *Philander*, and *Didelphis*). Many thylamyines occur in lowland tropical rainforest, but this group has also radiated in open habitats (e.g., savannas, deserts, and thorn scrub), and some species occur at high elevations in the Andes and at high latitudes in Patagonia.

## Chacodelphys

*Chacodelphys* is a monotypic genus of tiny (ca. 10 g) opossums. *Chacodelphys formosa* is only known from two complete specimens, but numerous cranial fragments have been recovered from owl pellets. Both of the complete specimens and all of the owl pellets containing cranial fragments were collected in subtropical mosaics of grassland and dry forest in northern Argentina (Voss et al., 2004; Teta et al., 2006; Teta and Pardiñas, 2007), where one specimen is definitely known to have been trapped in tall grass (P. Teta, personal commun.). Remarkably mouse-like in external appearance, *Chacodelphys formosa* has a short and apparently nonprehensile tail, suggesting terrestrial habits, whereas its carnassialized dentition suggests insectivory (Voss et al., 2004). Unfortunately, nothing definite is known about its behavior or diet from direct observation, and neither stomach contents nor feces are available for analysis.

## Cryptonanus

*Cryptonanus* is a recently described genus that includes at least four species known from scattered collection localities from French Guiana (Fig. 4.13) to northern Argentina (Voss et al., 2005; D'Elía and Martínez, 2006; Baglan and Catzeflis, 2016). These are small (<50 g) pouchless opossums with long prehensile tails, superficially similar to *Gracilinanus* (see below) but differing from members of that genus in several aspects of craniodental morphology (Voss et al., 2005). Judging from vegetation maps, habitat information recorded on specimen tags or in field notes, and a few published observations, species of *Cryptonanus* live in savannas, marshes, dry forests, scrub formations, and other more or less open lowland habitats, but (apparently) never in closed-canopy rainforest (Voss et al., 2005). Published behavioral observations of *C. chacoensis* suggest that it is an agile climber that constructs aboveground nests in low vegetation (Massoia and Fornes, 1972), despite the fact that most specimens of this and several other congeneric species have been trapped on the ground, sometimes in almost treeless habitats. Another species, *C. guahybae*, inhabits edges and gaps in subtropical restinga forest, where it has been trapped both on the ground and in low trees (Quintela et al., 2011). Nothing appears to be known about the natural diet of any species of *Cryptonanus*, although *C. chacoensis* is said to eat both raw meat and fruit in captivity (Massoia and Fornes, 1972).

## Gracilinanus

The six currently recognized species of *Gracilinanus* (Fig. 4.14) collectively range from near the Panamanian border of Colombia southward throughout most of

*Fig. 4.13.* An undescribed species of *Cryptonanus* captured in coastal savanna near Sinnamary, French Guiana (photo by Antoine Baglan)

the forested landscapes of tropical and subtropical South America to northeastern Argentina; published collection localities range from near sea level to about 4000 m (Teta et al., 2007; Creighton and Gardner, 2008; Voss et al., 2009a). Although the genus was once thought to be absent in Amazonia (e.g., by Costa et al., 2003), several specimens have recently been collected there (Solari et al., 2006; Voss et al., 2009a), and it seems likely that at least one species (*G. emiliae*) is widespread in the region. Recorded habitats for the genus include lowland rainforest, lowland dry forest, and montane ("cloud") forest. Based on known distributions, most species seem to be allopatric, but *G. agilis* and *G. microtarsus* are known to occur sympatrically at several southeastern Brazilian localities (Geise and Astúa, 2009).

*Fig. 4.14. Gracilinanus emiliae* captured on the road to Petit Saut, south of Sinnamary, French Guiana (photo by Sebastien Barrioz, courtesy of François Catzeflis)

Like many other didelphids that were once classified in the genus *Marmosa*, species of *Gracilinanus* are small (ca. 10–50 g), pouchless, black-masked opossums with long prehensile tails. The results of trapping studies and inventory fieldwork (e.g., Handley, 1976; Voss et al., 2001, 2009a; Vieira and Monteiro-Filho, 2003; Smith et al., 2012) suggest that species of *Gracilinanus* are primarily arboreal, and laboratory analyses of locomotion (Delciellos et al., 2006; Delciellos and Vieira, 2009a, 2009b) indicate that *G. microtarsus* is an agile climber and leaper, especially among small-diameter substrates. Apparently, this species nests in trees (Pires et al., 2010b).

Dietary information for *Gracilinanus* is only available from fecal analyses of two Brazilian species, *G. agilis* and *G. microtarsus*. Invertebrates (primarily insects) were the most frequently encountered items in all studies that quantified both animal and plant components of analyzed feces (Martins and Bonato, 2004; Martins et al., 2006a; Bocchiglieri et al., 2010; Lessa and da Costa, 2010; Camargo et al., 2014; Lessa and Geise, 2014), but frequent evidence of frugivory was also found. In the most extensive analysis published to date (based on 422 fecal samples; Camargo et al., 2014), the frequency of fruit consumption by *G. agilis* was estimated to be about 86%, and the same study found remains of small vertebrates (feathers, bird bones, and eggshell fragments) in just 4% of examined samples.

## Lestodelphys

*Lestodelphys* is a monotypic genus (Fig. 4.15) that occurs in deserts and cold Patagonian grasslands in Argentina, approximately from 32° to 49° S. It is the only didelphid genus that does not occur at tropical latitudes, and it occurs farther south than any other living marsupial (Formoso et al., 2015). *Lestodelphys halli* is small (ca. 50–90 g) but stoutly built; by comparison with tropical forest opossums,

*Fig. 4.15. Lestodelphys halli*, a captive individual from Chubut province, Argentina (photos by Darío Podestá)

it is short-faced, with small eyes and small ears; the body is compact, with short limbs and tail. The tail is incrassate (carrot-shaped and seasonally swollen with stored fat), an unusual trait that—among other small mammals—is correlated with seasonally restricted food resources (Morton, 1980). Among didelphids, only *Lestodelphys* and its sister taxon *Thylamys* have incrassate tails.

*Lestodelphys halli* is a terrestrial species that might climb in low bushes, but which must spend most of its life on the ground because woody vegetation is sparse in habitats where it is known to occur (Chapter 8). Captive animals introduced to terraria with sandy substrates immediately begin to dig, and some captives build subterranean galleries lined with grass, suggesting that natural refugia are often underground (Martin and Udrizar Sauthier, 2011), possibly in tunnels of their own construction, but apparently also in burrows constructed by rodents (Pearson, 2008). Captives strongly prefer raw meat over insects or vegetable matter, and they competently attack, kill, and eat mice (Birney et al., 1996a; Martin and Udrizar Sauthier, 2011). Stomach contents of trapped specimens have been reported to contain both small-mammal fur and arthropods (Pearson, 2008; Martin and Udrizar Sauthier, 2011), but during the austral winter, when arthropods are inactive, *L. halli* must subsist for months at a time on endotherms, probably cricetid rodents, which are abundant in the habitats where it occurs and are sometimes found half-eaten in traps set by collectors (Birney et al., 1996b). Summer diets include invertebrates (especially scorpions and beetles), lizards, birds, and fruit (Zapata et al., 2013).

## *Marmosops*

*Marmosops* contains 19 currently recognized species (Fig. 4.16) that collectively range from Panama to southern Brazil; recorded habitats include lowland rainforest, dry forest, and montane ("cloud") forest from sea level to about 3000 m. Species of *Marmosops* are small (ca. 20–120 g) pouchless opossums with blackish masks and long prehensile tails. Although *Marmosops* and *Marmosa*—long confused by taxonomists—are superficially similar, the two genera are easily distinguished by several external and craniodental characters (Voss et al., 2004b), and they are not closely related.

Trapping studies suggest that species of *Marmosops* are active both on the forest floor and in understory vegetation, but they apparently never climb into the canopy or subcanopy of tall lowland forest (Malcolm, 1991; Patton et al., 2000; Voss et al., 2001; Vieira and Monteiro-Filho, 2003). Free-ranging individuals encountered at night are often seen clinging to small-diameter vertical stems or perching on horizontal branches within a meter or two of the ground (personal obs.), but

*Fig. 4.16. Marmosops parvidens*, a captive individual from French Guiana
(photo by Antoine Baglan)

quantitative data from spool-and-line tracking studies suggest that most locomotion is on the ground (Leiner and Silva, 2007a; Loretto and Vieira, 2008). Slightly more than half of 77 individuals of *M. noctivagus* released in a Peruvian live-trapping study chose a terrestrial escape route (Fleck and Harder, 1995). By contrast, in a Venezuelan live-trapping study, 93% of 64 released individuals of *M. carri* escaped on the ground rather than by climbing trees, and "even when placed in trees or on vines, individuals would climb down and then run along the ground" (O'Connell, 1979: 76). In the Atlantic Forest of southeastern Brazil, sympatric species of *Marmosops* may use different vertical strata (*M. paulensis* apparently being more terrestrial than *M. incanus*; Leiner et al., 2010), but in the Guiana region sympatric species seem to segregate by forest type (*M. pinheiroi* apparently favors swampy sites with waterlogged soil, whereas *M. parvidens* seems to prefer well-drained sites; Voss et al., 2001).

Dietary information about *Marmosops* is limited to the results of fecal analyses of just two Brazilian species, one studied at a dry-forest site (*M. incanus*; Lessa and da Costa, 2010) and the other in a premontane rainforest (*M. paulensis*; Leiner and Silva, 2007b). Both species were found to consume invertebrates (mostly insects) and fruit, although small vertebrates and flower parts were eaten occasionally by *M. paulensis*. Unfortunately, fecal analyses are biased (tending to overestimate the dietary importance of hard-shelled invertebrates and small-seeded fruits; Chapter 9). Nevertheless, fruits eaten by *Marmosops* species notably include those of understory treelets that produce small crops over a long period of time (only a few fruits ripening each night; Leiner and Silva, 2007a). The overall impression gained from available data about habitat use and diet is that species of *Marmosops* are semiarboreal denizens of the forest understory that move about on the ground but also climb in low vegetation to search for insects and small fruits.

## *Thylamys*

*Thylamys* is a genus of nine currently recognized species (Fig. 4.17) that occur from coastal Peru and central Brazil southward to central Patagonia—approximately from 9° to 46° S—in both lowland and montane landscapes from near sea level to about 4000 m (Giarla et al., 2010; Formoso et al., 2011). Most species occur in xeric or semi-xeric open habitats, including tropical tree savannas (Vieira and Palma, 1996; Cáceres et al., 2007), temperate thorn scrub (Meserve, 1981; Lima et al., 2001), high-altitude and high-latitude deserts (Ojeda and Tabeni, 2009; Braun et al., 2010), and Patagonian steppe (Birney et al., 1996b). Like other "murine" opossums (formerly placed in *Marmosa*), species of *Thylamys* lack pouches and have dark circumocular masks, large ears, and prominent eyes; however, they resemble *Lestodelphys* and differ from superficially similar species in other genera by having incrassate, carrot-shaped tails that are seasonally swollen with stored fat. Two species in the subgenus *Xerodelphys* have short tails that lack external modifications for prehension, and their hands and feet lack the well-developed friction pads that provide traction for climbing smooth vertical stems in other didelphids; by contrast, seven species in the subgenus *Thylamys* have longer tails with the usual prehensile surface near the tip, and their hands and hind feet have well-developed friction pads (Carmignotto and Montfort, 2006; Giarla et al., 2010).

*Fig. 4.17. Thylamys pallidior*, a captive individual from Bahía Cracker, Chubut, Argentina (photo by Darío Podestá)

Despite a substantial ecological literature for some species, surprisingly little basic natural history information is available for any member of the genus *Thylamys*. Most live in open habitats with widely spaced trees or shrubs, so much locomotion is necessarily terrestrial, but some species are known to climb, at least occasionally, in woody vegetation (Cáceres et al., 2007; Albanese et al., 2011). Nests of *T. elegans* are said to be constructed under rocks and in low trees or shrubs, where abandoned bird nests are also sometimes used as diurnal refugia (Mann, 1978). Anecdotal accounts of diets (unsupported by quantification or methodological details; e.g., Mann, 1978), as well as a few analyses of feces and stomach contents, suggest that species of *Thylamys* are mostly insectivorous, but fruits and small vertebrates are also said to be eaten on occasion (Meserve, 1981; Birney et al., 1996b; Vieira and Palma, 1996; Albanese et al., 2012). Captive individuals of *T. elegans* can be maintained on a pure insect diet, and when offered a choice between insects and fruit, they always choose insects (Sabat et al., 1995). However, the desert species *T. pallidior* is known to also eat leaves throughout the year (Albanese et al., 2012). Because leaves are absent or infrequent in the known diets of other opossums that inhabit less xeric environments than *T. pallidior*, it seems probable that this desert species eats leaves for their water content and not for any caloric value.

*Thylamys pallidior* is the only New World marsupial that does not appear to require free water, obtaining what it needs from what it eats and producing a highly concentrated urine when deprived of succulent vegetable matter (Díaz et al., 2001).[7]

## Discussion

The current classification of didelphids, together with salient natural history information, is summarized in Table 4.1. Altogether, 116 living species of opossums are now recognized as valid, whereas only 87 species were recognized by Wilson and Reeder (2005)—an increase of 33% in less than two decades. New species are still being discovered in the field on an almost yearly basis, so an accurate tally of living opossum species is still unavailable. The outcome of future fieldwork can confidently be predicted to increase the count, but the outcome of ongoing taxonomic research is less predictable. Although revisionary studies sometimes result in the discovery of new species, they can also result in new synonymies (e.g., Pavan et al., 2014). Overall, however, it seems likely that the number of opossum species will continue to increase, albeit perhaps more slowly, for at least another decade.

By contrast, generic-level diversity seems more likely to remain stable: only a single additional genus (*Cryptonanus*) has been described since 2005. Although it is often mooted that taxonomic categories above the species level are biologically arbitrary, taxon membership is constrained by monophyly, and groups of species are seldom named unless they share a substantial number of phenotypic traits that set them apart from other clades of the same rank. Didelphid genera, tribes, and subfamilies are nonarbitrary in this sense, although it is certainly true that there are no hard-and-fast criteria for deciding whether a clade should be ranked (for example) as a tribe or as a subfamily. In practice, however, it is often useful to refer to genera in discussing evolutionary and ecological phenomena. Besides being demonstrably monophyletic and morphologically diagnosable (Voss and Jansa, 2009), didelphid genera also tend to be ecobehaviorally distinctive. Members of the same genus (congeners) often occupy similar habitats or microhabitats, eat much the same food, and behave alike in other ways, whereas members of different genera are frequently divergent in size, habitat occupancy, diet, or behavior. Such similarities and differences suggest that ecological-niche occupancy often corresponds to generic membership, and that many of the phenotypic traits used by taxonomists to diagnose genera may have adaptive explanations.

Table 4.1  Summary of Didelphid Diversity

| | Size (g)[a] | Habitat(s)[b] | Activity[c] | Locomotion | Diet[d] |
|---|---|---|---|---|---|
| **CALUROMYINAE** | | | | | |
| *Caluromys* (3 spp.) | 190–500 | RF, DF | N | arboreal | Fr, I, N, SV |
| *Caluromysiops* (1 sp.) | 300–500? | RF | N | arboreal | Fr?, I?, N, SV? |
| **GLIRONIINAE** | | | | | |
| *Glironia* (1 sp.) | 100–200? | RF, DF | N | arboreal | ? |
| **HYLADELPHINAE** | | | | | |
| *Hyladelphys* (1 sp.) | 10–20 | RF | N | arboreal | ? |
| **DIDELPHINAE** | | | | | |
| Marmosini | | | | | |
| *Marmosa* (21 spp.) | 20–170 | RF, DF, MF, De | N | arboreal | I, Fr, SV |
| *Monodelphis* (24 spp.) | 15–150 | RF, DF, MF, Gr | N & D | terrestrial | I, SV |
| *Tlacuatzin* (5 spp.) | 30–70 | DF | N | scansorial? | I, SV, N |
| Metachirini | | | | | |
| *Metachirus* (2 spp.) | 250–500 | RF | N | terrestrial | I, Fr, SV |
| Didelphini | | | | | |
| *Chironectes* (1 sp.) | 500–800 | RF | N | semiaquatic | Fi/Cr |
| *Didelphis* (6 spp.) | 600–5000 | RF, DF, MF, TF | N | scansorial | SV, I, Fr |
| *Lutreolina* (2 spp.) | 300–800 | Gr, MF | N & D? | terrestrial | SV, I |
| *Philander* (8 spp.) | 300–700 | RF, DF | N | scansorial | SV, I, Fr |
| Thylamyini | | | | | |
| *Chacodelphys* (1 sp.) | 10–15? | Gr? | ? | terrestrial? | ? |
| *Cryptonanus* (4 spp.) | 15–40 | DF, Gr | N | scansorial | ? |
| *Gracilinanus* (7 spp.) | 10–50 | DF, RF, MF | N | arboreal | I, Fr |
| *Lestodelphys* (1 sp.) | 60–100? | De | N? | terrestrial | I/SV |
| *Marmosops* (19 spp.) | 20–140 | RF, MF, DF | N | scansorial | I, Fr, SV |
| *Thylamys* (9 spp.) | 10–65 | DF, De | N | scansorial | I, Fr |

[a] Observed range of adult size, rounded to the nearest 10 g.
[b] Abbreviations: De, desert; DF, dry forest; Gr, grassland; MF, montane forest; RF, lowland rainforest; TF, temperate forest.
[c] Abbreviations: D, diurnal; N, nocturnal.
[d] Foods are listed in order of presumed importance, separated by commas (diagonal slashes separate foods that might be equally important).
Abbreviations: Cr, crustaceans; Fi, fish; Fr, fruit; I, insects; N, nectar; SV, small (terrestrial) vertebrates.

NOTES

1. Tyndale-Biscoe and Renfree (1987) claimed that *Caluromys* is folivorous but cited no supporting evidence. As far as we are aware, there are no published observations to suggest that leaves are a normal part of woolly opossum diets.

2. Only three weights of fully adult specimens of *Glironia venusta* have been reported in the literature. Based on the variability seen in other taxa, it would be reasonable to expect adults of this species to weigh anywhere from 100 to 200 g.

3. We are not aware of any evidence to support the claim that *Metachirus* "builds dens at the base of trees, sometimes with multiple tunnel entrances in and around the roots" (Baker and Dickman, 2018: 39). Among other reasons to doubt this alleged behavior, the small-clawed forefeet of *Metachirus* are entirely unsuited to tunneling.

4. Baker and Dickman (2018: 36) claimed that water opossums supplement their animal diets "with fruits and semiaquatic plants," and that females "dig 60-cm-long tunnels in riverbanks," but no supporting citations were provided for either statement, nor are we aware of any evidence for such behaviors.

5. Several behaviors attributed to *Lutreolina crassicaudata* can be traced back to Lydekker (1894: 204), who quoted almost verbatim from W.H. Hudson's (1892) popular and highly romanticized natural history of the Pampas. According to Lydekker's paraphrased version of Hudson's account, *Lutreolina* is both terrestrial and aquatic in its habits and "dives and swims with ease . . . constructing a globular nest of grass suspended from the flags and rushes." Hudson did not trouble to explain how he made such observations, nor have these behaviors been verified by subsequent researchers. Nevertheless, Lydekker's text has been repeatedly cited in the technical literature about *Lutreolina*, and it is clearly the original source for unattributed statements about swimming ability and nest construction in influential semipopular accounts. Other implausible behaviors attributed to *Lutreolina*, such as Baker and Dickman's (2018) absurd claim that it uses its tail as a spring to launch itself "up to 60 cm" in attacks on human intruders, appear to be similarly untethered to factual accuracy.

6. Other studies, however, have estimated the diet of *Philander opossum* in French Guiana to consist of up to 50% fruit (Atramentowicz, 1988).

7. Remarkably, this species is one of the few mammals known to inhabit the Atacama, the oldest and driest desert in the world, where it has been found at sites receiving less than 1 mm of annual rainfall (Carmona and Rivadeneira, 2006).

# III

# OPOSSUM PHENOTYPES

# 5

# Morphology

Morphology is the most accessible aspect of the phenotype, and it is all that is usually preserved in the fossil record. Much more is known about opossum morphology, which is easily studied from museum specimens, than about opossum behavior and physiology, which require access to live animals for observation and experiment. On the assumption that morphology is shaped by natural selection to maximize the effectiveness of evolved behaviors (locomotion, for example) or physiological traits (sense perception, thermoregulation, reproduction), taxonomic variation in morphology is an important source of evidence for ecological adaptations. However, morphology can also be shaped by nonadaptive processes—such as sexual selection—and it is important to distinguish adaptive from nonadaptive trait variation in any evolutionary synthesis.

Rather than providing a comprehensive review of didelphid morphology, which would not serve the primary goals of our book, this chapter briefly describes the principal attributes that distinguish opossums from other mammals (especially other marsupials), corrects a variety of misconceptions about opossum morphology, and documents ecobehaviorally correlated trait variation that is useful for reconstructing opossum adaptive radiation. Among other organ systems neglected in this incomplete survey, we do not consider skin glands, the male and female reproductive tracts, or the respiratory, circulatory, and excretory systems. Additionally, descriptions of the main sensory organs are deferred to subsequent physiological accounts of mechanoreception, vision, hearing, and chemosensation (Chapter 6).

## Size and Size Dimorphism

Living opossums are small to medium-sized marsupials, ranging in adult weight from about 10 g in the smallest known species (e.g., *Chacodelphys formosa*; Voss et al., 2004a) to more than 3000 g in some populations of the largest species (*Didelphis virginiana*; Hamilton, 1958). Most didelphids, however, weigh between about

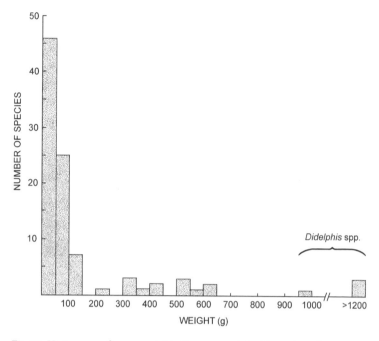

Fig. 5.1. Histogram of mean weights (in grams, g) for 95 species of opossums

20 and 500 g. Although histograms of species mean weights include several discontinuities (Fig. 5.1), most gaps are artifacts of averaging. The mean adult male weight of *Marmosa constantiae*, for example, is 103 g, but the observed range for this species (78–185 g; Voss et al., 2019) spans three size bins, including the gap between 150 and 200 g in our figure. Despite such intraspecific variation, there are noteworthy average size differences among certain clades. Members of the tribes Marmosini and Thylamyini, for example, are all small opossums (with mean adult weights <200 g), whereas members of the tribe Didelphini are consistently larger (with mean adult weights >300 g). The far-right end of the opossum size distribution is occupied exclusively by species of *Didelphis*—the largest living opossums, with mean adult weights >900 g—but these are not the largest opossums known to have ever lived. That distinction apparently belongs to the Pliocene species †*Thylophorops lorenzinii*, which may have weighed almost 9 kg (Zimicz, 2014).

Many didelphids exhibit obvious sexual size dimorphism, and in species with statistically significant sex differences, males are always larger, on average, than females (Table 5.1). In the most sexually dimorphic opossum for which reliable data are available (*Marmosa isthmica*), adult males can weigh almost 70% more

Table 5.1 Sexual Dimorphism in Adult Body Weights

| | $N_m/N_f$ | Males[a] | Females[a] | Dimorphism[b] | Source |
|---|---|---|---|---|---|
| *Caluromys lanatus* | 8/5 | 403 ± 81 g | 396 ± 32 g | 1.02 | Voss et al. (2019) |
| *Marmosa isthmica* | 35/26 | 99 ± 26 g | 59 ± 13 g | 1.68** | Rossi et al. (2010) |
| *Marmosa murina* | 177/160 | 51 ± 12 g | 38 ± 9 g | 1.34** | Rossi (2005) |
| *Marmosa rutteri* | 22/19 | 128 ± 19 g | 107 ± 18 g | 1.20** | J.L. Patton (unpublished) |
| *Monodelphis americana* | 59/36 | 37 ± 19 g | 25 ± 8 g | 1.48** | S.E. Pavan (unpublished) |
| *Monodelphis domestica* | 35/32 | 75 ± 26 g | 61 ± 21 g | 1.23* | S.E. Pavan (unpublished) |
| *Didelphis virginiana* | 20/20 | 2842 ± 869 g | 2004 ± 482 g | 1.42** | R.S. Voss (unpublished) |
| *Philander canus* | 21/23 | 340 ± 52 g | 299 ± 78 g | 1.14* | R.S. Voss (unpublished) |
| *Gracilinanus agilis* | 27/21 | 29 ± 5 g | 23 ± 6 g | 1.26** | Semedo et al. (2015) |
| *Marmosops incanus* | 127/82 | 66 ± 24 g | 42 ± 14 g | 1.57** | Mustrangi & Patton (1997)[c] |
| *Marmosops paulensis* | 25/19 | 43 ± 13 g | 28 ± 10 g | 1.54** | Mustrangi & Patton (1997)[c] |
| *Marmosops pinheiroi* | 12/11 | 27 ± 3 g | 22 ± 2 g | 1.23** | Díaz-Nieto & Voss (2016) |
| *Thylamys pallidior* | 22/19 | 24 ± 8 g | 24 ± 7 g | 1.00 | R.S. Voss (unpublished) |
| *Thylamys pusillus* | 16/15 | 22 ± 5 g | 21 ± 6 g | 1.05 | Voss et al. (2009b) |

[a] The sample mean plus or minus one standard deviation.
[b] Mean weight of males divided by mean weight of females.
* $p < 0.05$, ** $p < 0.001$.
[c] The "standard deviations" tabulated by Mustrangi and Patton (1997: Table 4) appear to be standard errors, so we multiplied each of their values by the square root of the corresponding sample size to obtain estimates of sample dispersion.

Asterisks indicate results of two-tailed $t$-tests for equality of male and female means:

than adult females, but sexual size dimorphism might be even more pronounced in certain other species (e.g., *Monodelphis dimidiata*; Baladrón et al., 2012). The smallest didelphids for which weight data are available (*Thylamys* spp.) are among the least sexually dimorphic in these comparisons, but large opossums are not consistently the most dimorphic, and analyses of osteological data (Astúa, 2010) suggest that opossums do not conform to Rensch's rule, which states that male-biased sexual size dimorphism is positively allometric (Abouheif and Fairbairn, 1997).[1]

Evidence from both laboratory and field studies (e.g., Barnes and Wolf, 1971; Cothran et al., 1985; Fernandes et al., 2010) suggests that the proximate cause of sexual size dimorphism in opossums is the faster postweaning growth rate of males (Chapter 6). Although sexual dimorphism might have adaptive explanations— e.g., reduced intersexual competition for food (Shine, 1989)—we are not aware of any compelling support for such hypotheses in mammals. Instead, male-biased sexual size dimorphism in mammals is significantly correlated with proxies for sexual selection (Lindenfors et al., 2007). Relevant information about opossum mating systems is sparse (Chapter 13), but in at least one sexually dimorphic species (*Didelphis virginiana*), males are known to compete directly with one another for access to estrous females, and male mating success is positively correlated with body weight (Ryser, 1992). Therefore, it seems reasonable to expect that taxonomic variation in opossum sexual size dimorphism is accompanied by corresponding variation in the strength of sexual selection for large males—a compelling topic for future research.

## External Morphology

Most opossums are superficially unremarkable mammals, with pointed muzzles, well-developed vibrissae, prominent eyes, membranous ears, soft pelage, subequal limbs, pentadactyl feet, and naked tails. In some or all of these respects, didelphids resemble other plesiomorphic marsupials (e.g., *Dromiciops* and dasyurids), as well as certain unspecialized placentals (e.g., solenodons, mouse lemurs, rice tenrecs, gymnures, and pen-tailed treeshrews). Indeed, it seems likely that many external morphological features of didelphids were inherited without substantial modification from the therian common ancestor and reflect ancient, widely shared mammalian adaptations.

### Body Pelage

All didelphids have soft (non-spinous) fur. Some arboreal taxa (e.g., *Caluromys*) have somewhat woollier fur than others, but texture is hard to define by objective criteria. Species of *Didelphis* appear shaggy (Fig. 4.10) due to the numerous long guard hairs that project far beyond the denser underfur, an odd trait for which no adaptive explanation has been suggested. Water opossum fur strongly resists wetting, resulting in a trapped layer of air when the animal is submerged, a phenomenon that is important for buoyancy in this species (Fish, 1993), as it also is in other small semiaquatic mammals (e.g., muskrats; Johansen, 1962).

The dorsal body pelage of most didelphids is uniformly colored and unpatterned, usually some dull shade of brown or gray. Among other small mammals, such drab coloration is usually attributed to selection for background-matching crypsis, and experimental studies with rodents support this hypothesis (e.g., Dice, 1947; Kaufman, 1974; Vignieri et al., 2010). However, theory suggests that the phenotypic outcome of selection for background-matching crypsis in any particular prey species depends on many factors, including the typical substrate on which it is found, the visual cues used by its predators, and the spectrum of ambient light in its microhabitat (Merilaita and Stevens, 2011). For arboreal didelphids, the relevant substrate is perhaps tree bark, whereas for terrestrial species it is probably soil or litter, and nocturnal opossums might more frequently be detected by predators with rod-dominated vision, whereas diurnal opossums might be subject to predation by species with cone-rich retinas. Careful measurements of ambient light in forests have discovered significant spectral differences from canopy to understory and from shade to gaps (Endler, 1993). Therefore, selection for background-matching camouflage could be expected to produce a range of dorsal pelage pigments, even among sympatric species, perhaps explaining how opossums with grayish fur and others with brownish or reddish fur can be cryptic in the same macrohabitat.

Among the few exceptions to the rule that opossum dorsal fur is drab and unpatterned are *Chironectes*—boldly banded dorsally with black and gray (Fig. 4.9)—and several colorful species of *Monodelphis*. The banded pattern seen in the water opossum does not, to our knowledge, resemble that of any other semiaquatic mammal, and it has no obvious explanation in terms of background-matching crypsis; possibly, it is an example of disruptive coloration (Caro, 2005), in which high-contrast color boundaries intersect the body outline, making it more difficult for predators to recognize. Several diurnal species of *Monodelphis* (including *M. americana, M. gardneri, M. iheringi,* and *M. scalops*; Pavan and Voss, 2016) have brownish dorsal fur that is longitudinally striped with black (Fig. 4.6), and clever experiments with plasticine models suggest that such striping reduces detectability by predators (Leone et al., 2019).[2]

Interestingly, two striped species of *Monodelphis* have evolved sexually dimorphic pelage. Adult male *M. americana* have completely reddish dorsal fur in which the striped pattern seen in females and juveniles is indistinct or absent, whereas adult male *M. scalops* develop a reddish head and rump separated by a grizzled-grayish midbody. In both *M. americana* and *M. scalops*, males are so divergently colored from females and juveniles that they were once considered different species (Pavan et al., 2014). Assuming that the striped phenotype of juveniles and females is optimally cryptic for the species, the brighter coloration of adult males in *M. americana* and *M. scalops* presumably incurs a fitness cost that must be overcompensated by some reproductive advantage, perhaps conspecific mate recognition mediated by female choice (Duda and Costa, 2015).

The hair bases of the dorsal fur are grayish in most opossums, but species of *Didelphis* uniquely exhibit white underfur (Fig. 4.10). Gray-based dorsal fur is widespread among nocturnal small mammals, but white underfur is highly unusual. The underfur of *Didelphis* spp. is prominently exposed when the animal is threatened and its long guard hairs are piloerected, resulting in a highly distinctive whitish appearance that might be aposematic, a possibility we consider elsewhere (Chapter 7).

The ventral pelage likewise exhibits noteworthy taxonomic variation. In some opossum species the ventral fur is partially or entirely gray-based because the individual hairs are grayish basally and abruptly paler (usually whitish, yellowish, orange, or brownish) distally; the overall color of the ventral fur then depends on the degree to which the dark basal pigmentation shows through the paler superficial coloration. In other species, the ventral fur is partially or entirely "self-colored," a traditional taxonomic descriptor that applies when the individual hairs have the same pigmentation (usually whitish or yellowish) from root to tip. Because differences in the patterning of gray-based versus self-colored ventral pelage are frequently conspicuous among closely related congeners that are otherwise difficult to distinguish by external characters (Patton et al., 2000: Fig. 41; Voss et al., 2018: Fig. 11), it seems likely that this character has often been under directional selection. Additionally, the ventral pelage of some species is characterized by bright fugitive pigments. The ventral fur of live *Monodelphis emiliae*, for example, is bright pink (Pavan and Voss, 2016: Fig. 14), but this color survives for just a few hours post mortem. Because the ventral pelage is exposed when threatened individuals stand on their hind legs during aggressive displays (Chapter 7), its pattern and coloration may serve some signaling function, perhaps for species recognition.

## Facial Markings

The fur surrounding the eye is not distinctively colored in *Caluromysiops, Lutreolina*, and *Monodelphis*, but most other didelphids have a more or less distinct circumocular mask of dark (usually blackish) fur that contrasts with the paler (usually brownish, whitish, or grayish) color of the crown and cheeks. In many dark-masked opossums (e.g., *Marmosa zeledoni*; Fig. 4.5), this marking extends anteriorly from the eye toward the bases of the mystacial vibrissae, and in a few species (e.g., *Glironia venusta*; Fig. 4.3) it also extends posteriorly to the base of the ear. The adaptive significance of dark masks, which are widespread in vertebrates, has been debated, and it seems unlikely that a common explanation applies to all of the taxa that exhibit such markings (Ficken et al., 1971; Ortolani, 1999; Caro, 2011). Among opossums, dark masks tend to be best developed in the smaller arboreal species, possibly to reduce light reflectance from the circumocular pelage in dark environments dappled with moonlight. However, in *Didelphis albiventris*, a large and aggressive savanna species (Fig. 4.10), bold black lines through the eyes are elements of a conspicuous black-and-white-striped facial pattern that has been interpreted as aposematic in ecobehaviorally similar carnivores (Newman et al., 2005).

Another distinctive facial marking is the pale spot above each eye that is consistently present in species of *Metachirus* and *Philander,* resulting in the "four-eyed" phenotype by which these animals are commonly known (Figs. 4.8, 4.12). Parsimony optimizations suggest that pale supraocular spots evolved independently in these genera (Voss and Jansa, 2009: Appendix 5), raising the question as to why this unusual trait evolved twice in opossums. Pale supraocular spots in didelphids are neither sexually dimorphic nor ontogenetically variable, so they seem unlikely to serve any social-signaling function. *Philander* and *Metachirus* are broadly sympatric, but if one taxon is mimicking the other to gain some fitness advantage, the nature of that advantage is unclear. To our knowledge, the adaptive significance of similar markings in other mammals (e.g., the pteropodid bat *Styloctenium wallacei*) is unknown, and we have no plausible hypothesis to offer here.

## Vibrissae

All didelphids are well-provided with vibrissae, long tactile hairs that are grouped into discrete tracts on the head and limbs (Lyne, 1959; Brown, 1971). The mystacial vibrissae (whiskers)—located just behind the nose on each side of the rostrum (Fig. 4.8)—are under voluntary muscular control and are used for active exploration of the physical environment. Associated behaviors include "whisking," in which these hairs are repetitively and rapidly swept back and forth, scanning for nearby physical objects, especially when the animal is navigating complex spaces in low-light conditions (Mitchinson et al., 2011).

Ecobehavioral correlations with vibrissal morphology have been noted at higher-taxonomic levels (e.g., when comparing marsupial orders or families; Lyne, 1959), and terrestrial opossums do seem to have somewhat shorter vibrissae than arboreal forms, but vibrissal contrasts are seldom striking in pairwise comparisons of similar-sized opossums with differing habits. For example, the mystacial vibrissae are only slightly shorter in terrestrial *Monodelphis domestica* (22 ± 2 mm) than they are in arboreal *Marmosa robinsoni* (27 ± 2 mm), and they are about the same length in semiaquatic *Chironectes minimus* as they are in scansorial *Didelphis marsupialis*. Although Marshall (1978b) alleged that the water opossum has "supernumerary facial bristles," only the usual mystacial, genal, supraorbital, and interramal vibrissae are present in the specimens we examined. Such conservatism suggests that vibrissal mechanoreception has not been under intense diversifying selection in the course of didelphid adaptive radiation. However, there is osteological and histological evidence that the mystacial vibrissae of *Chironectes* are more richly innervated—and, therefore, presumably more sensitive—than those of other didelphids (Sánchez-Villagra and Asher, 2002).

### Eyes and Ears

The eyes and external ears (pinnae) differ in size among opossums (Figs. 4.1–4.17), but this aspect of taxonomic variation has yet to be quantified or convincingly correlated with ecobehavioral traits. Sensory physiology (Chapter 6) suggests that nocturnal species should have larger eyes than diurnal species, and that arboreal species should have larger pinnae than terrestrial species, but measurements of eye diameter and pinna length have yet to be compiled across the family for relevant comparative analyses. The pinnae of many didelphids can be folded up (appearing crumpled) when the animal is alarmed or threatened; according to Barnes (1977), this folding is possible because the ear cartilage consists of many small pieces joined by striated muscle.

### Wrist

The wrists of males and females are externally similar in most didelphids, but striking sexual dimorphism is present in some small arboreal and scansorial taxa (Fig. 5.2). Prominent hairless tubercles supported internally by carpal ossifications are exhibited by adult males of *Cryptonanus*, *Gracilinanus*, *Marmosops*, *Tlacuatzin*,

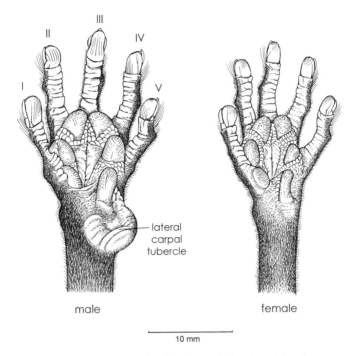

*Fig. 5.2.* Ventral view of left forefoot (manus) of male and female specimens of *Marmosops pinheiroi* illustrating external features of the plantar (volar) surface and the wrist. Digits are designated with Roman numerals from medial to lateral. In this sexually dimorphic species, males have larger hands than females, and they have a prominent carpal tubercle that is thought to be used for copulatory clasping.

and some species of *Marmosa*, whereas the wrist is externally featureless in conspecific females and immature males. Two distinct kinds of these sexually dimorphic tubercles can be distinguished: lateral ("ulnar") tubercles supported internally by the pisiform and medial ("radial") tubercles supported by the prepollex. Lunde and Schutt (1999) suggested that these unusual structures—apparently unknown among other mammals—function as clasping devices during copulation. By contrast, *Chironectes* has carpal tubercles that are neither sexually dimorphic nor ontogenetically variable (Fig. 5.4A, 5.4B). In this taxon, juveniles and adults of both sexes possess a fleshy process, supported internally by the pisiform, on the outside of the wrist. The function of this structure, which resembles a sixth digit, is unknown, but it perhaps assists the water opossum in detecting or securing fish or other aquatic prey.

## Hand

The didelphid forefoot (manus) is provided with five well-developed claw-bearing digits, and there is abundant behavioral evidence that opossums are capable of manual prehension (Chapter 7). Although didelphids are conservative in many aspects of manual morphology, they are not all alike, and some taxonomic differences can be correlated with microhabitat use and locomotor behavior. Arboreal opossums have relatively longer fingers than terrestrial taxa, an adaptation for grasping small-diameter supports that is more easily quantified osteologically than externally (see below), but arboreal and terrestrial species also differ in externally obvious digital proportions (Table 5.2). In terrestrial opossums—as well some scansorial species and the water opossum—the third manual digit is longer than the other fingers (Figs. 5.3A, 5.3B), proportions that correspond to the mesaxonic morphotype of comparative anatomists (e.g., Brown and Yalden, 1973; Patel

*Table 5.2*    Several External-Morphological Traits Associated with Locomotion

|  | Manus | Pes | Tail[a] | Caudal ratio[b] |
|---|---|---|---|---|
| *Caluromys* | dIV longest | dIV longest | prehensile | 1.53 (1.46–1.56) |
| *Caluromysiops* | dIV longest | dIV longest | prehensile | 1.10 |
| *Glironia* | dIII = dIV | dIV longest | prehensile | 1.17 |
| *Hyladelphys* | dIII = dIV | dIV longest | prehensile | 1.47 |
| *Marmosa* | dIII = dIV | dIV longest | prehensile | 1.40 (1.16–1.68) |
| *Monodelphis* | dIII longest | dIII longest | nonprehensile | 0.55 (0.46–0.65) |
| *Tlacuatzin* | dIII = dIV | dIV longest | prehensile | 1.23 (1.12–1.42) |
| *Metachirus* | dIII longest | dIV longest | nonprehensile | 1.24 (1.19–1.28) |
| *Chironectes* | dIII longest | webbed | nonprehensile | 1.32 |
| *Didelphis* | dIII longest | variable[c] | prehensile | 0.93 (0.77–1.00) |
| *Lutreolina* | dIII longest | dIII longest | nonprehensile | 1.00 (0.93–1.06) |
| *Philander* | dIII longest | dIV longest | prehensile | 1.05 (0.98–1.17) |
| *Chacodelphys* | dIII longest | dIV longest | nonprehensile | 0.81 |
| *Cryptonanus* | dIII = dIV | dIV longest | prehensile | 1.25 (1.21–1.36) |
| *Gracilinanus* | dIII = dIV | dIV longest | prehensile | 1.43 (1.31–1.64) |
| *Lestodelphys* | dIII longest | dIII longest | nonprehensile | 0.67 |
| *Marmosops* | dIII longest | dIV longest | prehensile | 1.34 (1.12–1.50) |
| *Thylamys (Thylamys)* | dIII longest | dIV longest | prehensile | 1.22 (1.12–1.41) |
| *Thylamys (Xerodelphys)* | dIII longest | dIV longest | nonprehensile | 0.78 (0.77–0.79) |

[a] "Prehensile" tails are those with external modifications for grasping (see text), whereas "nonprehensile" tails lack such modifications despite behavioral evidence for some grasping ability.
[b] Length of tail divided by length of head and body. Tabulated values are species means (for monotypic genera) or the mean-of-means and observed range of species means (for polytypic genera). Morphometric data from Voss and Jansa (2009) and the revisionary literature.
[c] Varies among species: dII = dIII = dIV (in *D. albiventris* and *D. virginiana*), or dIV longest (in *D. aurita* and *D. marsupialis*).

and Maiolino, 2016). In most arboreal species, by contrast, the third and fourth manual digits are subequal and longer than the other fingers, corresponding to the paraxonic morphotype (Fig. 5.3C); in opossums with a paraxonic manus, dV is also substantially longer than it is in species with mesaxonic forefeet. A third condition, the ectaxonic morphotype, is seen in *Caluromysiops* and *Caluromys* (Fig. 5.3D), whose fourth manual digit is slightly, but distinctly, the longest finger.

These taxonomic differences are easily understood in functional terms. Arboreal opossums use a zygodactylous grip, grasping small-diameter supports between the second and third digits (Youlatos, 2010; Antunes et al., 2016). In other words, digits I and II oppose digits III–V, and in species that support their weight on vertical stems with a friction grip, these sets of digits must subtend a sufficiently

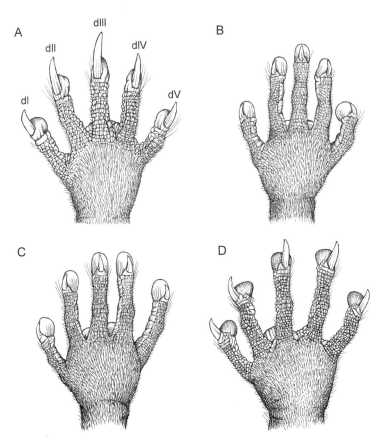

*Fig. 5.3.* Dorsal views of right forefoot: **A**, *Monodelphis brevicaudata*; **B**, *Marmosops incanus*; **C**, *Marmosa robinsoni*; **D**, *Caluromys philander*

wide arc for friction forces generated by muscular effort to resist the downward pull of gravity (Cartmill, 1974, 1985). Elongation of dIV and dV increases the arc subtended by the manus, thereby increasing the diameter of supports that can be effectively climbed by friction.

Taxonomic differences in manual claw size are also correlated with locomotor habits. Most scansorial and arboreal didelphids that climb in understory vegetation (e.g., on slender vertical stems, twigs, and vines) have small claws that do not extend much (if at all) beyond the large, fleshy apical pad of each manual digit (Figs. 5.3B, 5.3C), but some arboreal taxa known to frequent the canopy and subcanopy (where they must often negotiate trunks and limbs too wide to be climbed by friction alone) have claws that extend well beyond the fleshy apical pads (Fig. 5.3D). *Gironia venusta*, an arboreal species that exhibits squirrel-like agility on large-diameter vertical trunks (Chapter 4), has exceptionally large, laterally compressed, and strongly recurved claws. The longest claws in the family, however, belong to strictly terrestrial taxa such as *Monodelphis* (Fig. 5.3A) and *Lestodelphys*, whose documented behaviors include scratching or burrowing in soil. The vestigial manual claws of the water opossum (Fig. 5.4A) are embedded in the fleshy apical pad of each elongated finger.

Like many other plantigrade mammals (Brown and Yalden, 1973), most didelphids have 11 fleshy pads on the hairless ventral (plantar or volar) surface of the manus: two near the wrist, four among the bases of the digits, and one at the apex of each digit (Fig. 5.2). The thickened epidermis that covers these plantar pads is provided with dermatoglyphs (friction or papillary ridges resembling those on human fingertips; Hamrick, 2001). Plantar pads tend to be larger and to have more pronounced dermatoglyphs in arboreal opossums than in terrestrial species (Argot, 2001), but the water opossum uniquely lacks any trace of plantar pads or dermatoglyphs; instead, its hand is provided with integumentary structures unlike those of any other mammal.

*Chironectes* uses its hands to locate and grasp aquatic prey (Chapter 4), derived functions for which these appendages are uniquely specialized. In addition to the previously mentioned carpal tubercle and vestigial claws, special features of the water opossum manus include two kinds of papillae on its naked skin. First, the entire plantar surface (including the palm and the undersides of the digits) is densely covered with microscopically dentate papillae—the "Hornpapiller" (horn papillae) of Brinkmann (1911). These unusual structures have fleshy bases, but they are divided distally into multiple sharp, cornified tips, giving the skin a rough, sandpapery texture that is probably useful for securing slippery prey. Second, scattered among the horn papillae on the plantar surface, but also present on the

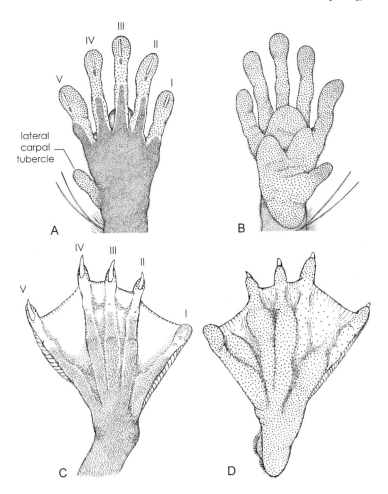

*Fig. 5.4.* Dorsal and ventral views of left forefoot (**A, B**) and hind foot
(**C, D**) of the water opossum, *Chironectes minimus.* Stipple indicates the
approximate distribution of Brinkmann's organs (see text).

naked dorsal surface of the distal phalanges and on the dorsal surface of the carpal
tubercle (Fig. 5.4), are numerous smoothly hemispherical papillae that are richly
innervated and appear to have a tactile function; their histological structure is unlike that of any other cutaneous sensory organ known in mammals, but it strikingly
resembles that of sensory papillae found on the tongues of ducks (Brinkmann, 1911).
Since these papillae lack a proper name and deserve future attention from sensory physiologists, we propose that they be called Brinkmann's organs after the
Danish morphologist who first described them in substantive detail.[3]

## Hind Foot

Most didelphids have a hind foot (pes) with five separate, well-developed digits, of which the fully opposable and clawless hallux (dI) is set off at a wide angle from the others (Fig. 5.5). The exception is *Chironectes*, whose pedal digits are loosely bound together by webs of skin to form a paddle-like swimming organ.[4] Remarkably, however, the hallux of web-footed *Chironectes* remains fully opposable, and this extraordinary animal is fully capable of pedal prehension. Although the hallux tends to be maximally developed in arboreal species, it is always long enough to contact the tips of the other digits when they are flexed, even in terrestrial taxa such as *Lestodelphys*, *Lutreolina*, and *Monodelphis*. The second digit bears an asymmetrical grooming claw with a rounded apex, but the remaining pedal digits are armed

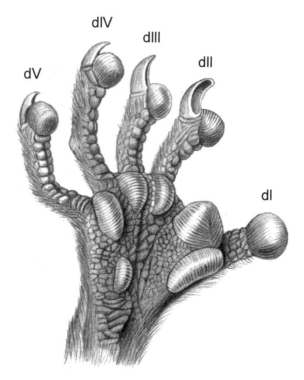

Fig. 5.5. Ventral view of right hind foot of *Gracilinanus marica* (after Boas, 1918)

with claws that are sharp, laterally compressed, and dorsoventrally recurved like those of most other plantigrade mammals (Brown and Yalden, 1973). In strictly terrestrial opossums (*Lestodelphys*, *Lutreolina*, and *Monodelphis*), the third pedal digit is longer than the adjacent second and fourth digits. By contrast, dII, dIII, and dIV are subequal (none distinctly longer than the others) in *Didelphis albiventris* and *D. virginiana*. Among all other didelphids (including *Didelphis marsupialis*), the second, third, and fourth pedal digits progressively increase in length, such that dIV is the longest toe opposing the hallux. It is noteworthy that the relative lengths of pedal digits do not strictly covary with those of the manual digits, as they might be expected to do if transformations of the pes and manus were functionally or developmentally determined by the same factors. For example, whereas dIV is the longest pedal digit in *Marmosops*, dIII is the longest manual digit in that genus.

The hind foot of most didelphids has dermatoglyph-bearing plantar pads resembling those previously described for the forefoot, and both the pads and their dermatoglyphs are better developed in arboreal than in terrestrial species (Argot, 2002). The conspicuous exception is *Chironectes*, whose webbed hind feet (Figs. 5.4C, 5.4D) lack any trace of plantar pads or dermatoglyphs. Instead, the plantar surface of the hind foot of the water opossum is covered by flat scales, among which are scattered Brinkmann's organs—the small, hemispherical sensory papillae previously described on the skin of the hands.

## *Pouch and Mammae*

Although possession of a pouch (marsupium) that contains and protects the nursing young is inextricably associated with the word "opossum" in the minds of many people, most didelphids lack a pouch. Pouchless opossums include species of *Cryptonanus*, *Glironia*, *Gracilinanus*, *Hyladelphys*, *Lestodelphys*, *Marmosa*, *Marmosops*, *Metachirus*, *Monodelphis*, *Thylamys*, and *Tlacuatzin*. In these genera the nursing young are not enclosed by any protective structure (Fig. 4.8). By contrast, well-developed pouches are consistently present in adult females of *Caluromys*, *Chironectes*, *Lutreolina*, *Didelphis*, and *Philander*. The presence or absence of a pouch is uncertain for two genera: *Caluromysiops* is said to have one (Izor and Pine, 1987), but no actual description of it has been published, whereas the external morphology of *Chacodelphys* is known from just a single male specimen.

The only known phenotypic trait that correlates with presence of a pouch in opossums is size: most pouchless species are small (with mean adult weights <200 g) and all pouched species are substantially larger (with mean adult weights >300 g).

Species of *Metachirus* (with adult weights of 250–500 g) are the only large opossums known to lack a pouch. Phylogenetic reconstructions (Voss and Jansa, 2009) suggest that a marsupium evolved independently at least twice, once in the common ancestor of Caluromyinae (including *Caluromys* and *Caluromysiops*) and again in the common ancestor of Didelphini (*Chironectes, Lutreolina, Philander,* and *Didelphis*), but taxonomic differences in pouch morphology do not correlate with phylogenetic relationships. For example, the marsupium of *Caluromys philander* uniquely consists of deep lateral skin folds that enclose the nursing young and open in the midline—resembling the morphology that Tyndale-Biscoe and Renfree (1987: Fig. 2.8) incorrectly attributed to didelphids in general—whereas in *Caluromys lanatus, Didelphis,* and *Philander* the lateral pockets are joined posteriorly, forming a more extensive enclosure that opens anteriorly. Yet another condition is seen in *Chironectes* and *Lutreolina*, in which the lateral pockets are connected anteriorly, forming a marsupium that opens posteriorly. In *Chironectes* the pouch is said to be closeable by muscular contractions, such that the nursing young are protected from drowning while the mother is submerged (Enders, 1937).

In all opossums that possess a pouch, the mammae are contained inside it, but the mammae of pouchless taxa are variously distributed. The taxonomic details need not concern us here (for a full account, see Voss and Jansa, 2009), but in species with only a few (up to about seven) mammae, the teats are confined to the same more or less inguinal position that the pouch would occupy if it were present. Species with larger numbers of mammae have additional teats positioned anteriorly, usually in pairs, along the upper abdomen and lower thorax. The total number of teats exhibits considerable taxonomic variation—from four (e.g., in *Hyladelphys kalinowskii*; Voss et al., 2001) to 27 (in *Monodelphis dimidiata*; Pavan and Voss, 2016)—suggesting a corresponding variation in litter size that remains to be convincingly correlated with latitude, habitat, and other life history traits (Chapter 13).

## Scrotum

Like many placental mammals, marsupials carry their testes in a pendulous scrotum, but in marsupials the scrotum is anterior to the penis, whereas the penis is anterior to the scrotum in most placentals. Opossum scrota are often sparsely furred, so the scrotal skin is usually exposed. The scrotal integument appears whitish or pinkish in some species, but in others it is bright blue, resembling the vividly colored scrotal skin of vervet monkeys (Cramer et al., 2013). Although this trait has not been comprehensively surveyed in the family, didelphids with blue scrota include species of *Caluromys, Hyladelphys, Gracilinanus, Marmosa, Marmosops*, and *Thylamys*. However, some members of some of these genera do not have blue scrota, and scrotal coloration may be ontogenetically variable among conspecific individuals (Barnes, 1977).

The blue coloration of opossum scrotal skin is structural (not pigmental) and results from the coherent scattering of reflected light by quasi-ordered arrays of parallel dermal collagen fibers (Prum and Torres, 2004). Mammalian structural colors are present only in opossums and cercopithecid primates, and they are thought to have evolved in the context of social signaling by taxa with trichromatic color vision (Prum and Torres, 2004). However, although this hypothesis makes sense for structurally colored primates (species of *Chlorocebus* and *Mandrillus*)—diurnal, highly social, and presumably trichromatic—it seems problematic for opossums, which are nocturnal, solitary, and probably lack trichromatic color vision (Chapter 6). Therefore, the blue scrota of opossums merit additional research.

## Tail

The mammalian tail is a multipurpose organ, frequently used for balance, but often modified in different lineages for grasping, propulsion, social displays, thermoregulation, defense, and other functions (Hickman, 1979). The tail of most didelphids is a slender, muscular organ, approximately round in cross section and tapering evenly from base to tip, but two thylamyine genera (*Thylamys* and *Lestodelphys*) have carrot-shaped, incrassate tails that are seasonally distended by stored fat (Chapter 6). Surprisingly, the tail of the water opossum (*Chironectes*) is not externally modified for swimming and lacks the specialized caudal features—such as lateral flattening or ventral keels of stiff hairs—seen in other semiaquatic mammals such as otter shrews (Tenrecidae: *Potamogale*), desmans (Talpidae: *Desmana, Galemys*), and muskrats (Cricetidae: *Ondatra*).

Most didelphids have a tail that is substantially longer than the combined length of the head and body, but some taxa are much shorter-tailed (Table 5.2). The known range of relative tail length in the family is bracketed on the one hand by *Gracilinanus emiliae* and *Marmosa lepida*—arboreal species with tails that are >60% longer than the combined length of the head and body, on average—and on the other by species of *Monodelphis*—terrestrial opossums with tails that can be less than half as long as the head and body. Although arboreal taxa are not always longer-tailed than terrestrial forms, the caudal ratio (tail length divided by head-and-body length) is correlated with arboreality. All very long-tailed opossums (with caudal ratios >1.50) are arboreal or scansorial, and all very short-tailed opossums (with caudal ratios <0.80) are probably terrestrial.

Body pelage (soft fur composed of ordinary coat hairs that are not associated with epidermal scales) extends to a variable extent onto didelphid tails. Body fur extends onto the dorsal surface of the tail all the way to the tip in *Caluromysiops* (Fig. 4.2) and *Glironia* (Fig. 4.3), and it extends onto the tail base for a centimeter or more in several other taxa (e.g., *Caluromys derbianus*; Fig. 4.1), but in most other opossums the tail appears macroscopically naked from base to tip. The exposed caudal skin in all didelphids is visibly scaly except on the ventral surface near the tip, where the epidermis is variously modified for prehension in arboreal and scansorial species.

All didelphids are probably capable of caudal prehension to some extent (e.g., using their tails to carry nesting material; Chapter 7), but external morphological features associated with grasping ability are variably developed in the family. The unfurred caudal surfaces of the water opossum (*Chironectes*), several strictly terrestrial taxa (*Lestodelphys*, *Lutreolina*, *Metachirus*, and *Monodelphis*), and two others with unknown locomotor behaviors (*Chacodelphys* and species in the subgenus *Xerodelphys* of *Thylamys*) are covered with unmodified scales from base to tip. By contrast, the ventral surface of the tail of all other opossums is provided with a distal prehensile surface that is transversely creased and scaleless, and the tip of the tail has a fleshy pad bearing dermatoglyphs. Some strictly arboreal taxa additionally exhibit basal modifications for grasping. Whereas the unfurred ventral surface of the tail base in most didelphids is covered by smooth scales, the scales on the underside of the base of the tail are heavily cornified, forming hard raised tubercles in *Glironia* (da Silva and Langguth, 1989: Fig. 1), *Caluromysiops*, and some species of *Caluromys*.

The exposed skin of didelphid tails is variously pigmented. Some species have uniformly colored tails that are usually grayish or brownish, but many species have tails that are distinctly paler distally. In most taxa with pale caudal markings, the basal color is grayish or brownish, and the transition is either gradual, the basal color fading to whitish distally (e.g., in *Metachirus*; Fig. 4.8), or irregularly mottled (e.g., as in *Caluromys*; Fig. 4.1). However, in most species of Didelphini (*Chironectes*, *Didelphis*, *Lutreolina*, and *Philander*), the tail is blackish basally with an abruptly white tip (Figs. 4.9, 4.10). The adaptive significance of contrastingly colored tail tips has been debated, and it seems probable that no general explanation (e.g., predator deflection or caudal luring; Ortolani, 1999) applies to all of the mammalian species that exhibit such markings. However, differences in tail-tip coloration among three sympatric species of *Philander* in southwestern Amazonia—where specimens of *P. pebas* have tails that are usually less than ¼ white, *P. canus* have tails that are about to ⅓ to ½ white, and *P. mcilhennyi* have tails that are usually >½ white (Voss et al., 2018)—suggests a possible role in species recognition.

## Skeleton

Because only mineralized tissues are normally preserved in the fossil record, vertebrate morphologists have lavished more attention on the skeleton than on any other organ system. Mammalian skeletal features of particular interest for evolutionary inference are teeth, cranial foramina, and limb bones, which are often informative about trophic, sensory, and locomotor adaptations, respectively. To interpret the adaptive significance of skeletal features of extinct mammals, the ecological and behavioral traits of Recent species with similar morphologies is key. In this context, opossum skeletal morphology has been studied for clues about the ecology and behavior of basal therians, stem metatherians, and early primates (e.g., Crompton and Hiiemae, 1969, 1970; Lemelin, 1999; Argot, 2001, 2002, 2003). However, conserved aspects of skeletal morphology among opossum species known to differ in adaptively important ecobehavioral traits suggest that there are also limits to what can be inferred from bones and teeth.

### Skull and Mandible

The opossum head skeleton (Fig. 5.6) exhibits many primitive therian features and closely resembles the inferred phenotype of the marsupial common ancestor. All of the usual cranial bones are present in much the same gross configuration as in basal therians, and the sutures that separate them are ontogenetically persistent. There is an essential sameness to opossum skulls because their basic architecture is similar. All opossums have the same number and kinds of teeth (see below), so the bony structures that support the dentition and surround the oral cavity are similar from genus to genus. Among other shared features, didelphids have a prominent rostrum enclosing a large nasal cavity that contains a well-developed array of turbinals (Rowe et al., 2005; Macrini, 2012). Likewise, all opossums have convergent orbits (Pilatti and Astúa, 2016), and the ear region is osteologically similar from genus to genus, despite minor differences of interest primarily to taxonomists (e.g., Reig et al., 1987). Opossum mandibles have all the usual mammalian processes for articulation and muscular insertion, and the symphysis is always unfused.

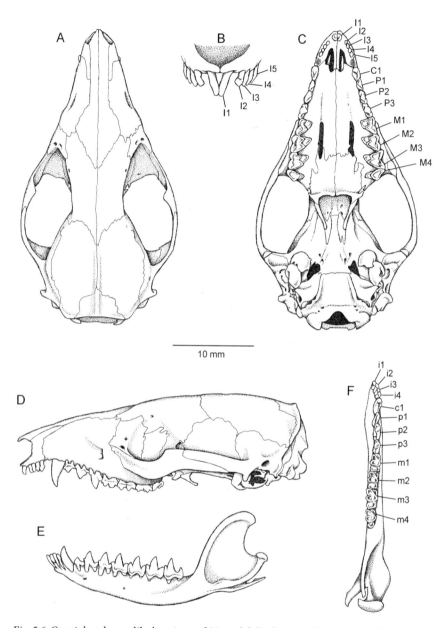

*Fig. 5.6.* Cranial and mandibular views of *Monodelphis brevicaudata*: **A**, dorsal view of skull; **B**, frontal view of upper incisor arcade; **C**, ventral view of skull; **D**, lateral view of skull; **E**, lateral view of mandible; **F**, dorsal view of mandible. Dental abbreviations are explained in the text

Opossum skulls exhibit taxonomic variation in size and shape (see illustrations in Voss and Jansa, 2009), but size and shape are not independent aspects of cranial form. The developmental correlation of size and shape—ontogenetic allometry—follows the same general pattern in didelphids that it does in other therian mammals (Flores et al., 2018). In general, bony structures that enclose precociously maturing soft tissues, such as the brain and the eyes, tend to scale with negative allometry, whereas bones that contain the dentition as well as the crests and processes associated with the masticatory musculature tend to be isometric or to scale with positive allometry. Therefore, younger (smaller) individuals have relatively larger braincases and orbits but smaller rostrums and less prominent muscular sculpturing than larger (older) individuals. Similar proportional differences can be observed when comparing the adult skulls of smaller versus larger species, a related phenomenon known as static allometry. Multivariate analyses suggest that size (including static-allometric shape) accounts for most of the interspecific variance in cranial measurements of adult opossums (Chemisquy et al., 2020; Jansa et al., unpublished).

Nevertheless, a few size-independent cranial differences—differences in shape that cannot be explained by allometry—hint at ecological adaptations. The short-faced, somewhat mustelid-like skull of *Lutreolina crassicaudata*, for example, is often noted in connection with that species' predatory habits (e.g., by Astúa de Moraes et al., 2000), whereas the enormous orbital fossae of *Glironia venusta* suggest a correspondingly prominent sensory role for vision. For some cranial trait variation there is supporting evidence from soft-tissue anatomy to support functional interpretation (Sánchez-Villagra and Asher, 2002), but the functional cranial morphology of didelphids has not received the attention that has been devoted to this topic in other mammals (e.g., Kay and Kirk, 2000; Crumpton and Thompson, 2013; Dumont et al., 2014), so the adaptive significance of most interspecific differences is obscure.

### Dentition

Like most heterodont mammals, opossums possess four morphologically and functionally distinct kinds of teeth, including (from anterior to posterior) incisors, canines, premolars, and molars. Like other marsupials, but unlike placental mammals, opossums have only two dental loci—the last premolar locus in the upper and lower jaws—with deciduous (milk) precursors. All adult didelphids normally have 50 teeth, of which there are five incisors (I1–I5), one canine (C1), three premolars (P1–P3), and four molars (M1–M4) in each upper jaw and four incisors (i1–i4), one canine (c1), three premolars (p1–p3), and four molars (m1–m4) in each lower jaw.

Opossum incisors are small teeth that are obviously used for mechanically undemanding tasks; they are conspicuously unsuited for forceful gnawing, gouging, stabbing, or any of the other vigorous dental activities reported for mammals with enlarged chisel-shaped or blade-like incisors (Ungar, 2010). *Didelphis virginiana* uses these teeth for scooping up soft food (Hiiemae and Crompton, 1971), and *Lestodelphys halli* is reported to use them for skinning mice (Martin and Udrizar Sauthier, 2011), but opossum incisors are presumably also used for nipping and grasping in many other contexts, including grooming and ectoparasite removal.

The upper incisor dentition is arrayed in a more or less parabolic arcade in front of the canines. These teeth consist of a styliform I1 that is always separated from I2–4 by a distinct gap, whereas the posterior incisors have buccolingually flattened, sharp-edged, and closely spaced crowns. The lower incisors are procumbent teeth, undifferentiated in crown morphology, with no gaps among them. Cineradiography (Bhullar et al., 2019) and manipulation of crania and mandibles suggest that the lower incisors occlude with I2–4 by simple edgewise contact. The first upper incisor (I1) does not occlude with any lower tooth, and its morphology does not suggest any obvious trophic function, but the much-worn crowns of older specimens suggest that it is often used for something, possibly grooming.

The canines of most opossums are unicuspid fangs like those of many other mammals, but the canines of small opossums sometimes have accessory cusps, and the lower canines of some small species are procumbent and premolariform (Voss and Jansa, 2009). The canines are the only teeth in opossum dentitions that are often sexually dimorphic, and upper-canine height is consistently the most dimorphic craniodental measurement in size-dimorphic species (Table 5.3). Although canine sexual dimorphism has yet to be systematically surveyed in the family, visual comparisons suggest that it is greatest in certain high-latitude species such as *Monodelphis dimidiata* (see Pine et al., 1985; Chemisquy and Prevosti, 2014) and *Didelphis virginiana* (Fig. 5.7), although not all opossums that live at high latitudes are sexually dimorphic.

Canines are obviously important trophic structures (opossums invariably kill their prey by biting; Chapter 7), but they may also function as advertisements of fighting ability. Canine weaponry is prominently exposed in the gape display that all opossums use when threatened by predators or rivals (Fig. 4.12), and this behavior—like the tooth-exposing snarls of carnivorans—doubtless conveys information about the owner's readiness and capacity to inflict fitness costs on

*Table 5.3*    Sexual Dimorphism in Selected Craniodental Dimensions
of *Didelphis virginiana*[a]

|  | Males | Females | Dimorphism[b] |
|---|---|---|---|
| Condylobasal length | 117.3 ± 6.9 mm | 106. 7 ± 5.3 mm | 1.10** |
| Least interorbital breadth | 24.2 ± 2.3 mm | 20.6 ± 1.4 mm | 1.17** |
| Zygomatic breadth | 63.6 ± 5.1 mm | 55.2 ± 3.7 mm | 1.15** |
| Palatal length | 70.2 ± 3.6 mm | 64.9 ± 3.4 mm | 1.08** |
| Palatal breadth | 34.1 ± 1.0 mm | 33.2 ± 1.2 mm | 1.03* |
| Length of upper molar series | 23.1 ± 1.2 mm | 22.4 ± 0.7 mm | 1.03* |
| Width of M3 | 6.8 ± 0.3 mm | 6.6 ± 0.3 mm | 1.03 |
| Height of upper canine | 19.3 ± 1.7 mm | 12.6 ± 1.2 mm | 1.54** |

[a] Based on measurements of 20 adult males and 20 adult females from Highlands County, Florida.
[b] Male mean divided by female mean. Asterisks indicate results of two-tailed Student's $t$-tests for equality of male and female means: * $p < 0.05$, ** $p < 0.001$.

opponents (Parker, 1974). Defense against predators is not a plausible explanation for didelphid canine dimorphism because predation probably affects males and females equally. However, males of some opossum species are known to compete with one another for reproductive access to females (Ryser, 1992), and larger canines might confer a decisive advantage in such contests with or without actual fighting, so sexual selection seems like a plausible explanation (Rico-Guevara and Hurme, 2019). Although it has been hypothesized that opossum canine dimorphism is an allometric byproduct of sexual selection for larger male body size (Chemisquy and Prevosti, 2014),[5] research on other mammals suggests that canine weaponry can be the primary target of sexual selection (Thorén et al., 2006), and we suspect that this may also be true of didelphid species with large-fanged males.

Unlike the multifunctional anterior dentition, the cheek teeth—premolars and molars—are used exclusively for food reduction. The permanent premolars of most opossums are laterally flattened blade-like teeth, each with a single main cusp that may or may not be flanked by small accessory cusps. This taxonomically widespread (and undoubtedly primitive) therian morphology has received little attention from functional morphologists, but descriptions of opossum feeding behavior and close inspection of photographs (e.g., Figs. 4.5 and 4.15) suggest that the primary function of these teeth is to cut large food items into pieces that are small enough for subsequent molar chewing. Their knife-like morphology seems suited for this purpose, and cineradiographic imaging has shown that the two posterior pairs of upper and lower teeth (P2/p2 and P3/p3) occlude edge-to-edge in a simple slicing stroke when the working-side jaws are closed (Bhullar et al., 2019).

A

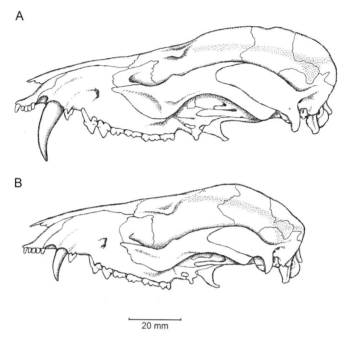

B

20 mm

*Fig. 5.7.* Skulls of male (**A**) and female (**B**) specimens of *Didelphis virginiana* (from Highlands County, Florida) illustrating sexual dimorphism in canine size

Premolar morphology exhibits modest taxonomic variation in the family (Voss and Jansa, 2009). The first upper and lower teeth (P1/p1) are always small and non-occluding, and p2 is usually the tallest lower premolar, but P2 and P3 vary in relative size from genus to genus. In opossums known to have mixed diets of invertebrates and fruit (e.g., species of *Gracilinanus*, *Marmosa*, and *Marmosops*; Chapter 9), P2 and P3 are subequal in size, but in predominantly frugivorous opossums (*Caluromys* spp.), P2 is larger than P3, whereas in all primarily faunivorous opossums (e.g., species of *Monodelphis*, Didelphini, *Thylamys*, and *Lestodelphys halli*), P3 is consistently larger. This makes functional sense because soft items like fruit are presumably easier to fit in the mouth and cut up with the more anterior tooth, whereas greater mechanical advantage—from a tooth closer to the jaw joint—is probably needed to subdivide more resistant items such as insect exoskeletons or vertebrate bones and ligaments.

Didelphid molars are good examples of the tribosphenic morphotype that characterized basal therians, but which is now retained by only a few other living taxa

(Ungar, 2010). When the jaws are closed, the principal cusps and interconnecting crests of tribosphenic molars form a series of interlocking triangles and basins. Such teeth effectively process a wide range of diets because their cusps, crests, and basins work together to puncture, cut, crush, and grind many different kinds of food.

The details by which food is reduced by opossum molars—first worked out by careful analysis of wear facets and later extended by cineradiography (Crompton and Hiiemae, 1970; Bhullar et al., 2019)—are beyond the scope of this book, but the essential features of tribosphenic functional morphology are easily understood. In the course of a single unilateral chewing sequence (chewing is done on one side of the mouth at a time), food is first punctured and pulped by the main cusps—without any tooth-to-tooth contact—as the jaws are repeatedly and forcefully brought together by muscular contractions. Once the bolus is broken down to the point where dental contact is possible, it is cut up into fragments as the lower- and upper-molar crests shear past one another, and it is simultaneously crushed and ground up in the basins of the upper and lower teeth.[6]

The initial cuspal pulping of food followed by cutting, crushing, and grinding between occluding dental surfaces are hallmark features of tribosphenic mastication, but these mechanical functions are not equally important for different kinds of food. The soft fruits known to be eaten by opossums (Chapter 9) require minimal processing: merely pulping such items is probably sufficient to burst fragile cell walls and release the easily digested sugary fluids they contain. Animal tissues, by contrast, are more biomechanically challenging. Insects are protected by an indigestible chitinous exoskeleton that must be broken open to expose soft tissues to digestive enzymes, and in vitro experiments have shown that protein digestibility increases with decreasing fragment size (Moore and Sanson, 1995). Dismembering vertebrate prey (cutting though tendons, ligaments, and bones) is the task of the premolar dentition (see above), but vertebrate skin, muscle, and viscera are tough, resistant tissues that must still be reduced by molar chewing to facilitate rapid digestion.

Consistent with such expectations, the molars of frugivorous opossums (e.g., *Caluromys*) have less occlusal relief (lower cusps with shorter crests) and wider basins than the molars of faunivorous taxa (e.g., *Lutreolina* and *Lestodelphys*), whose molars typically have taller cusps, longer crests, and narrower basins (Voss and Jansa, 2009). Additionally, more faunivorous taxa have relatively larger posterior molars than frugivorous taxa (Fig. 5.8), presumably because mechanical advantage is gained by chewing more resistant food with teeth closer to the jaw joint. Unfortunately, few of these functionally relevant features of opossum molar dentitions

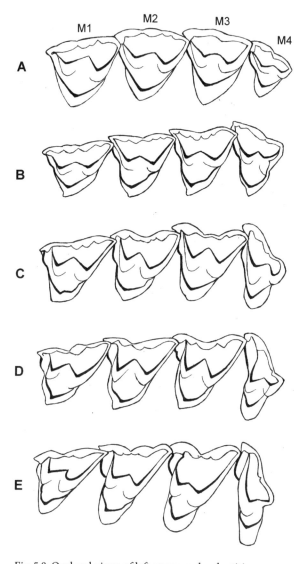

*Fig. 5.8.* Occlusal views of left upper molar dentitions:
**A**, *Caluromys philander*; **B**, *Glironia venusta*; **C**, *Marmosa rutteri*; **D**, *Marmosops soinii*; **E**, *Lestodelphys halli*. This sequence illustrates the progressive enlargement of transverse shearing crests (especially of the posteriormost teeth) from frugivorous *C. philander* to faunivorous *L. halli*.

are captured in two-dimensional projections of individual teeth, which appear to show a strong phylogenetic effect but no significant correlation with diet (Chemisquy et al., 2015). Although more sophisticated methods of tooth-shape quantification (e.g., Boyer, 2008; Santana et al., 2011) might provide more interpretable results, we agree with Chemisquy et al. (2015) that the functional versatility of tribosphenic molars and the mixed diets of most species plausibly explain the essential conservatism of opossum molar teeth.

*Axial Skeleton*

Numerous features of didelphid vertebrae and ribs are correlated with locomotor behavior (Argot, 2003), but only a few are noteworthy here. The thoracic and lumbar elements of arboreal opossums are clearly designed to increase the stability of the trunk, which must remain rigid (resisting sagittal bending) when the animal spans canopy gaps by bridging and cantilevering (Chapter 7). Most remarkably, the ribs of *Caluromys* (and *Caluromysiops*; Flores, 2009) are anteroposteriorly expanded, appearing very wide in lateral view.[7] By contrast, the axial morphology of terrestrial opossums such as *Metachirus* allows much greater sagittal flexibility, an

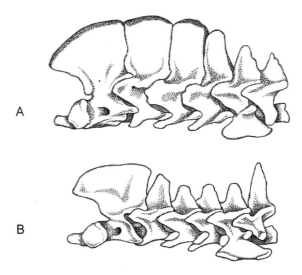

*Fig. 5.9.* Left lateral views of cervical vertebrae 2–7 of *Didelphis marsupialis* (**A**) and *Philander opossum* (**B**) illustrating the expanded and ankylosed neural processes in the former species, an adaptation that may prevent cervical dislocation by predators (see text)

attribute that is probably important for cursorial locomotion. Most scansorial didelphids exhibit intermediate phenotypes that are plausibly explained by locomotor behaviors that place less extreme mechanical demands on the axial skeleton.

A striking exception is *Didelphis*, whose unusually rigid vertebral column defies functional explanation in terms of locomotion (Argot, 2003). In particular, sagittal and lateral bending of the neck is prevented by the tight articulation of hypertrophied neural processes of the cervical vertebrae (Fig. 5.9). Immobilization of the cervical vertebrae, whether by actual fusion of adjacent bones or by tight articulation of hypertrophied processes, is an uncommon trait in mammals, for which various adaptive explanations have been proposed (VanBuren and Evans, 2017). Argot (2003) plausibly suggested that the vertebral reinforcements seen in *Didelphis* spp. are defensive, protecting these physically awkward and slow-moving opossums from predatory attacks, a hypothesis subsequently explored by Giannini et al. (2011).

## Appendicular Skeleton

Opossum girdle and limb elements exhibit numerous locomotor adaptations, some of which are common to all examined taxa, whereas others reflect functional divergence among genera with distinctive locomotor habits (Argot, 2001, 2002). Like other noncursorial small mammals, most didelphids have a crouching posture in which the proximal long bones (humeri and femora) are more frequently positioned horizontally than vertically, and limb movements are three-dimensional rather than parasagittal (Jenkins, 1971).[8] Didelphid appendicular elements are correspondingly well designed to permit a wide range of motion, which is obviously advantageous for negotiating the spatially complex environments that most species inhabit. Most remarkably, the didelphid upper ankle (cruroastragalar) joint allows full reversal of the hind foot, such that the animal can descend vertical surfaces head-first (Jenkins and McClearn, 1984).

Adaptive divergence among didelphids is reflected in appendicular proportions, joint morphology, and the configuration of bony processes for muscular origin and insertion (Hildebrand, 1961; Argot, 2001, 2002). Appendicular traits that can be quantified by measurement are of special interest for evolutionary analyses, but relevant data are currently available from just a few taxa. Among several morphometric outliers, *Chironectes* resembles semiaquatic placental mammals by having proximal limb bones (humeri and femora) that are short in proportion to more distal limb elements, whereas *Metachirus* converges on cursorial forms by having forelimbs that are short in proportion to its hindlimbs. The clearest

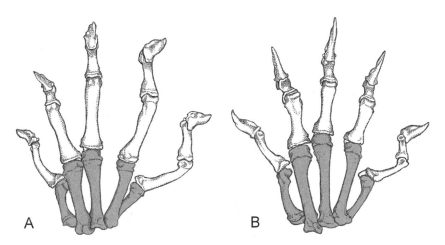

*Fig. 5.10.* Right manual skeletons of *Marmosa murina* (**A**) and *Monodelphis domestica* (**B**) illustrating the relatively short metacarpals (shaded) and long phalanges of *Mar. murina* (an arboreal species) by comparison with metacarpal/phalangeal proportions in *Mon. domestica* (a terrestrial species)

correlations with locomotor behavior, however, are provided by measurements of hand and foot bones.

The hand skeleton includes wrist elements (carpals), the long bones of the palm (metacarpals), and the finger bones (manual phalanges). The foot skeleton includes ankle elements (tarsals), the long bones of the sole (metatarsals), and the toe bones (pedal phalanges). As noted earlier, arboreal opossums have relatively longer fingers and toes than terrestrial opossums, and these proportions can be quantified by computing ratios between measurements of the phalanges and the metapodials with which they articulate. Such ratios (phalangeal indices) are positively rank-correlated with arboreality in both marsupials and primates and appear to reflect adaptations for manual and pedal prehensility in both groups (Lemelin, 1999; Weisbecker and Warton, 2006). Opossum sister taxa with divergent locomotor habits (e.g., *Marmosa* and *Monodelphis*) have correspondingly disparate phalangeal indices, and their osteological phenotypes are visibly distinct (Fig. 5.10).

## Viscera

The abdominal and thoracic viscera of opossums have received little attention from comparative morphologists, but most of the principal nonreproductive organs (heart, lungs, liver, kidneys, etc.) seem unremarkable in the few species that have been examined to date (e.g., *Marmosa robinsoni*; Barnes, 1977). The digestive tract, however, is notable for its simplicity: the stomach is simple and undivided, and the intestines are very short. According to Barnes (1977), the combined length of the small intestine, caecum, and colon was just 176 mm for a specimen of *Marmosa robinsoni* by comparison with 653 mm for a rodent of similar size (*Meriones unguiculatus*). This anatomical simplicity is consistent with opossum diets, which provide an abundance of easily digestible nutrients: protein from animal prey and simple carbohydrates from fruit (Chapter 9). By contrast, herbivores have more compartmentalized and much longer digestive tracts because complex carbohydrates in foliage and other plant structural tissues must be broken down by microbial fermentation, either in a multichambered stomach, an elaborate caecum, or a greatly expanded proximal colon. (All three anatomical arrangements for microbial fermentation are seen in herbivorous Australasian marsupials; Hume, 1999).

Modest taxonomic variation in the relative lengths of opossum gut segments hints at differences in nutritional physiology, but in the absence of accurate dietary information, specific trophic adaptations are hard to identify (Santori et al., 2004). The most that can be said at present is that no opossum is known to exhibit a digestive morphology compatible with herbivory. The only anatomical evidence that complex carbohydrates might be important in the diet of any taxon is the sacculated caecum and proximal colon of *Caluromys philander*, which differ strikingly from the tubular structure of these organs in other didelphids (Charles-Dominique et al., 1981; Santori et al., 2004). As remarked by Charles-Dominique et al. (1981), this might have something to do with gum consumption, because plant exudates contain polysaccharides that are resistant to mammalian digestive enzymes and must be broken down by microbial activity (Chapter 9).

## Central Nervous System

The central nervous system of *Didelphis virginiana* has been studied in considerable detail (Johnson, 1977), but little comparative information is available from other opossums. The neuroanatomical details are of no compelling interest here, but researchers have emphasized that few of the observed differences between marsupial and placental brains seem to have real functional consequences. The well-known absence of a corpus callosum in marsupials, for example, is compensated by alternative pathways of neocortical hemispheric connection (e.g., via an enlarged anterior commissure; Ashwell, 2010).

Of all the measures of central nervous system functionality that could be imagined, none, perhaps, is cruder than brain weight, and yet weighing brains (or, equivalently, measuring brain volume) has long been popular among comparative morphologists. Because, on average, larger animals have larger brains than smaller animals, brain-size comparisons are usually quantified in terms of encephalization: the residual for a species from the value predicted by a regression of brain weight on body weight. A considerable literature has long assumed a close relationship between encephalization and intelligence (Jerison, 1985), so the notion that marsupials are relatively smaller-brained than placental mammals (Jerison, 1973; Striedter, 2005) may have contributed to the widespread belief that marsupials are not very smart.

Unfortunately, early studies of mammalian encephalization included brain and body weights from just a few marsupials and were misleading on this point. The only marsupial represented in Jerison's (1973) influential dataset, for example, was *Didelphis virginiana*, which is, in fact, rather small-brained (Weisbecker and Goswami, 2014). Recent analyses that include data from a wider range of taxa suggest that marsupials and placentals are similarly encephalized, and that opossums, in particular, are not statistically distinguishable from nonprimate placentals in relative brain size (Weisbecker and Goswami, 2010, 2014). In effect, opossums and placentals seem to be equivalently endowed with neural tissue, despite considerable taxonomic variation about the central tendency in both groups.

In a much-cited article on opossum brain size, Eisenberg and Wilson (1981) claimed that arboreality and an associated "demographic strategy" of longer life span and smaller litter size are positively correlated with encephalization. Although subsequent analyses of larger marsupial datasets with more sophisticated methods (e.g., Weisbecker et al., 2015) have found no statistically significant correlation of encephalization with arboreality, they do confirm a significant negative correlation between encephalization and litter size. These results—and others reported by Weisbecker et al. (2015)—suggest that brain size is energetically constrained, but that encephalization seems to lack consistent behavioral correlates.

## Discussion

Didelphids resemble early therians in many morphological features, and they retain numerous traits that are probably ancestral for marsupials. No opossum is—or, insofar as known, ever was—very large; none is digitigrade or syndactylous; none has traits associated with fossorial locomotion, gliding, or bipedal hopping; none is diprotodont or has bilophodont or selenodont molars; and none has a digestive tract suitable for processing foliage or other plant structural tissues. The contrasting morphological features of many Australasian marsupials are correlated with major ecobehavioral transformations associated with an impressive adaptive radiation that occurred in the virtual absence of competing therian clades (Black et al., 2012). By contrast, taxonomic variation in morphology among living opossums is on a much more modest scale and reflects a correspondingly limited range of adaptations.

In general, adaptive morphological divergence among opossums is more apparent in traits correlated with locomotion than in traits correlated with other aspects of behavior or ecology. In particular, a large suite of morphological characters distinguishes arboreal taxa on the one hand from terrestrial taxa on the other, with scansorial taxa exhibiting intermediate morphologies. Arboreal opossums, of which *Caluromys* is the paradigmatic example, possess many locomotor traits classically associated with tetrapod climbing behavior (Cartmill, 1974, 1985), notably including strongly prehensile appendages (hands, feet, and tails) and trunk skeletons designed to resist sagittal bending. Although terrestrial opossums (such as *Monodelphis* and *Metachirus*) retain some ability for manual, pedal, and caudal prehension, they are demonstrably less endowed in these respects than arboreal taxa, and their trunk skeletons allow much more dorsoventral flexibility. The only opossum with a locomotor phenotype that substantially diverges from this spectrum of arboreal-to-terrestrial morphologies is *Chironectes*, which exhibits several adaptations (e.g., water-resistant pelage, webbed hind feet, and postcranial skeletal proportions) that are widely shared with semiaquatic placental mammals.

Adaptive variation among opossums is also apparent in morphological traits related to diet, among which dental features (especially premolar and molar proportions) are more readily interpretable than seemingly minor taxonomic differences in gut morphology. Opossums also differ in morphological features that are probably adaptations for avoiding detection by predators (e.g., concealing or disruptive dorsal pelage coloration), for surviving predatory attacks (fused cervical vertebrae in several species of *Didelphis*), for reproduction (presence/absence of a pouch, number of mammae), for energy storage (incrassate tails in *Lestodelphys* and *Thylamys*), and, possibly, for species recognition (ventral pelage coloration, tail markings). That we cannot plausibly identify many additional morphological adaptations is perhaps more an artifact of our ignorance than an indication that they do not exist. However, it will become clear in subsequent chapters that some adaptively important physiological traits (e.g., snake-venom resistance) and behaviors (ophiophagy) have no morphological correlates, such that morphology alone provides an incomplete record of opossum phenotypic evolution.

Lastly, there remain several taxonomically variable traits that are probably the result of sexual selection. This topic has previously received little attention in the didelphid literature, but information summarized in the present chapter suggests that sexual selection has shaped several aspects of opossum morphology including size dimorphism (widespread in the family, but not present in all taxa), dimorphic pelage coloration (in at least two species of *Monodelphis*), osseous structures of the wrist (in several arboreal or scansorial genera), and dental weaponry. In particular, sexually dimorphic opossum canine teeth fit the recognition criteria for intrasexually selected weapons (Rico-Guevara and Hurme, 2019) and merit comparisons with better-known and more thoroughly researched examples of dental weapons in other mammals (Plavcan and van Schaik, 1992; Thorén et al., 2006). Because canine height is easily measured from museum specimens, dimorphism in this dimension might serve as a useful proxy for sexual selection in future comparative studies.

NOTES

1. Note that weight is proportional to volume, which scales as the cube of linear dimensions, so sexual dimorphism in taxonomic measurements seems less impressive than sexual dimorphism in weight. For example, sexual dimorphism in adult head-and-body length of *Marmosa isthmica* is only 1.15 (males 15% larger than females), by comparison with 1.68 (males 68% larger) for sexual dimorphism in weight of the same species (data from Rossi et al., 2010).

2. Although it seems probable that longitudinal striping is an adaptation for concealment from predators, it is not clear how it should be classified among the several commonly recognized optical tricks by which this effect is achieved. Leone et al. (2019) refer to striping as an example of disruptive coloration, but Caro (2005) interpreted striping as pattern-blending crypsis. However, this is perhaps a distinction without a difference.

3. Hamrick (2001) described the horn papillae of *Chironectes*, but he did not mention Brinkmann's organs. Hamrick thought that the horn papillae might have a tactile function, but their heavily cornified tips and lack of special innervation is more consistent with a mechanical role in securing prey.

4. By contrast, digits II and III are tightly bound together by skin—a condition known as syndactyly—in many Australasian marsupials. Statements to the contrary notwithstanding, no didelphid is even incipiently syndactylous (Weisbecker and Nilsson, 2008; Voss and Jansa, 2009).

5. Unlike most teeth, which do not increase in size after the crown is fully erupted, the canines of didelphids (and other faunivorous marsupials; Aplin et al., 2010) are larger in older adults than in younger animals because the root is continuously extruded from the alveolus throughout life (Voss and Jansa, 2009).

6. For readers familiar with tribosphenic nomenclature, the principal shearing action in marsupial molars is accomplished when the postmetacristae of the upper molars occlude with the paracristids of the lower molars, and the crushing action occurs by a double mortar-and-pestle effect as the protocone of each upper molar is driven into the talonid basin of each occluding lower molar, and each lower-molar hypoconid is simultaneous driven into the trigon basin of each occluding

upper tooth. From this position (known as centric occlusion), the food contained in the talonid and trigon basins is then ground up as the lower jaw rotates on its long axis such that the inner (lingual) surface of the hypoconid is drawn across the outer (labial) surface of the protocone. This mandibular rotation corresponds to the so-called lingual phase of the masticatory cycle (Mills, 1967), which Crompton and Hiiemae (1969, 1970) claimed does not occur in opossums, but which has recently been observed using more sophisticated methods of cineradiographic analysis (Bhullar et al., 2019).

7. Anteroposteriorly expanded ribs increase thoracic rigidity by reducing the spaces occupied by intercostal musculature, an adaptation that is also seen in other arboreal mammals such as anteaters, slow-climbing primates, and some tree shrews (Granatosky et al., 2014).

8. This generalization is based on cineradiographic observations of *Didelphis virginiana*, but photographs (e.g., in Chapter 4) suggest that most opossums have similar postures. A probable exception is *Metachirus*, whose erect stance (Fig. 4.8) and limb movements more closely resemble those of cursorial mammals (L.H. Emmons, personal commun.; R.S.V., personal obs.).

# 6

# Physiology

Physiology is a broad field, and it is far beyond the scope of this book to review all of the physiological processes that have been investigated with didelphid models. Instead, we focus on topics that are relevant for understanding ecological adaptations, including some that have been alleged to constrain marsupial evolution in various ways. In addition, some early misconceptions about opossum physiology have been corrected by subsequent studies, and it is useful to summarize that research here.

## Metabolism, Energy Reserves, and Water Balance
### Metabolism and Thermoregulation

Measurements of basal metabolic rate (BMR, the mass-specific maintenance rate of energy use at thermoneutrality)[1] are about 25% to 30% lower, on average, for marsupials than they are for placental mammals of equivalent body size (MacMillen and Nelson, 1969; Dawson and Hulbert, 1970; Hayssen and Lacy, 1985), but there are several noteworthy and ecologically relevant exceptions. Xenarthrans, for example, have significantly lower BMRs than marsupials of equivalent size, whereas BMRs for marsupials are indistinguishable from those of similar-sized primates and bats (Hayssen and Lacy, 1985). Didelphids were originally thought to have somewhat higher BMRs than Australian marsupials of the same size (McNab, 1978), but subsequent studies suggest that this is not the case; in fact, didelphids seem to be metabolically typical marsupials (Cooper et al., 2009, 2010).

Although McNab (1978, 1986) proposed several ad hoc adaptive explanations for taxonomic variation in marsupial BMRs, some of the anomalous values that seemed to need explaining in his data were erroneous (Thompson, 1988), and recent analyses of larger marsupial datasets (with corrected values for some species) have not discovered many statistically significant correlations between BMRs

and ecological variables. For example, marsupial BMRs are not significantly correlated with diet, as was once believed. In fact, there is really very little residual
variation to explain because body weight accounts for >97% of the taxonomic variation in marsupial basal metabolic rates (Withers et al., 2006).

Below thermoneutrality many didelphids lower their body temperature ($T_b$) in
response to cooler ambient temperatures ($T_a$); such thermolability conserves energy that would otherwise be spent in maintaining a steeper gradient between $T_b$
and $T_a$ (Ribeiro and Bicudo, 2007; Cooper et al., 2009, 2010). Some small didelphids
also conserve energy by entering torpor, a spontaneously reversible hypometabolic
state accompanied by reductions of body temperature, heart rate, and ventilation
(breathing) rate. Although torpor has only been reported in the literature from
seven species in five genera (*Gracilinanus, Lestodelphys, Marmosa, Monodelphis,*
and *Thylamys*; Geiser and Martin, 2013; Riek and Geiser, 2014), it might be taxonomically widespread. Among those species studied to date, torpor reduces body
temperature by 6°C to 24°C, with correspondingly depressed metabolic rates that
range from 17% to 81% of normal BMR (Riek and Geiser, 2014). Torpor is entered
spontaneously by some species provided with ad libitum food and maintained at
relatively high ambient temperature, apparently as part of their normal circadian
cycle (Fig. 6.1), but other species are only reported to enter torpor when deprived
of food or when subjected to low ambient temperatures (Bozinovic et al., 2005;
Cooper et al., 2009; Busse et al., 2014). Because torpor has been reported from both
tropical species (e.g., *Marmosa robinsoni*; McNab, 1978) and temperate forms (e.g.,
*Lestodelphys halli*; Geiser and Martin, 2013), it is clearly not just an adaptation for
survival in cold climates. Instead, torpor seems to be an ecologically widespread
mechanism for energy conservation (or for reducing evaporative water loss; see
below) in small marsupials, and its essential physiological characteristics do not
seem to differ between didelphids and Australian taxa (Riek and Geiser, 2014).

Like other marsupials, opossums lack brown fat (Barnes, 1977; Hayward and
Lisson, 1992), a special kind of adipose tissue that is important for nonshivering
thermogenesis in small-bodied placental mammals (Oelkrug et al., 2015). Nevertheless, opossums (and other marsupials) can defend high body temperatures
against prolonged cold exposure by other mechanisms (Dawson and Olson, 1988),
including increased liver oxidative capacity (Villarin et al., 2003) and, probably,
muscular nonshivering thermogenesis (Nowack et al., 2017). Because other vertebrate endotherms lacking brown fat (e.g., birds) have radiated spectacularly, it
seems unlikely that the absence of this tissue in opossums and other marsupials
is a significant physiological handicap.

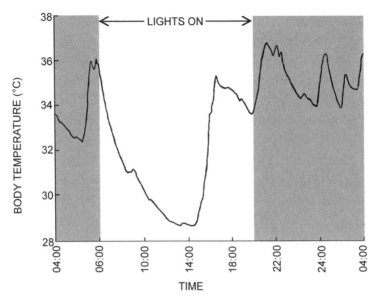

*Fig. 6.1.* Daily pattern of heterothermy in captive *Monodelphis domestica* (redrawn from Busse et al., 2014). The diurnal decrease in body temperature ($T_b$) often starts around the beginning of the light phase, and subsequent rewarming typically starts a few hours before the dark phase begins. Lowered light-phase $T_b$ in this nocturnal species is accompanied by spontaneous torpor, even at high ambient temperature and without food deprivation.

## Energy Reserves

Small mammals that live in seasonally hostile environments survive predictable episodes of resource scarcity by storing food, entering prolonged dormancy, and/ or by catabolizing energy reserves (ordinary white fat). Neither food storage nor seasonal dormancy (hibernation) has been reported for any didelphid species, and substantial fat reserves are only known to occur in a few taxa. Free-living tropical opossums are typically lean and lack substantial amounts of adipose tissue, but captive *Marmosa robinsoni* are said to accumulate subcutaneous fat (Barnes, 1977), and wild individuals presumably do so as well when food is abundant. The temperate North American opossum *Didelphis virginiana* was long prized by rural hunters for its succulent fat, which made for a toothsome (if greasy) meal on many a farmhouse table when other meat was scarce (Hartman, 1952). According to two unpublished theses cited by Harder and Fleck (1997), as much as one-third of the winter energy requirements of *D. virginiana* are met through catabolism of fat and muscle. The anatomical distribution of stored fat in *D. virginiana* has not been described in the literature, but large quantities of fat do not appear to be stored in the tail, which remains a slender, muscular organ even in visually obese individuals at the onset of winter.

By contrast, species of *Thylamys* and *Lestodelphys* have incrassate tails that are periodically distended with stored fat (Figs. 4.15, 4.17). Anecdotal observations that the caudal fat of captive animals is visibly diminished or replenished in just a few days of fasting or food abundance, respectively, suggest that this tissue is primarily useful as a short-term energy reserve (Martin and Udrizar Sauthier, 2011). Incrassate tails are known to occur in a wide range of other small mammals, including some Australian marsupials; according to Morton (1980), this trait is typically associated either with seasonal dormancy or with insectivory in arid habitats. Because no opossum is known to hibernate, the latter circumstance better fits the didelphid examples of caudal fat storage.[2]

*Water Balance*

Rainforest mammals have little need to conserve water under normal circumstances, but species that live in arid habitats must avoid dehydration by various mechanisms (Degen, 1997). Torpor, which reduces evaporative water loss by decreasing the ventilation rate, has been suggested as one means of water economy by arid-adapted mammals (Geiser and Brigham, 2012), including some didelphids (Busse et al., 2014), but a more frequently studied mechanism is renal-concentrating ability. To date, the urinary characteristics of only a few arid-adapted opossums have been studied, but the results suggest an impressive ability to conserve water by this means. Observed urine osmotic concentrations of *Monodelphis domestica* (which inhabits the semiarid Caatinga) are said to approach or exceed those of many Australian desert marsupials (Christian, 1983), whereas *Thylamys "pusillus"* (probably *T. pallidior*) of the Monte Desert is reported to produce an even more highly concentrated urine (Díaz et al., 2001). Few as they are, these studies suggest that at least some didelphids have evolved effective adaptations to arid conditions despite their rainforest ancestry.

# Reproduction, Growth, and Longevity
## *Reproduction*

The essential features of marsupial reproduction—short uterine embryogenesis, extremely altricial neonates, and prolonged lactation—have been admirably reviewed elsewhere (Tyndale-Biscoe and Renfree, 1987) and need not be described in detail here. Based on the few opossum species that have been reasonably well studied, didelphids do not seem to differ from Australian taxa in any of these key reproductive traits (Table 6.1). On average, the period of lactation (birth to weaning) is about five times as long as gestation. Among the taxa studied to date, *Caluromys philander* is noteworthy for its substantially longer gestation and longer interval from birth to weaning by comparison with members of the subfamily Didelphinae. Ovulation is spontaneous in all studied opossum species except *Monodelphis domestica*, in which pubescent females do not ovulate in the absence of males (Fadem, 1987). Because estrus is delayed by lactation, another litter is not conceived while the young from a previous litter are still nursing. The age of females at first estrus ranges from 127 days (in *M. domestica*) to about 280 days (in *Marmosa robinsoni*), but it is noteworthy that age of female puberty is unknown for any caluromyine.

The energetics of reproduction have only been investigated in detail for a single opossum species, *Monodelphis domestica*. Reproductively active female *M. domestica* assimilate about twice as much energy in the interval from conception to weaning (ca. 1260 Kcal) than nonreproductive females (about 590 Kcal) over an equivalent time period (Hsu et al., 1999). Because daily energy assimilation by female *M. domestica* during gestation and early lactation does not differ significantly from that of nonreproductive females, almost the entire energetic cost of reproduction is incurred late in lactation, especially in the weeks immediately before weaning, when energy assimilation rises sharply (Fig. 6.2). These results clearly indicate that late lactation is the most energy-demanding period for reproductive female opossums, and that it is likely to be physiologically stressful unless food is abundant in the habitat.

The same study (Hsu et al., 1999) compared the energetics of reproduction in *Monodelphis domestica* with the energetics of reproduction in a similar-sized placental mammal, the hamster *Mesocricetus auratus*. Maternal mass-specific assimilated energy (kilocalories per gram maternal body weight) was significantly greater in hamsters than in opossums during gestation, but not during lactation nor over the entire interval from conception to weaning. Although the energetic cost of reproduction (kilocalories per young per day) was higher in hamsters than in opossums, the reproductive interval (conception to weaning) is correspondingly shorter in hamsters. As estimated in this comparative study, opossums appear to produce young at lower cost per offspring, but over a longer reproductive interval, than do like-sized placental mammals.

Because young marsupials are continuously attached to the mammae (nipples) for the first few weeks of postnatal life, the number of mammae sets an absolute

Table 6.1   Key Reproductive Features from Captive Studies of Five Didelphid Species

| | Litter size[a] | Gestation | Lactation[b] | Sexual maturity[c] | References |
|---|---|---|---|---|---|
| Caluromys philander | 4 (F & C) | 24 days | 120 days | unknown | Perret & M'Barek (1991) Atramentowicz (1995) |
| Didelphis virginiana | 7–9 (F) | 13 days | 96–104 days | ca. 180 days | Reynolds (1952) |
| Marmosa robinsoni | 8–9 (C) | 14 days | 60–70 days | ca. 280 days | Barnes & Wolf (1971) Godfrey (1975) |
| Monodelphis domestica | 7 (C) | 15 days | 56 days | 127 days | Stonerook & Harder (1992) Harder & Jackson (1999) |
| Philander quica | 4 (C) | 13–14 days | 70–80 days | unknown[d] | Hingst et al. (1998) |

[a] Mean number of nursing young recorded from field (F) or captive (C) studies, rounded to the nearest integer; ranges indicate variation in mean values among field populations or laboratory generations.
[b] The interval from birth to weaning.
[c] Female age at first estrus, usually based on small sample sizes and therefore only approximate in most cases. When only age at birth of first litter was known, we subtracted gestation length to obtain age at first estrus.
[d] Not clear from information provided by authors.

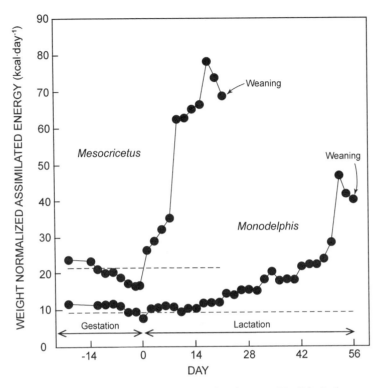

*Fig. 6.2.* Average weight-normalized assimilated energy (Kcal/day) of reproductively active short-tailed opossums (*Monodelphis domestica*) and hamsters (*Mesocricetus auratus*). Dashed lines indicate the average assimilated energy of nonreproductive conspecific females maintained under the same ambient conditions (redrawn from Hsu et al., 1999).

upper limit on litter size (Russell, 1982; Tyndale-Biscoe and Renfree, 1987). However, not all nipples are functional, and not all nursing young survive to weaning, so the number of weaned young per litter is often less than the number of nipples (Reynolds, 1952; Barnes and Barthold, 1969; Russell, 1982). Field estimates of litter size are usually based on numbers of attached young or (for females captured without attached offspring) on numbers of obviously used nipples. Mean litter sizes determined by such criteria vary more than threefold among species: from about four in *Caluromys philander* (in French Guiana; Julien-Laferrière and Atramentowicz, 1990) to about 14 in *Marmosa robinsoni* (in Venezuela; O'Connell, 1989). Statistical analyses suggest that opossum litters are, on average, smaller in species that live in wet tropical habitats and larger in species that live in seasonally dry habitats or at high latitudes (Battistella et al., 2019), but litter size can also vary among conspecific populations. Mean litter size varies geographically in widespread species of common opossums, apparently in response to length of the breeding season; populations with shorter breeding seasons have larger litters than populations with longer breeding seasons, presumably due to selection for higher per-litter reproductive effort when the probability of successfully raising multiple litters is lower (Rademaker and Cerqueira, 2006).

### Postweaning Growth

The only extensive growth data available from known-age captives of any didelphid species were obtained in the course of nutritional research with *Monodelphis domestica*. In this study (Cothran et al., 1985), growth was modeled by sigmoidal functions fitted separately to weight data for males and females from birth to 550 days. The results (Fig. 6.3) suggest that growth rates are substantially higher for males than for females, resulting in ever-increasing sexual size dimorphism with advancing age. Although the fitted growth curves seem to show accelerated male growth even before weaning, weanling weights for males (ca. $23 \pm 4$ g) and females (ca. $22 \pm 4$ g) are not significantly different. Therefore, sexual size dimorphism in this species appears to result almost entirely from the higher postweaning growth rate of males. Unfortunately, the authors did not plot the actual weights they obtained, so it is difficult to assess how well these functions fit their data. The growth curves for both sexes seem to reach asymptotic values (ca. 91 g for females and 136 g for males), but asymptotes estimated by curve-fitting can be artifactual (Zullinger et al., 1984). Cothran et al. (1985) noted that some wild-caught males are much larger (170–180 g) than any raised in their laboratory, and although there might be several alternative explanations for this, one possibility is that male *M. domestica* continue to grow indefinitely. Indeterminate growth has been suggested for some opossums (e.g., by Gardner, 1973), but other analyses seem to indicate that adults do not increase in size beyond a certain age (e.g., Tyndale-Biscoe and Mackenzie, 1976; Bergallo and Cerqueira, 1994). Because none of these studies appear to be conclusive, and because it is possible that growth may be asymptotic in some species but not in others, this topic merits further investigation.

### Longevity

Marsupials are not, in general, long-lived mammals (Austad and Fischer, 1991), and opossums are no exception. Field studies of several species indicate that wild populations experience almost complete annual turnover, and that few individuals

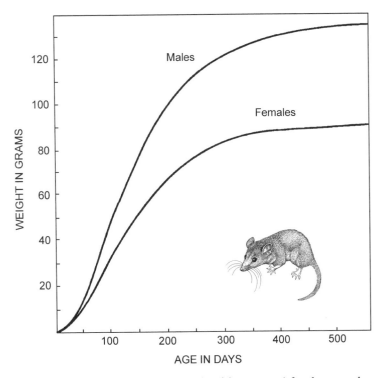

*Fig. 6.3.* Sigmoidal functions (von Bertalanffy's equations) fitted to growth data from captive-bred cohorts of male and female *Monodelphis domestica* (redrawn from Cothran et al., 1985)

survive for more than a single breeding season (Chapter 13). Captive animals, of course, can be expected to live longer than free-living conspecifics exposed to predators and parasites, but the interpretation of captive lifespans (Collins, 1973) is complicated by the fact that the maintenance requirements of most opossums are poorly understood.

In laboratory colonies of *Monodelphis domestica*, a species that thrives in captivity, natural death usually occurs between 36 and 42 months of age; the maximum recorded lifespan is 55 months, but evidence of senescence appears as early as 18 months, with the onset of reproductive decline in females (VandeBerg, 1989). An early onset of physiological senescence has also been reported for *Didelphis virginiana*—another species that is easy to maintain in captivity—in which third-year females (that is, females entering their second breeding season) exhibit markedly decreased fertility (Jurgelski and Porter, 1974). Other opossums that have been successfully maintained in captivity seem to have similar lifespans to those reported for *M. domestica* (e.g., "about three years" for *Philander quica*; Hingst et al., 1998), but caluromyines may be an exception. Some individuals of *Caluromysiops irrupta* captured in the wild are known to have lived for almost seven years in zoos (Izor and Pine, 1987). Because arboreality is correlated with increased longevity in statistical analyses of mammalian datasets (e.g., Shattuck and Williams, 2010), even such slender evidence of a similar correlation within Didelphidae is intriguing.

Theory predicts that senescence will occur earlier in species subject to higher extrinsic mortality than in species subject to lower extrinsic mortality (Williams, 1957). This expectation has been abundantly supported by laboratory and field research with a variety of animal taxa, notably including Austad's (1993) comparison of aging phenomena in wild populations of *Didelphis virginiana* exposed to different levels of natural predation (Chapter 11). Information about senescence in relation to predation in additional opossum species would be welcome, as would information about the demographic impact of parasitism (Chapter 10) on wild populations of both arboreal and terrestrial taxa.

## Sensory Physiology

The size of sensory organs tends to be positively correlated with their functional importance for the organism, so the large and well-developed eyes, pinnae, turbinal bones (supporting the olfactory epithelium), and vibrissae of opossums suggest that vision, hearing, smell, and touch are correspondingly acute. Indeed, direct observations of foraging behavior of free-ranging and captive animals have shown that opossums locate prey by sight, sound, and odor (McManus, 1970; Charles-Dominique et al., 1981; Tuttle et al., 1981; Rasmussen, 1990). All three senses are surely also important for predator avoidance, and vibrissal mechanoreception doubtless plays a key role in navigating dark, cluttered spaces. Although it seems indisputable that opossum sensory physiology is crucial for survival in these and other contexts, there are few relevant experimental studies, so sensory function must often be inferred from morphological or genomic clues.

### Vision

The opossum eye and its associated neural structures are comparable to those of placental mammals with excellent vision (e.g., felids and primates; Oswaldo-Cruz et al., 1979; Volchan et al., 2004; Lutz et al., 2018) and seem well designed to provide good nocturnal imaging ability. Optokinetic experiments with *Monodelphis domes-*

*tica* suggest that visual acuity in this species equals or exceeds that of laboratory rodents measured by the same method (Dooley et al., 2012), and larger opossums could be expected to have even more acute vision, as could like-sized species with larger eyes (Kiltie, 2000; Veilleux and Kirk, 2014). Like other nocturnal mammals (Peichl, 2005), opossums have strongly rod-dominated retinas. In *Didelphis virginiana*, the ratio of cone cells (useful primarily in strong light) to rod cells (which provide most visual input at low light levels) ranges from 1:50 at the center of the retina to 1:120 at the retinal periphery (Kolb and Wang, 1985); in *Thylamys elegans*, cones comprise just 0.4% to 1.2% of retinal photoreceptors (Palacios et al., 2010). A tapetum lucidum—the reflective retinal layer that serves to amplify photoreception in many nocturnal vertebrates—is present in some opossums (e.g., *Didelphis*) but apparently not in others (e.g., *Marmosa*; Walls, 1939; Oswaldo-Cruz et al., 1979).

Whether or not opossums have color vision is the topic of some controversy. Vertebrate color vision is based on cone photopigments encoded by opsin genes (Jacobs and Rowe, 2004; Jacobs, 2009, 2010), which can be classified as either short-wave-sensitive (SWS) or long-wave-sensitive (LWS). Although it is assumed that color vision is absent in taxa with only one expressed cone opsin gene, color vision is not necessarily present in species with two or more cone opsins, because color vision also requires neural mechanisms for comparing the output of cone photoreceptors with different spectral sensitivities; therefore, behavioral evidence of color discrimination is important. Most mammals have only two cone opsin photopigments (one each from the SWS1 and LWS gene families; Peichl, 2005) and are thought to be capable of only dichromatic color vision (equivalent to red-green color blindness in humans). Fully trichromatic mammalian color vision is known to be present only in some primates (by duplication of the LWS opsin gene) and a few Australian marsupials (in which the pigmental basis for color perception is still unknown).

Early behavioral experiments with *Didelphis virginiana* suggested that this species has trichromatic color vision (Friedman, 1967), but those experiments may have been inadequately controlled for confounding brightness cues (Jacobs and Williams, 2010; Gutierrez et al., 2011). Two cone opsins (SWS1 and LWS) have been found in didelphid retinas (Hunt et al., 2009; Palacios et al., 2010), suggesting that opossums are biochemically equipped for dichromatic color vision, but cones are so sparsely represented in opossum retinas that the importance of cone-based vision seems doubtful (Walls, 1942). Electroretinographic measurements of *Didelphis virginiana* suggest that this species has only a single functional class of cone photoreceptors with a peak spectral sensitivity corresponding to that of an LWS opsin (Jacobs and Williams, 2010). By contrast, a recent behavioral study suggests that *D. albiventris* has dichromatic color vision (Gutierrez et al., 2011), and electroretinography suggests that *Thylamys elegans* has two classes of cone photoreceptors with spectral peaks corresponding to SWS1 and LWS opsins (Palacios et al., 2010). In effect, there is evidence that some didelphids are functionally monochromatic (as are nocturnal procyonids and some nocturnal primates), whereas others might be dichromats.

## Hearing

Hearing is important for communication in some social species (e.g., humans, marmots, and meerkats) and for navigation by echolocating taxa (bats and cetaceans), but for most mammals the primary function of hearing is sound localization: the ability to quickly locate the source of unexpected sounds—such as those made by predators or prey—so that the eyes can be brought to bear for additional sensory input. Sound localization can be based on binaural cues (obtained by comparing the auditory input from left and right ears) or monaural cues (obtained from comparing auditory input from a single ear in different positions; Heffner and Heffner, 1992).

Binaural cues include the time difference between sounds impinging on the left and right ears. A function of the interaural distance, this time difference ($\Delta t$) is obviously greater, and therefore more reliably detected, for large species (with widely spaced ears) than for small species (with close-set ears). For sound originating at an azimuth of 90 degrees (directly to the right or left), $\Delta t$ is about 786 microseconds for a human (with an interaural distance of 180 mm) versus about 39 microseconds for a mouse (with an interaural distance of 9 mm; Heffner and Heffner, 1992). Another binaural cue is the difference in the frequency-intensity spectra ($\Delta fi$) incident on the two ears, which is caused by the sound shadow cast by the head itself. This spectral difference is frequency dependent, because high-frequency sounds are more attenuated by the head than low-frequency sounds, but $\Delta fi$ is also size dependent, because large heads cast more effective sound shadows than small heads. Therefore, all else being equal, large mammals are better at binaural sound localization than small mammals, but small mammals can do better at using binaural spectral cues if their ears are tuned to higher frequencies.

Monaural sound localization is largely dependent on the morphology and mobility of the pinnae. In effect, each pinna acts like an auditory funnel that selectively admits and sometimes amplifies high frequencies along its directional axis but attenuates high frequencies from other directions (low frequencies are much less affected by pinna position; Heffner and Heffner, 1992). Because species with small heads and close-set ears are relatively ineffective at binaural sound localization, monaural cues are expected to be correspondingly more important for small mammals. Additionally, monaural cues are crucial for three-dimensional sound localization because binaural cues are not effective at locating sounds in the vertical plane, which impinge simultaneously on both ears. Therefore, all else being equal, arboreal or scansorial species of small mammals would be expected to have larger pinnae than strictly terrestrial species.

Audiograms are only available for three didelphid species (Ravizza et al., 1969; Frost and Masterton, 1994), all of which exhibit a range of best sensitivity that extends well into the ultrasonic spectrum (above the upper limit of human hearing, ca. 20 kHz): about 4–32 kHz in *Didelphis virginiana* and about 8–64 kHz in *Monodelphis domestica* and *Thylamys elegans*. These results are consistent with the expectations of sound-localization theory, which predicts a greater reliance on higher frequencies in species with small heads and close-set ears for the reasons explained above. However, all three species are much less sensitive at all tested frequencies than most placental mammals. For example, the average lowest auditory threshold of the three opossum species is 24 decibels (dB), whereas that of *Rattus norvegicus* is 0.4 dB (Frost and Masterton, 1994).[3] Nevertheless, *D. virginiana* exhibits excellent sound-localization acuity by comparison with a wide range of tested placental exemplars (Heffner, 1997).

*Olfaction*

Mammals have several distinct chemosensory organs (Breer et al., 2006), but the detection of airborne odorant molecules is primarily accomplished by the main olfactory epithelium of the nasal cavity.[4] This specialized mucosa is supported by the turbinals, paper-thin scrolls of bone that are constantly ventilated by inspired air (Rowe et al., 2005). Airborne molecules—typically volatile organic compounds—are detected by millions of primary olfactory neurons with dendrites embedded in the main olfactory epithelium. The axons of these receptor cells pass through perforations in the cribriform plate at the back of the nasal cavity to synapse with secondary neurons in the olfactory bulb of the forebrain. Remarkably, each primary olfactory neuron expresses just one olfactory receptor (OR) gene and fires only in response to the odorant molecule or molecules that bind to the protein it encodes. Many hundreds of OR genes are present in most mammalian genomes, plausibly accounting for the very wide range of odors that humans and other tested species can distinguish (Firestein, 2001; Reed, 2004).

*Table 6.2*    Number of Functional Olfactory Receptor (OR) Genes in
12 Mammalian Genomes[a]

| Species (& coverage[b]) | Order | Family | OR genes |
|---|---|---|---|
| *Monodelphis domestica* (7.3×) | Didelphimorphia | Didelphidae | 1273 |
| *Dasypus novemcinctus* (2×) | Xenarthra | Dasypodidae | 1146 |
| *Echinops telfairi* (2×) | Afrosoricida | Tenrecidae | 585 |
| *Tupaia belangeri* (2×) | Scandentia | Tupaiidae | 985 |
| *Microcebus murinus* (1.9×) | Primates | Cheirogaleidae | 573 |
| *Homo sapiens* (10×) | Primates | Hominidae | 581 |
| *Erinaceus europaeus* (1.9×) | Erinaceomorpha | Erinaceidae | 295 |
| *Sorex araneus* (1.9×) | Soricomorpha | Soricidae | 1028 |
| *Myotis lucifugus* (1.7×) | Chiroptera | Vespertilionidae | 381 |
| *Canis familiaris* (7.6×) | Carnivora | Canidae | 883 |
| *Rattus norvegicus* (7×) | Rodentia | Muridae | 1420 |
| *Cavia porcellus* (1.9×) | Rodentia | Caviidae | 561 |

[a] From Hayden et al. (2010: Supplementary Information, Table 2).
[b] Average number of reads per nucleotide.

Among other aspects of olfactory performance that might be subject to natural selection, sensitivity and discriminatory ability seem likely to be paramount. Opossums have yet to be included in any experimental study of olfactory sensitivity (e.g., Wackermannová et al., 2016), and no morphological or genetic proxies for olfactory sensitivity are known. The size of a species' OR gene repertoire, however, is often assumed to be correlated with the organism's ability to discriminate odors. Data on the number of functional OR loci are available from genomic sequencing of several dozen mammalian species, among which *Monodelphis domestica* stands out as among the best endowed for olfactory discrimination (Table 6.2).

*Touch*

Mechanoreception is an anatomically dispersed function, but the most exquisitely sensitive mammalian touch receptors are the vibrissae: large sinus hairs typically found on the head and limbs. To oversimplify the function of a surprisingly versatile and taxonomically diversified structure, each vibrissa acts as a lever that transmits and multiplies the pressure caused by contact with objects in the environment to a bundle of sensory cells wrapped around its base (Patrizi and Munger, 1966; Loo and Halata, 1991; Brecht et al., 1997). Opossums are provided with a full complement of mammalian vibrissae (Lyne, 1959), of which the mystacial hairs—the large "whiskers" found on either side of the rostrum just behind the nose (Fig. 4.8)—are the most prominent and best studied. Mammalian mystacial vibrissae are richly innervated, with primary sensory neurons forming large bundles of afferent fibers that pass through the infraorbital foramen to synapse with secondary sensory neurons in the trigeminal ganglion.

Mystacial vibrissae are associated in some mammals with a characteristic behavior, whisking, that has been extensively studied in laboratory rodents. Whisking is caused by rapid contractions and relaxations of intrinsic and extrinsic mystacial muscles that cause the whiskers to be swept backward and forward in wide arcs that intercept any object within reach of the longest hairs (Mitchinson et al., 2011; Grant et al., 2013; Sofroniew and Svoboda, 2015). Whisking behavior in opossums seems to be essentially similar to that of laboratory rodents, suggesting a similar role for their mystacial vibrissae in actively sensing the spatial configuration of adjacent space during exploratory behavior (Mitchinson et al., 2011). Additionally, the mystacial vibrissae of *Monodelphis domestica* and laboratory rodents project to equally large areas of the primary somatosensory cortex, where stimuli from individual vibrissae are similarly processed (Ramamurthy and Krubitzer, 2016). These observations suggest that opossums and laboratory rodents are likely to obtain equivalently complex information from mystacial mechanoreception; in the latter, such information includes the distance, relative motion, size, shape, and surface texture of nearby objects (Sofroniew and Svoboda, 2015).

As discussed earlier, the sensory papillae found on the plantar skin of the hands and feet of *Chironectes minimus* (Brinkmann's organs; Chapter 5) also seem likely to be specialized mechanoreceptors. Although these remarkable structures are plausibly useful for underwater foraging, their sensory physiology remains completely unknown. Because *C. minimus* is neither difficult to capture (Voss et al., 2001) nor unusually challenging to maintain in captivity (Oliver, 1976), there is an obvious opportunity here for innovative sensory research on this ecobehaviorally unique marsupial.

## Toxin Resistance

By contrast with the topics discussed above, which concern physiological functions common to most mammals, toxin resistance is an unusual trait that only a few taxa are known to possess. Although the distinction is chemically arbitrary, biological toxins are commonly classified by delivery method: as venoms, if delivered by injection, or as poisons, if delivered by ingestion. Examples of toxin resistance in opossums include resistance to snake venoms and to the poisonous skin secretions of amphibians. Because both traits enable opossums to consume prey that would otherwise be unavailable, toxin-resistant phenotypes can validly be considered trophic adaptations.

## Snake-Venom Resistance

Snake-venom resistance is known to occur in some hedgehogs, some mongooses, a few mustelids, and members of the opossum tribe Didelphini. What these unrelated mammals have in common is ophiophagy, a diet that includes venomous snakes (Voss and Jansa, 2012). Among didelphids, snake-venom resistance and/or ophiophagy has been documented for species of *Didelphis*, *Lutreolina*, and *Philander* (Vellard, 1945, 1949; Werner and Vick, 1977; Sazima, 1992; Oliveira and Santori, 1999; Almeida-Santos et al., 2000; Rodrigues, 2005; Gómez-Martínez et al., 2008).

Snake venoms are complex mixtures of toxic molecules. Recent proteomic research suggests that most venoms contain several dozen toxic proteins and peptides that act synergistically to shut down vital physiological functions, such as breathing or hemostasis (Fox and Serrano, 2008; Calvete, 2010; Mackessy, 2010). Whereas neurotoxic venoms typically cause victims to suffocate by poisoning membrane function at neuromuscular synapses, hemotoxic venoms cause massive bleeding by attacking hemostatic proteins. Although some opossums are known to attack and eat snakes with neurotoxic venom (Gómez-Martínez et al., 2008), laboratory studies of venom resistance in didelphids have focused primarily on snake venoms with hemotoxic effects.

Two primary mechanisms of snake-venom resistance have been identified in opossums and other ophiophagous mammals: toxin-neutralizing serum factors and adaptive changes in venom-targeted substrate or ligand molecules. Toxin-neutralizing serum factors were the first to be discovered and characterized (e.g., by Vellard, 1945; Perales et al., 1986, 1994; Catanese and Kress, 1993; Jurgilas et al., 2003). These factors confer "transferable immunity," so called because serum from venom-resistant opossums affords some protection to experimentally envenomed laboratory species. Insofar as known, all toxin-neutralizing serum factors are non-enzymatic glycoproteins paralogous with human plasma alpha 1B-immunoglobulins (Neves-Ferreira et al., 2010). They are not antibodies; in fact, no antibodies to snake venoms are apparently produced by ophiophagous opossums, presumably because endogenous antitoxins clear venom antigens from the bloodstream too rapidly for an antibody-mediated immune response to occur.

Adaptive changes in venom-targeted substrate or ligand molecules are less biomedically interesting than toxin-neutralizing serum factors because the resulting immunity is nontransferable, so this mechanism of venom resistance has received correspondingly less attention. The best-known examples among placental mammals are amino-acid substitutions in a synaptic membrane protein that inhibit binding by snake-venom neurotoxins (Barchan et al., 1992; Drabeck et al., 2015). Another example, now understood in comparable detail, consists of amino-acid substitutions in opossum von Willebrand factor (vWF) that inhibit binding by C-type lectins found in pitviper venom (Jansa and Voss, 2011). A crucial blood protein, vWF initiates primary hemostasis (formation of a platelet plug), and the substitutions that inhibit binding with snake-venom toxins are present in the group of large opossums known to attack and eat pitvipers and to resist the toxic effects of pitviper venom (*Didelphis, Lutreolina,* and *Philander*; see above).

Although multiple mechanisms of snake-venom resistance are known to occur in opossums that eat venomous snakes (Voss and Jansa, 2012), the molecular basis of resistance has been determined for just a few toxins: metalloproteinases and A$_2$ phospholipases (both neutralized by serum factors) and C-type lectins (neutralized by adaptive changes in ligand molecules). Because snake venoms contain dozens of toxic compounds, venom resistance must be a correspondingly complex trait, of which we are just beginning to understand the molecular details.

*Resistance to Amphibian Toxins*

Opossum resistance to ingested amphibian poisons is less well documented than snake-venom resistance, but *Didelphis marsupialis* and *D. virginiana* are known to eat toads (Anura: Bufonidae), which are protected by bufadienolides (Hayes et al., 2009), and *D. virginiana* also eats newts (Urodela: Salamandridae), which are protected by tetrodotoxin and related compounds (Mebs et al., 2010). Evidence that common opossums consume such potently toxic prey with impunity comes from analyses of stomach contents (reviewed by Gardner, 1982), feeding experiments (Hart et al., 2019), and anecdotal but compelling field observations (Garrett and Boyer, 1993; Laurance and Laurance, 2007). Unfortunately, most studies of opossum food habits are based on analyses of feces (Chapter 9), which seldom preserve morphological evidence of amphibian consumption (Hamilton, 1958), so the dietary importance of poisonous amphibians is hard to assess from the literature.

Bufadienolides belong to a class of steroid compounds known as cardiac glycosides that produce their toxic effects by inhibiting the sodium-potassium pump (Na$^+$/K$^+$-ATPase) in cardiac muscle cells. Cardiac glycosides are produced as secondary metabolites by numerous plants (e.g., milkweeds) and by toads, but some herbivorous insects (e.g., milkweed bugs) and a few toad-eating vertebrates (including the hedgehog *Erinaceus europaeus*) have evolved convergent amino-acid substitutions in $Na^+/K^+$-ATPase that confer physiological resistance to these defensive toxins (Ujvari et al., 2015).

Tetrodotoxin, by contrast, produces its toxic effects by selectively binding with voltage-gated sodium ($Na_v$) channels in muscle and nerve membranes. Because no biosynthetic pathway for tetrodotoxin has yet been identified in metazoans, tetrodotoxin-defended vertebrates (e.g., newts; Yotsu-Yamashita et al., 2012) are thought to acquire toxicity either via symbiotic bacteria or through the food chain. Resistance to tetrodotoxin has evolved in several taxa—notably in newt-eating snakes—by amino-acid substitutions in muscular and neuronal $Na_v$ channels (Soong and Venkatesh, 2006).

Although the mechanisms by which common opossums resist the toxic effects of amphibian chemical defenses have yet to be investigated, repeated discoveries of molecular convergence among bufadienolide- and tetrodotoxin-resistant taxa (Soong and Venkatesh, 2006; Ujvari et al., 2015) suggest that similar amino-acid substitutions in $Na^+/K^+$-ATPase and $Na_v$ channels could be expected among toad- and newt-eating didelphids. Interestingly, research on other vertebrates that eat poisonous amphibians suggests that the evolution of toxin resistance incurs physiological costs, some of which have been demonstrated experimentally (Brodie and Brodie, 1999; Lee et al., 2011), whereas others have been inferred from phylogenetic reversals (Ujvari et al., 2015). Given the molecular targets of bufadienolides and tetrodotoxin, it seems reasonable to assume that the physiological costs of toxin-resistant molecular phenotypes include reduced cardiac and neuromuscular performance, which would be interesting physiological traits to compare in toxin-resistant and -nonresistant species of opossums.

## Discussion

Although it is often claimed that marsupials are evolutionarily and ecologically "constrained" by various aspects of their metabolic and reproductive physiology (e.g., McNab, 1978, 1986; Goin et al., 2016), it is not clear that any metabolic or reproductive trait has inhibited the adaptive radiation of opossums. To be sure, didelphids are maximally diverse in lowland tropical rainforests—uniquely permissive environments with stable temperature regimes, abundant rainfall, and diverse food resources—but opossums occur at higher latitudes (to 49° N and S), at higher elevations (to at least 4000 m in the Andes), and in drier habitats (with <1 mm of annual rainfall) than those occupied by placental clades of comparable crown age that are seldom described as physiologically "constrained" (e.g., platyrrhine primates and phyllostomid bats; Rojas et al., 2016; Silvestro et al., 2019). In effect, opossums as a group do not appear to be strikingly handicapped metabolically or reproductively by comparison with sympatric placental radiations.

Available information likewise suggests that opossums are provided with sensory abilities comparable to those of other nocturnal mammals, although possibly with somewhat less sensitive hearing and perhaps with better-than-average olfactory acuity. Unfortunately, most relevant observations derive from research on just a few species, which might not be typical of the family in all respects. It does seem safe to conclude, however, that opossums as a group are neither obviously deficient nor unusually endowed with respect to known aspects of visual, auditory, olfactory, or tactile sensation. However, future research on species with unique morphologies (e.g., *Chironectes*) is certain to enhance understanding of opossum sensory adaptations.

The only known physiological trait that would seem to confer an ecological advantage on some opossums is their ability to resist the toxic effects of snake venom and, apparently, amphibian poisons. Such phenotypes are surely advantageous, because they allow toxin-resistant species to exploit otherwise dangerous prey that might be key for survival during periodic episodes of resource scarcity (Chapter 9). These, however, are trophic adaptations possessed by a single clade of opossums and do nothing to explain the persistence of the group as a whole in competitor-rich Neotropical habitats. Thus, if there were some general explanation for the evolutionary success of didelphids, it would not appear to reside in any known aspect of their physiology.

NOTES

1. The "maintenance rate of energy use" implies that metabolic measurements are obtained from subjects that are inactive, nongrowing, nonreproductive, and postabsorptive; thus, there are abundant sources of experimental error. Metabolic rates are typically quantified in terms of oxygen consumption (ml $O_2$/hr), and for comparative purposes the mass-specific basal rate (ml $O_2$/g·hr) is the quantity of interest. Because metabolic rates scale with negative allometry, larger endotherms have, on average, lower (mass-specific) BMRs than smaller endotherms (Withers et al., 2016).

2. Contra McNab (1978), no tropical rainforest marsupial is known to store fat in its tail. All opossums with incrassate tails inhabit arid or semiarid habitats subject to seasonal drought and resource scarcity (Chapters 4 and 8).

3. The relative insensitivity of opossum hearing is plausibly attributed by Ravizza et al. (1969) to energy loss at the tympanum, which is attached at its circumference to a thin bony ring

(the tympanic annulus), whereas the tympanum in rodents and many other placental mammals is attached to a completely ossified auditory bulla.

4. Another olfactory structure, the vomeronasal organ, is crucially important for the detection of pheromones and other chemical signals that mediate reproduction and social behavior in mammals. This accessory chemosensory system has received much research attention with laboratory colonies of *Monodelphis domestica* (see Harder and Jackson, 2010), but like other aspects of the social phenotype (e.g., scent glands and scent-marking behavior), we do not treat it here.

# 7

# Behavior

Behavior is the principal interface between morphology and physiology on the one hand and the environment on the other, so the adaptive significance of morphological and physiological traits is often difficult to understand in the absence of behavioral context. To date, only a few didelphid species (e.g., *Caluromys philander*, *Didelphis virginiana*, and *Monodelphis domestica*) have been studied in any behavioral detail, but much of what we know about them is based on observations of captive individuals. Unfortunately, because captives are seldom exposed to the full range of stimuli encountered in their native habitat, some adaptively important behaviors might escape the attention of laboratory researchers.

Didelphid behavioral studies were last reviewed by Kimble (1997), who discussed several topics that are not considered here (e.g., vocalizations, scent marking, and learning). Instead, we focus on behaviors that are relevant for understanding taxonomic variation in morphological traits, or that contribute to our understanding of ecological adaptations. Additionally, we emphasize observations of behavior by unrestrained animals in their natural habitats, a category of information that has expanded rapidly in the recent literature.

## Diel Activity

Most didelphids are nocturnal (Fig. 7.1). Nocturnal activity has been reported for many species based on direct observations of free-ranging individuals in the wild (e.g., Enders, 1935; Charles-Dominique et al., 1981; Janson et al., 1981; Tuttle et al., 1981; Rasmussen, 1990; Ryser, 1992), on the results of trapping studies and automated photography (Vieira and Baumgarten, 1995; Oliveira-Santos et al., 2008; Norris et al., 2010; Gregory et al., 2014), and on telemetry of radio-collared animals (Sunquist et al., 1987; Ryser, 1995; Galliez et al., 2009). Nocturnal activity can also be inferred from the abundant remains of some species in owl pellets (Carmona and Rivadeneira, 2006; Motta, 2006; Teta et al., 2009). Observations of nocturnal

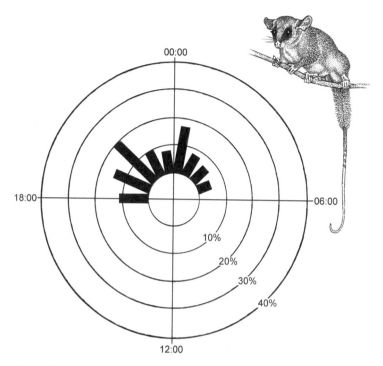

*Fig. 7.1.* Histogram of diel activity by free-ranging *Marmosa paraguayana* documented by automated photography in southeastern Brazil. Bars represent hourly frequencies of 90 time-stamped camera-trap photographs (redrawn from Oliveira-Santos et al., 2008)

activity by some opossum species has also been reported from captive studies (e.g., Park et al., 1940; McManus, 1971; Bucher and Fritz, 1977).[1] By contrast (and with exceptions as noted below), most didelphids are not normally active in the daytime, as indicated by the methods mentioned previously, by the absence of diurnal observations at sites where opossums are known to be abundant (e.g., Voss et al., 2001), and by diurnal nest occupancy (e.g., Loretto and Vieira, 2011).

More or less compelling evidence of nocturnality from such sources is available for members of most didelphid genera (Table 4.1), but relevant data are deficient or equivocal for *Lestodelphys*, *Lutreolina*, and *Chacodelphys*. Only captive observations of diel activity are available for *Lestodelphys*, which seems to be predominantly nocturnal but is sometimes active in the daytime (Martin and Udrizar Sauthier, 2011). Although *Lutreolina crassicaudata* is said to be nocturnal (Marshall, 1978a), we have found no explicit description of nocturnal activity by this mink-like species—which often inhabits dense grass that probably conceals it from visually oriented predators—and there is at least one report of diurnal activity in the wild (Rodrigues, 2005). Very little is known about the diel activity of *Chacodelphys*, whose nocturnal habits are suggested only by remains found in owl pellets (Teta et al., 2006; Teta and Pardiñas, 2007).

By contrast, several species of *Monodelphis* are known to be diurnally active (Chapter 4). These species are among the most colorful and boldly marked members of the family Didelphidae, suggesting that their pelage pigmentation (presumably adaptive either for concealment or species recognition; Chapter 5) has evolved under different light conditions than those encountered by other opossums. Because diurnally active species of *Monodelphis* include members of all five subgenera (Pavan and Voss, 2016), diurnal activity might be ancestral for the genus. In fact, the only species of *Monodelphis* definitely known to be nocturnal is *M. domestica*, whose relevant behaviors and metabolic rhythms are attested by both field observations (Streilein, 1982a) and laboratory experiments (e.g., Busse et al., 2014; Seelke et al., 2014). With such taxonomic variation, and with a high-coverage genome available (Samollow, 2008), species of this genus would seem to be ideal subjects for researching the genetic basis of diel activity patterns in mammals.

## Lunar Phobia

The activity of nocturnal mammals is often affected by bright moonlight, but lunar effects on mammalian behavior vary considerably from taxon to taxon (Prugh and Golden, 2014). The woolly opossum *Caluromys philander* exhibits classic lunar phobia, reducing its nightly activity when the moon is near full (Julien-Laferrière, 1997), but the activities of several other nocturnal opossums (e.g., *Didelphis marsupialis*, *Metachirus nudicaudatus*; Pratas-Santiago et al., 2016) seem unaffected by lunar phase, and at least one species (*Didelphis albiventris*; Huck et al., 2017b) is said to be more active in bright moonlight than when the moon is dark. Risk of detection by visually oriented predators is often suggested as an explanation for lunar phobia, and it seems plausible that *C. philander* is more vulnerable to owls (Charles-Dominique et al., 1981) as it forages for fruit in the brightly moonlit canopy by comparison with terrestrial and scansorial opossums, which might be less vulnerable to avian predation while foraging in the heavily shaded and densely cluttered forest understory. Additional information about the effects of moonlight on opossum behavior would be useful for evaluating this and other hypotheses (besides predation risk) that might account for species differences (Prugh and Golden, 2014).

## Locomotion

Like other tetrapods, opossums use a variety of locomotor behaviors to traverse different substrates. Although distinct suites of behavioral traits seem to be associated with arboreal, terrestrial, and semiaquatic locomotion, only a few species have been studied in this context, so generalizations are still elusive. Nonetheless, some opossum behaviors associated with locomotion on different substrates strikingly resemble those previously reported among other mammals with convergent habits, lending credibility to adaptive interpretations.

Species of *Caluromys* are often active on small-diameter substrates in the forest canopy and subcanopy (Charles-Dominique et al., 1981; Rasmussen, 1990). Locomotion on small-diameter arboreal substrates is a biomechanically challenging activity where falling is an omnipresent hazard and gaps among adjacent supports of unknown reliability must often be negotiated. Field and laboratory studies of locomotion by *C. derbianus* and *C. philander* (e.g., Rasmussen, 1990; Schmitt and Lemelin, 2002; Lemelin et al., 2003; Lemelin and Schmitt, 2007; Youlatos, 2008, 2010; Dalloz et al., 2012) have reported numerous primate-like behaviors in these species, including diagonal-sequence walking (placement of a hind foot is followed by placement of the contralateral forefoot), powerful hallucal grasping with the hind feet, and zygodactylous grasping with the forefeet. Other characteristic arboreal behaviors of *Caluromys* described by these studies include below-branch quadrupedal locomotion, foot-hanging (a below-branch suspensory posture with only the hind feet supporting the body), tail-hanging (a below-branch suspensory posture with only the tail supporting the body), bridging (spanning gaps with at least three feet anchored), and cantilevering (spanning gaps with only the hind feet anchored).

Locomotor behavior in *Monodelphis domestica*, a terrestrial species, has only been studied in the laboratory, primarily for the purpose of comparisons with *Caluromys* (Lemelin et al., 2003; Lemelin and Schmitt, 2007). Not surprisingly, *M. domestica* is slow and unsteady on small-diameter substrates (dowels and branches), often slipping or falling from supports that *C. philander* negotiates with ease. None of the primate-like arboreal behaviors reported for *Caluromys* (see above) seem to have been observed for *M. domestica*, which (among other differences) primarily uses lateral-sequence rather than diagonal-sequence gaits.

Studies of gait mechanics provide a good example of how biomechanical analyses can support adaptive explanations for observed behavioral differences. Theoretically, the lateral-sequence walking gaits favored by *Monodelphis domestica* and most other terrestrial mammals improve stability on level substrates by max-imizing the support provided by large tripods that enclose the animal's center of gravity (Hildebrand, 1980). By contrast, the diagonal-sequence walking gaits pre-ferred by *Caluromys philander* and many primates appear to enhance stability dur-ing arboreal locomotion by ensuring that a grasping hind foot is placed on a tested support below the animal's center of gravity just before the contralateral fore-foot strikes down on an untested support (Cartmill et al., 2002). Interestingly, *Didelphis virginiana*—a scansorial species—uses both lateral- and diagonal-sequence walking gaits (White, 1990).

Jumping or leaping has seldom been reported in field studies of free-ranging opossums, but it must often be necessary in order for arboreal animals to cross gaps in canopy or subcanopy vegetation that are too wide to be bridged or spanned by cantilevering.[2] Consistent with this expectation, arboreal opossums (e.g., *Caluromys philander*, *Gracilinanus microtarsus*, and *Marmosa paraguayana*) can be induced to jump longer distances between elevated supports than can scansorial species (*Didelphis marsupialis* and *Philander quica*), and at least one terrestrial spe-cies (*Metachirus myosuros*) cannot be induced to jump at all (Delciellos and Vie-ira, 2009b).

Like many other mammals, terrestrial and scansorial opossums (e.g., *Lutreo-lina* and *Didelphis*) swim competently by quadrupedal paddling, using both lat-eral- and diagonal-sequence strokes of the forelimbs and hindlimbs (Fish, 1993; Santori et al., 2005). By contrast, the water opossum *Chironectes* swims with alter-nate strokes of its webbed hind feet only, as do many semiaquatic placental mam-mals (e.g., desmans [*Desmana*], beavers [*Castor*], and muskrats [*Ondatra*]). The forefeet of *Chironectes* are not used for swimming but are held outstretched under the chin, with the digits unflexed; apparently, the tail is not used for rectilinear swimming (Mondolfi and Padilla, 1957; Fish, 1993).

Digging is perhaps not appropriately considered a mode of locomotion (except for fossorial taxa), but it seems relevant to remark in this section that most opos-sums do not dig at all. Indeed, the small-clawed forefeet of many species (Figs. 5.3B, 5.3C) are obviously unsuited for digging, especially in root-filled forest soils. Spe-cies of *Monodelphis* (Fig. 5.3A) are much longer-clawed and have been reported to scratch around in loose soil and leaf litter, but they appear to dig infrequently (Pheasey et al., 2018). By contrast, desert-dwelling *Lestodelphys*, whose long claws resemble those of *Monodelphis*, is said to dig burrows in captive enclosures and has been found occupying burrows in the wild.

## Feeding

Many mammals use only the mouth for food procurement, but didelphids often use their prehensile hands to secure and manipulate edible objects (Figs. 4.5, 4.15). Potentially dangerous prey—such as scorpions—are sometimes pinned to the ground with the forefeet before being bitten (Streilein, 1982a; Pheasey et al., 2018), but in other cases biting seems to precede forefoot involvement (Oliveira and Santori, 1999; Almeida-Santos et al., 2000), and intraspecific variation in premasticatory forefoot use suggests that this behavior may depend on prey size and prey defenses; fuzzy caterpillars, for example, are often scratched or rolled with the forefeet to remove irritating hairs before being eaten (González and Claramunt, 2000). Both terrestrial opossums (e.g., *Monodelphis domestica*; Streilein, 1982a) and arboreal species (e.g., *Caluromys derbianus*; Rasmussen, 1990) have been observed to snatch flying insects from the air with their forefeet. As previously noted, the water opossum (*Chironectes*) uses its hands to search for and grasp underwater prey prior to oral processing (Mondolfi and Padilla, 1957; Rosenthal, 1975; Oliver, 1976). Most descriptions of didelphid feeding behavior (including those cited above and many others) describe the use of the hands to transfer food to the mouth or to position large items in the mouth for biting.

Some aspects of feeding behavior are relevant for dietary inferences based on indirect evidence, such as fecal analysis. Small opossums, for example, preferentially consume the head, thorax, and abdomen of large insects but often discard the legs and wings (Enders, 1935; Pheasey, et al., 2018). Because legs and wings can be important for prey identification, the importance of large insects in the diets of small opossums might be consistently underestimated from scat samples. Similarly, mice are partially or completely skinned before they are eaten by small opossums (González and Claramunt, 2000; Martin and Udrizar Sauthier, 2011), so taxonomically diagnostic hairs might be absent or infrequent in scat. Detailed observations about fruit eating are scarce, but some opossums are reported to swallow only the pulp of large-seeded fruits (Charles-Dominique et al., 1981; Atramentowicz, 1988).

Like most other mammals, opossums first orally grasp food items with the anterior dentition, but as previously noted (Chapter 5), the incisors are unlikely to play a major role in food procurement or food reduction. Instead, killing bites, when necessary, are delivered with the canines. In all described cases of opossums feeding on mechanically resistant objects (e.g., vertebrate prey too large ingest whole), the item is next positioned with one or both forefeet for unilateral biting by the postcanine dentition until it is separated into sufficiently small fragments for intraoral chewing (Hiiemae and Crompton, 1971; González and Claramunt, 2000; Pheasey et al., 2018). Careful examination of photographs suggests that food separation prior to intraoral chewing is primarily accomplished with the premolars, which seem appropriately designed for this purpose (Chapter 5). Intraoral chewing involves complex, precise movements of the tongue and jaws that have been closely observed with cineradiography in functional-morphological studies of *Didelphis virginiana* and *Monodelphis domestica* (Hiiemae and Crompton, 1971; Bhullar et al., 2019). In this stage of mastication, food is processed exclusively by the molar dentition.

## Social Behavior

Like most nocturnal mammals, didelphids are normally solitary (Charles-Dominique, 1983). Apart from agonistic encounters, matings, and brief interactions of unknown significance, adult opossums do not usually associate with one another in the wild, and nests are almost always found to be occupied either by single adult individuals or by a single lactating female and her nursing offspring.[3] Apparently, didelphids do not exhibit allogrooming, a behavior known to reduce social tension in other mammals (McManus, 1970; Charles-Dominique, 1983; Russell, 1984; González and Claramunt, 2000). No compelling evidence of pair-bonding or even of prolonged postweaning associations of mother and offspring has been reported for any opossum species. Indeed, captive studies of several species have reported that cannibalism can occur if mother and offspring are not promptly separated at weaning (Barnes and Wolf, 1971; Charles-Dominique, 1983; Atramentowicz, 1995; Hingst et al., 1998).

## Agonistic Behavior, Fighting, and Passive Defense

Agonistic encounters among conspecifics and reactions to threats from other species are always accompanied by oral gaping, the so-called open-mouth display (McManus, 1970; Streilein, 1982c; González and Claramunt, 2000; Martin and Udrizar Sauthier, 2011). This behavior (Fig. 4.12) exposes the upper and lower canines—the only effective morphological weaponry that opossums possess—and doubtless signals the animal's capacity and willingness to bite. Another noteworthy agonistic behavior is the semierect or fully bipedal stance that many small opossums assume in both conspecific and interspecific confrontations (Enders, 1935; Beach, 1939; Streilein, 1982b; González and Claramunt, 2000; Martin and Udrizar Sauthier, 2011). Apparently, large opossums (e.g., species of *Didelphis*; McManus, 1970; Streilein, 1982c) do not adopt such upright postures.

Actual fighting among conspecifics has seldom been described and is presumably rare, except, perhaps, when food resources are very concentrated (Charles-Dominique, 1983), and in the context of sexual competition. According to Ryser (1992), who studied radio-collared *Didelphis virginiana* during the mating season in Florida, several males typically compete for access to an estrous female, and males were often observed to have wounds, suggesting that severe fighting occurs during these episodes. Staged encounters between captive adult males usually involve fighting, during which most attacks are directed at the head, throat, and shoulder; typical wounds sustained in such fights are canine punctures (McManus, 1970); remarkably, prior residency in the enclosure where fights were staged did not appear to confer any advantage, nor were any effective submission postures observed. Therefore, this species seems to lack some of the behavioral cues commonly used by other mammals to avoid or de-escalate combat.

Some species of common opossums use passive defense when attacked by predators. The best known of such behaviors is death feigning—sometimes called "playing possum" or "thanatosis." Death-feigning appears to be better developed in *Didelphis virginiana* than in other congeneric species, but in the absence of any detailed ethological studies of Neotropical common opossums, this impression might be misleading. The subject of considerable folklore (Hartman, 1952), feigned death in *D. virginiana* is a highly stereotyped behavior that is consistently exhibited only when the animal is seized by the neck and violently shaken, typically by a dog (Norton et al., 1964; Franq, 1969; Gabrielsen and Smith, 1985). Unlike sleeping animals, opossums feigning death are fully conscious; the eyes remain open, and the behavior is always preceded by defecation and release of fluid from the anal glands. "Recovery" occurs quickly, typically just as soon as the attacker leaves the scene.

Most researchers have concluded that feigning death helps *Didelphis* avoid predation, but it is not clear why a hungry predator should lose interest in a dead victim. One plausible explanation involves the noxious substances that some death-feigning species secrete when they are attacked: feigning death might inhibit the predator's killing behavior and gain time for defensive secretions to work their deterrent effects (Rogers and Simpson, 2014).[4] As far as we are aware, the chemistry of anal gland secretions by common opossums has not been investigated, but the paste or fluid secreted by these glands is sometimes described as foul smelling, and it might also have physiological effects on predators if ingested. Other aspects of the morphology and behavior of common opossums support the notion that death feigning has evolved as part of a complex defensive adaptation, key aspects of which remain obscure. To the best of our knowledge, death feigning has not been documented as a consistently exhibited behavior in any other mammal.[5]

## Nesting

Sleeping places must be chosen carefully if animals are to be concealed from predators and sheltered from the elements while unconscious. The fauna of disease-bearing hematophagous insects to which the animal is exposed while asleep can also be affected by nest location (Gaunt and Miles, 2000). Thus, nesting behavior (including site choice and nest construction) has potentially important fitness consequences. For the purposes of this review, "nests" will refer to the diurnal refugia of nocturnal opossums because nothing is known about the nocturnal refugia of diurnal species.

Not surprisingly, arboreal and scansorial opossums tend to nest in trees, whereas the nest sites of terrestrial opossums are at or near ground level (Miles et al., 1981; Loretto et al., 2005; Moraes and Chiarello, 2005b; Delciellos et al., 2006; Lira et al., 2018). The most commonly reported nesting places of arboreal and scansorial species include tree cavities, tangled lianas, and (especially) leaf accumulations in palm crowns (Miles et al., 1981; Moraes and Chiarello, 2005b; Lira et al., 2018). Bird nests, however, are sometimes appropriated by small opossums (Enders, 1935; Cáceres and Pichorim, 2003), and *Didelphis aurita* is known to occupy the abandoned tree nests of coatis (Monticelli-Almada and Gasco, 2018). Two species, *Philander quica* and *Hyladelphys kalinowskii*, are known to occupy aboveground nests built entirely of leaves and possibly made by the animals themselves; however, whereas the leaf nests of *P. quica* are said to simply be woven together (Lira et al., 2018), those of *H. kalinowskii* are cemented together and fastened to supporting understory vegetation with a white substance of unknown origin (Catzeflis, 2018).

Ground-level nesting sites are also diverse. *Didelphis virginiana* occupies a wide range of refugia, including hollow logs and the abandoned burrows of other mammals (skunks, armadillos, woodchucks, foxes, etc.; Gardner, 1982). In the dry Caatinga habitats of northeastern Brazil, *D. albiventris* is said to favor cavities in rocky outcrops (Streilein, 1982a). The ground-level and subsurface leaf-litter nests of *Metachirus myosuros* are said to be carefully camouflaged (almost impossible to detect visually) and defended by unique behaviors (Loretto et al., 2005). The Patagonian opossum *Lestodelphys halli* may excavate subterranean galleries or use abandoned rodent burrows for nesting (Martin and Udrizar Sauthier, 2011).

Most opossums provision their sleeping places with dry leaves, grass, or other suitable bedding. This nesting material is transported using the tail, a remarkable behavior that has been reported for at least 11 species in seven genera (Delgado-V. et al., 2014).[6] Automated video cameras have also recorded at least one female *Didelphis marsupialis* using its pouch to transport nesting material. Such investments of time and effort are presumably advantageous when the same nest is occupied continuously (or repeatedly) over an extended period of time, and radiotelemetry suggest that this is often the case (e.g., Fitch and Shirer, 1970; Lira et al., 2018). Research with captive *Monodelphis domestica* suggests that females build nests more consistently and use more nesting material than males, and that both sexes build better quality nests at lower ambient temperatures than they do at warmer temperatures (Fadem et al., 1986).

## Mating

Copulatory behavior has only been described for four didelphid species and appears to differ between terrestrial and arboreal taxa. Whereas two terrestrial taxa (*Didelphis virginiana* and *Monodelphis domestica*; Reynolds, 1952; Trupin and Fadem, 1982) copulate while lying on their sides on the ground, at least two arboreal species (*Marmosa robinsoni* and *Tlacuatzin canescens*; Barnes and Barthold, 1969; Valtierra-Azotla and García, 1998) copulate while one or both partners are suspended from elevated supports by their tails. According to Barnes and Barthold (1969), male *M. robinsoni* are unable to achieve erection and intromission without overhead purchase for the tail.

## Care of Young

Care of the young is exclusively provided by the female opossum (Russell, 1984). In early postnatal life, opossum maternal behavior is rudimentary. Neonates climb from the cloaca to the nipples unassisted, and any neonates that fail to attach to a nipple are ignored and left to die (Reynolds, 1952; Barnes and Barthold, 1969). Female *Didelphis* are said to pay little or no attention to nursing young during the weeks-long interval that they are continuously attached to the nipples, but females of *Caluromys* and *Philander* have been observed to lick nursing young (presumably to clean them; Charles-Dominique, 1983), and a female *Marmosa* was observed to retrieve experimentally detached young and to help them reattach (Beach, 1939). Once the young have opened their eyes and are not constantly attached, the female leaves them in the nest as she forages or carries them about on her back until they are weaned; this interval ranges from about a week to over six

*Fig. 7.2.* Female *Marmosa murina* carrying unweaned young in French Guiana (photo by Antoine Baglan)

weeks, depending on the species (Reynolds, 1952; Julien-Laferrière and Atramentowicz, 1990; Atramentowicz, 1995). The spectacle of female opossums heavily burdened with offspring at this stage of development (Fig. 7.2) is compelling evidence that leaving young behind in the nest must sometimes incur an unacceptably high risk of loss due to predation or other accidents.

## Discussion

Many of the didelphid behaviors described above and in other reviews (e.g., Kimble, 1997) are unexceptionally mammalian. Most mammals are nocturnal, and a solitary lifestyle is almost universal among nonflying nocturnal species. The absence of any male involvement in care of the young, and the termination of maternal care at weaning are also widespread (and possibly ancestral) mammalian traits. Additional aspects of opossum behavior mentioned in other chapters—in connection with sensory physiology (Chapter 6), dietary habits (Chapter 9), and population biology (Chapter 13)—are likewise familiar from ethological research with a wide range of other taxa. Therefore, although some opossum species have eccentric habits, opossums as a group do not appear to be behaviorally unusual mammals.

As might be expected, didelphids exhibit a much narrower range of behaviors than Australasian marsupials, which include taxa of widely differing autecologies. No opossums glide, hop bipedally, or travel underground; none forms pair bonds, engages in allogrooming, or lives in herds or family groups. However, didelphid variation in behavioral traits does not appear notably impoverished by comparison with behavioral variation within ecologically similar Australasian clades of roughly equivalent crown age, such as dasyurids or peramelids. In fact, didelphids and dasyurids are behaviorally similar in many respects (Croft, 2003).

Among opossums, the main axis of ethological variation concerns arboreal versus terrestrial adaptations. This pattern of behavioral disparity is clearest in locomotor traits—obviously because tree-living and ground-dwelling species encounter distinctly different substrates as they go about their nightly rounds—but nest-site selection, lunar phobia, and mating behavior might be other behavioral correlates of vertical stratification. *Chironectes* is clearly an outlier due to its semiaquatic habits, *Lestodelphys* seems to be the only opossum that digs, and *Didelphis virginiana* is the only species definitely known known to feign death, but opossum behavioral research is still in its infancy, and it seems probable that future research will discover behavioral phenotypes unique to other taxa that currently seem to be ethologically indistinguishable.

NOTES

1. Note, however, that behavioral observations in captivity do not always agree with field data. Captive water opossums (*Chironectes*), for example, are said to often be active in the daytime (Oliver, 1976), whereas field studies consistently indicate nocturnal habits (Voss et al., 2001; Galliez et al., 2009; Leite et al., 2013). The reason for such discrepancies is not certainly known, but as Croft (2003: 333) remarks, "In captivity, feeding times are quickly learnt and thus may confound the timing and expression of foraging behavior."

2. The only reported observations of unforced leaping by opossums seem to be Enders' (1935) account of locomotion by *Marmosa isthmica* and Charles-Dominique et al.'s (1981) description of *M. demerarae* ("On peut facilement le reconnaître dans le haut des arbres, à sa course rapide et saccadée, et aux nombreux sauts qu'il fait entre les branches").

3. Of 346 records of five species of opossums using artificial nest boxes in southeastern Brazil, 89% were of solitary individuals, 8% were of females with nursing young, and just 3% were of "multiple individuals" (Astúa et al., 2015). The latter category is, unfortunately, uninterpretable because no information was provided about age or relatedness. The same authors report a single (equally uninterpretable) observation of communal denning by *Didelphis albiventris* in an urban setting.

4. Curiously, this hypothesis is ignored in Humphreys and Ruxton's (2018) review of death feigning, which does not mention opossums at all.

5. Other phenotypic attributes of *Didelphis* that seem relevant in this context include white underfur (Fig. 4.10), which, when exposed by piloerection, makes the animal unmistakable for any

other, and might be an aposematic signal (Chapter 5). Common opossums are also slow for their size, which is also true of many aposematic species (e.g., porcupines and skunks) that rely on defenses other than speed to escape predators (Garland, 1983). Lastly, the peculiar vertebral reinforcements (including ankylosed cervical elements; Fig. 5.9), which have no plausible explanation in the context of locomotion (Argot, 2003; Giannini et al., 2011), might make the animal harder to kill and (like feigned death) might buy time for a chemical defense to take effect. According to Coulson (1996), death feigning is unknown among Australian marsupials. Ewer (1968) claimed that African ground squirrels (*Xerus erythropus*) and "the fox" (presumably *Vulpes vulpes*) feign death, but we have not been able to locate any other reference to such behavior in either species.

6. Caudal transportation of nesting material has been observed in taxa with (e.g., *Caluromys* and *Marmosa*) and without (e.g., *Metachirus* and *Monodelphis*) external evidence of tail prehensility. Illustrations of opossums carrying nesting material with their tails are in Oliver (1976), González and Claramunt (2000), Dalloz et al. (2012), and Delgado-V. et al. (2014). Similar behaviors have been reported for a few Australian marsupials but not for any placental mammal.

# IV

# OPOSSUM NATURAL HISTORY

# 8

# Habitats

The ecological context of opossum evolution consists of a hierarchy of environments—biome, habitat, and microhabitat—that are important for understanding locomotory and dietary adaptations and much else besides. Biomes (identified in our text with capital letters) are large contiguous regions of more or less homogeneous climate that typically contain mosaics of floristically and structurally distinct habitats. The Cerrado biome of central Brazil, for example, includes a wide range of grassland and dry-forest formations (Eiten, 1972; Oliveira-Filho and Ratter, 2002), whereas Amazonia includes lowland rainforest as well as isolated patches of savanna vegetation (Pires and Prance, 1985). However, what most zoologists would call a habitat (e.g., lowland rainforest) can also include a patchwork of local vegetation types (e.g., well-drained primary forest, palm swamps, etc.) that differ in ecologically significant ways. Additionally, most habitats contain distinct microhabitats (canopy, subcanopy, understory, etc.), each with important properties for its animal inhabitants.

The important but sometimes misused terms sympatry and syntopy merit brief discussion. The former denotes two or more species living at the same locality (within dispersal range of one another), whereas the latter is used for multiple species living in the same habitat. Species cannot be syntopic unless they are sympatric, but sympatric species are not necessarily syntopic. Because biologists can reasonably differ about what constitutes a habitat versus a microhabitat (primary and secondary rainforest, for example), syntopy is necessarily a somewhat subjective concept, but the distinction from sympatry is nevertheless important.

Below we review the principal kinds of habitats occupied by opossums, noting their geographic distribution, climate, and other ecological characteristics, especially physical structure and food resources. For the purpose of this review, we use "habitat" to designate the predominant vegetation type in which a species is usually found. Often (but not always), this corresponds to the local climax formation, but human disturbance, fire, poor drainage, and other factors often result in quite distinct plant communities that coexist side-by-side at localities where only a single "habitat" is indicated on large-scale vegetation maps. For specimens unaccompanied by ecologically informative field notes, such maps provide the only available information about habitat occupancy, and for some species that is all we have to go by.

## Habitat Occupancy

Opossums collectively occur in a wide range of habitats from about 49° N (in southern Canada) to about 49° S (in southern Argentina), and from sea level to about 4000 m in the Andes. Although some species are geographically widespread, most are consistently associated with just one kind of habitat, with the result that each species is normally exposed to a limited range of physical and biotic conditions to which it is adapted in one way or another. In particular, habitat occupancy largely determines the range of available food resources and the physical structure of sheltering vegetation. Because opossum diversity varies dramatically among habitats (Table 8.1), it is important to understand why some habitats support larger numbers of species than others.

### Lowland Rainforest

Most opossums (about 55% of currently recognized species; Appendix 1) live in tropical lowland rainforest, the dominant climax vegetation below about 1500 m elevation in areas that receive about 2000 mm or more of annual rainfall, from southern Mexico to southeastern Brazil (Fig. 8.1). Tropical lowland rainforest is replaced by montane forest ("cloud forest") at elevations above about 1500 m in the Andes, it grades into dry forest across isopleths of decreasing rainfall in both Central and South America, and it merges imperceptibly with moist subtropical forests in southern Brazil, eastern Paraguay, and northeastern Argentina. In Central America, lowland rainforest principally occurs along the Caribbean coast, which receives substantially more rainfall than the Pacific side. In South America, lowland rainforest occurs along the Pacific coast of western Colombia and northwestern Ecuador (a very wet region known as the Chocó), in discontinuous patches along the Caribbean littoral of Colombia and northern Venezuela, throughout most of the Guianas and the Amazon Basin (collectively known as Amazonia), and along the coastline of southeastern Brazil (the so-called Atlantic Forest, or Mata Atlântica). Rainforest also extends as a riparian formation into drier landscapes,

Table 8.1 Numbers of Syntopic Opossum Species at Seven South American Localities

| Locality | Latitude | Elevation[a] | Rainfall[b] | Habitat[c] | Didelphids[d] | References |
|---|---|---|---|---|---|---|
| Nuevo San Juan, Peru | 5° S | <200 m | 2900 mm | RF | 16 spp. | Voss et al. (2019) |
| Paracou, French Guiana | 5° N | <50 m | 3200 mm | RF | 14 spp. | Voss et al. (2001)[e] |
| Noel Kempff Mercado, Bolivia | 15° S | ca. 200 m | 1500 mm | DF | 12 spp. | Emmons et al. (2006) |
| Emas, Brazil | 18° S | ca. 800 m | 1700 mm | DF | 7 spp. | Rodrigues et al. (2002) |
| Mar Chiquita, Argentina | 38° S | sea level | 1000 mm | Gr | 3 spp. | Bó et al. (2002) |
| Ñacuñán, Argentina | 34° S | 520 m | 300 mm | De | 2 spp. | Ojeda (n.d.) |
| Carpanta, Colombia | 5° N | 3000–3100 m | >3000 mm | MF | 1 sp. | López-Arévalo et al. (1993) |

[a] Above sea level, rounded to the nearest 10 m.
[b] Annual average, rounded to the nearest 100 mm.
[c] De, desert; DF, dry forest; Gr, grassland; MF, montane forest; RF, rainforest.
[d] Counts include only species found in main habitat.
[e] Includes two species reported after Voss et al.'s (2001) survey (F. Catzeflis, personal commun.; Adler et al., 2012).

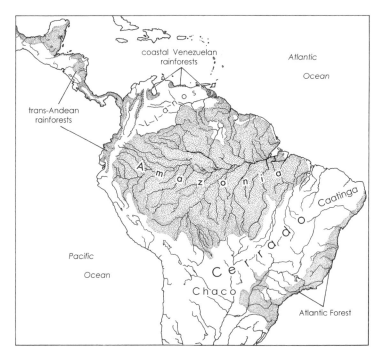

*Fig. 8.1.* Geographic distribution of lowland rainforest (dense stipple) in relation to biomes dominated by dry forests, montane forests, savannas, and deserts in Central and South America. High elevations in the Andes separate the trans-Andean rainforests from Amazonia, whereas the "Dry Diagonal" of Chaco, Cerrado, and Caatinga separates Amazonia from the Atlantic Forest

thus accounting for the presence of Amazonian and Atlantic Forest species along major rivers in the Cerrado (Johnson et al., 1999).

Primary lowland tropical rainforest has a closed canopy—typically some 25–45 m high in the Neotropics—that is persistently leafy throughout the year (Richards, 1952). Emergent trees are massively buttressed (Fig. 8.2), and palms are usually abundant in the subcanopy and understory; woody climbers (lianas), stranglers, and hemiepiphytes are other characteristic growth forms, and the undergrowth is typically dominated by saplings, treelets, and giant herbs. Due to rapid decomposition of fallen leaves and other debris, the litter layer is usually sparse, barely concealing the mineral topsoil, which is often exposed on hill slopes and ridgetops.

At upland sites (far from river floodplains), lowland rainforest can extend monotonously as a more or less homogeneous formation for many miles, interrupted

*Fig. 8.2.* Understory of primary lowland rainforest at Paracou, French Guiana (photo by Robert Voss). Fourteen species are sympatric at this well-sampled locality, of which eight (*Didelphis imperfecta, D. marsupialis, Hyladelphys kalinowskii, Marmosops parvidens, Mps. pinheiroi, Metachirus nudicaudatus, Monodelphis touan,* and *Philander opossum*) are known to occur in the primary forest understory.

only by treefalls, blowdowns, and narrow breaks along the larger streams. By contrast, floodplain landscapes in western Amazonia are complex mosaics of successional vegetation that can include monospecific stands of fast-growing pioneer trees, canebrakes, swamps, and old-growth forest (Terborgh, 1983; Puhakka and Kalliola, 1995). Secondary growth—successional vegetation regenerating in abandoned agricultural clearings (Guariguata and Ostertag, 2001)—is also present throughout the rainforested Neotropics, and a variety of seasonally flooded formations adds additional habitat complexity to the riparian landscapes of central Amazonia (Prance, 1979).

The distinguishing trophic characteristic of lowland tropical rainforest for resident vertebrate faunas is the high productivity and prolonged availability of diverse food resources. In particular, insects and other terrestrial arthropods are active throughout the year, as are numerous small vertebrates (notably amphibians and reptiles). Additionally, most woody rainforest plant species (70% to 94% of those at representative Neotropical sites; Jordano, 2000) produce fleshy fruits to attract vertebrate seed dispersers. Because fleshy-fruit-bearing rainforest plants include a wide range of growth forms (canopy trees, lianas, understory treelets, hemiepiphytes, etc.) that produce fruits differing in size, presentation, and other physical characteristics, they afford ample opportunity for niche-partitioning by vertebrate frugivores (Smythe, 1986; Fleming et al., 1987; Levey et al., 1994). However, even rainforests experience seasonal fluctuations in resource abundance, such that both primary and secondary production tend to be greatest in the rainy season and least available to vertebrate consumers in the dry season (Janson and Emmons, 1990). Rainforest primary productivity (e.g., of fruit and new leaves) is also spatially heterogeneous, being highest in the sunlit canopy, where many frugivorous and folivorous animals are also found. By contrast, productivity is substantially lower in the shaded understory, the fauna of which is largely sustained—directly or indirectly—by a steady rain of detritus (fruit, flowers, leaves, and wood) from overhead. Nevertheless, primary production can be high at or near ground level in gaps—treefalls, blowdowns (Nelson et al., 1994), river margins, and so forth—where understory frugivores and other consumers (e.g., folivorous insects) are often concentrated (Hallé et al., 1978; Levey, 1990; Wunderle et al., 2005).

The distinguishing structural features of lowland tropical rainforest are the wide separation of the canopy from the ground and the complex woody architecture of the intervening space. Whether or not distinct strata are recognizable (Grubb et al., 1963), vegetation profiles of primary Neotropical rainforest (e.g., in Richards, 1952) illustrate an almost-continuous volume of tree crowns extending from the understory to the canopy. Although large-diameter vertical trunks are a conspicuous feature of lowland rainforest, lianas offer alternative routes of canopy access for understory species, or of understory access for canopy dwellers. Another relevant aspect of Neotropical rainforest architecture is that adjacent tree crowns in the canopy and subcanopy are often interconnected by lianas or are close enough to be bridged or leaped across. By contrast, the crowns of treelets and giant herbs in the heavily shaded understory of primary rainforest are sometimes widely separated, such that small climbing animals often cannot move from one understory plant to another without descending to the ground.

### Lowland Dry Forest

Many of the larger opossums that live in rainforest are also found in geographically contiguous dry forests, but some small species (e.g., *Gracilinanus agilis*, *Monodelphis domestica*, *Marmosops ocellatus*, *Thylamys pusillus*, and *Tlacuatzin canescens*)

*Fig. 8.3.* Dry forest in the Serra do Amolar, Mato Grosso do Sul, Brazil (photo by Rogério Rossi). Only an incomplete species list is available from this locality, but as many as 12 didelphid species occur sympatrically in similar vegetation elsewhere in southwestern Brazil and eastern Bolivia.

are primarily or exclusively dry-forest species. The climatic conditions that support dry-forest vegetation vary widely, but several months of severe drought seem to be key (Murphy and Lugo, 1986). At most dry-forested Neotropical study sites (Fig. 8.3), the dry season is from four to eight months long, and annual rainfall varies from about 700 to 1600 mm (Gentry, 1995). Dry forest is the dominant climax vegetation on the Yucatan Peninsula, along most of the Pacific coast of Mexico and Central America, along parts of the Caribbean coast of Colombia and Venezuela, in some rain-shadowed Andean valleys, along the Pacific coast of southwestern Ecuador and northwestern Peru, and across the so-called Arid Diagonal (of Chaco, Cerrado, and Caatinga) that separates Amazonia from the Atlantic Forest (Gentry, 1995). However, savannas can occur under the same climatic conditions that support dry forests, and more or less extensive grasslands are present as fire-maintained subclimax vegetation, or where soils are so poor that they cannot support woody plants, in many dry-forested regions, especially the Cerrado (Oliveira-Filho and Ratter, 2002). Dry forests are deciduous formations in which the woody vegetation is partially to almost completely leafless during the dry season, but they vary considerably in other characteristics (Murphy and Lugo, 1986). Whereas some dry forests—usually those with >1000 mm of annual rainfall—are almost as tall and structurally complex as rainforest, others—typically in drier regions or on poorer soils—are much reduced in canopy height; in even drier climates, dry forest grades into desert thornscrub. In general, dry forest can be characterized as less tall, structurally simpler, less species-rich (especially in large trees), with a more open canopy, and with fewer epiphytes than rainforest (Murphy and Lugo, 1986; Gentry, 1995). Characteristic morphological traits of dry-forest woody plants include thorns, photosynthetic bark, and conspicuous flowers; among other floristic features, cacti (especially arborescent species) become increasingly prominent in dry-forest communities along gradients of increasing aridity (Gentry, 1995; Killeen et al., 1998).

Due to their shorter growing seasons, dry forests are generally less productive than rainforest (Murphy and Lugo, 1986), and fewer species of woody plants (only about 30% to 60% in representative dry-forest floras; Jordano, 2000) produce fleshy fruits to attract vertebrate seed dispersers. The defining ecological characteristic of dry forest, however, at least insofar as resident vertebrate faunas are concerned, is the pronounced seasonality of primary and secondary production. In particular, fruiting and leaf flush are concentrated at the onset of the rainy season (van Schaik et al., 1993), which is also when insect prey are most abundant (Janzen, 1973a; Ceballos, 1995). By contrast, the dry season is famine time for active resident (nondormant and nonmigratory) dry-forest vertebrates, when few dietary options are reliably available.

### Montane (Cloud) Forest

Many didelphids that inhabit lowland rainforest also occur on the lower slopes of the Andes and other mountains, but some species (e.g., *Gracilinanus aceramarcae, Marmosa tyleriana, Marmosops creightoni*, and *Monodelphis gardneri*) are montane-forest endemics. Montane forests (Fig. 8.4) occur discontinuously in the highlands of Mexico and Central America, more or less continuously along the Andes (especially on the eastern slopes) from Colombia to northwestern Argentina, and as islands in a sea of lowland rainforest or dry forest on many small mountain ranges and isolated peaks along the Caribbean coast of Venezuela, across the Guiana highlands, and in southeastern Brazil.

Several distinct montane-forest formations have traditionally been recognized by tropical plant ecologists (e.g., Lower Montane Rainforest, Upper Montane Rainforest, Subalpine Rainforest; Grubb, 1977; Webster, 1995), whereas other researchers have suggested that forest structure and plant communities change

*Fig. 8.4.* Montane forest at 2700 m above sea level in Parque Nacional Río Abiseo, San Martín, Peru (photo by Pedro Peloso). Undescribed species of *Gracilinanus, Marmosa,* and *Marmosops* were trapped at this locality and at nearby sites with similar vegetation (Silvia Pavan, personal commun.).

continuously with elevation on wet tropical mountains, such that discrete life zones cannot be recognized (Lieberman et al., 1996). However, there is universal consensus that montane forest differs from lowland rainforest in many important respects, and that the differences are greater in proportion to the elevational separation of the forests being compared. Directly or indirectly, such differences result from the adiabatic lapse rate of air temperature with elevation (about 0.6°C per 100 m in the humid tropics) and the presence of low-lying clouds (fog). In general, the lower limit of montane forest approximately coincides with the fog line, which is lower on isolated peaks, on small mountain ranges, and on outlying spurs than it is along the main range of major cordilleras. In the Andes, the transition from lowland rainforest to cloud forest often occurs in the interval from 1400 to 1600 m above sea level.

With increasing elevation on humid tropical mountains, forests dwindle in canopy height (from ca. 25–45 m in the lowlands to only a few meters near treeline); palms, lianas, and giant herbs become less abundant and eventually drop out completely; vascular and nonvascular epiphytes become more abundant; and plant species richness decreases (Richards, 1952; Grubb, 1977; Webster, 1995; Lieberman et al., 1996). The ground-litter layer increases in depth at higher elevations (e.g., from about 3 cm at 300 m above sea level to >10 cm at 2020 m in Panama; Olson, 1994), and the epiphyte-laden trunks and horizontal branches of upper-montane forest trees accumulate their own litter. Near treeline, an almost-continuous carpet of wet debris and moss extends from ground level into the treetops, effectively homogenizing the substrate traversed by arboreal and terrestrial animals.

Whereas lower-montane forests are perhaps nearly as productive as their lowland counterparts,[1] primary productivity at higher elevations is reduced by persistently cold temperatures, fog (which reduces incident solar radiation), and possibly also by a shortage of mineral nutrients (Grubb, 1977). Anecdotal evidence of extremely slow plant growth at high elevations on wet tropical mountains (Terborgh, 1971; Janzen, 1973b) suggests that upper-montane forest plants are unlikely ever to produce substantial fruit crops. Reduced primary productivity and persistent cold (which directly affects insect growth rates) are almost certainly responsible for the reduced diversity and abundance of arthropods at high elevations in the tropics, as well as for the very marked scarcity of large insects in alpine habitats (Mani, 1968; Janzen et al., 1976; Olson, 1994).

### Other Habitats

Only a few didelphids live in arid landscapes, among which we include the tropical thornscrub surrounding the northern part of the Gulf of Venezuela (inhabited by *Marmosa xerophila*; Fig. 8.5), the temperate thornscrub of northern Chile (inhabited by *Thylamys elegans*; Meserve, 1981), the Atacama and Monte deserts (inhabited by *Thylamys pallidior*; Carmona and Rivadeneira, 2006; Albanese and Ojeda, 2012), and the Patagonian shrub-steppe (inhabited by *Lestodelphys halli*; Zapata et al., 2013). All of these habitats receive substantially less than 500 mm of annual rainfall and have sparse woody vegetation consisting of only a few dozen drought-resistant species of spiny shrubs and small trees (Sarmiento, 1976; Paruelo et al., 2007). Although only a small fraction of woody plant species bear fleshy fruits in most desert floras (Bronstein et al., 2007), the thornscrub surrounding the Gulf of Venezuela includes abundant cacti (whose fruits are a seasonally important source of calories); additionally, insects are active throughout the year at this latitude (Thielen et al., 1997b). By contrast, cacti are less abundant in high-latitude South American deserts (Mourelle and Ezcurra, 1996), where insects are inactive during the austral winter (Pol and Casenave, 2004), a time of resource scarcity for small nonhibernating vertebrate secondary consumers.

*Fig. 8.5.* Tropical thornscrub on the Paraguaná Peninsula, Falcón, Venezuela (photo by José Ochoa). Apparently, *Marmosa xerophila* is the only species of small nonvolant mammal that occurs in this habitat, which receives less than 500 mm of rainfall annually.

Grasslands are another habitat where few opossums are found. In fact, only *Didelphis albiventris, Lutreolina crassicaudata,* and *Monodelphis dimidiata* are definitely known to occur in extensive areas of treeless grassy vegetation (e.g., the pampas and campos of Argentina and Uruguay; Paruelo et al., 2007; Bó et al., 2002). Consisting of a single stratum of low herbaceous plants, few (if any) of which produce fleshy fruits, grasslands typically support a limited range of nonflying vertebrates, mostly herbivores, granivores, and predators. Significantly, *D. albiventris, L. crassicaudata,* and *M. dimidiata* all belong in the latter category. Other grasslands sparsely inhabited by opossums occur in subtropical habitat mosaics of the Chaco (Fig. 8.6) and in the once-extensive tropical savannas of the Cerrado.

Only a single didelphid (*Didelphis virginiana*) occurs in the deciduous forests of eastern North America. These forests are distinguished from other habitats treated in this review by experiencing marked seasonal fluctuations in temperature but not in precipitation; leaf loss in such habitats is therefore a vegetative adaptation to temperatures too low to sustain primary production, not to seasonal

*Fig. 8.6.* Subtropical Chacoan grassland at Reserva El Bagual, Formosa, Argentina (photo by Pablo Teta). *Lutreolina crassicaudata* and *Chacodelphys formosa* occur locally in this habitat, and several other species (e.g., *Cryptonanus chacoensis, Thylamys pusillus*) occur along its brushy margins.

water deficit (Barnes, 1991). Fewer than half of the woody plant species in most temperate deciduous forests produce fleshy fruits (Jordano, 2000), which are generally available to vertebrate consumers only in the summer and early fall. Because most insects are dormant in cold weather, winter is a time of food scarcity as well as cold stress for most temperate deciduous-forest vertebrates, many of which migrate to warmer latitudes or hibernate until the spring (Kitchings and Walton, 1991). For nonhibernating and nonmigratory vertebrate secondary consumers large enough to eat them, small endotherms (e.g., insectivores, rodents, rabbits, and birds) are among the few reliable winter food resources in cold-temperate habitats.

## Microhabitat Use

Animals are not, in general, randomly dispersed in the habitats they occupy, tending to occur more frequently in some parts than in others for a variety of reasons. For small nonflying vertebrates, microhabitat use can be motivated by efficiency of locomotion, resource distribution, protection from predators, or interference from competing species. Without considerable research effort, it is seldom possible to explain observed patterns of microhabitat use, but the patterns themselves are often obvious and easily quantified.

Most research on microhabitat use by sympatric opossums has been carried out at forested study sites. Evidence for species differences in forest microhabitat use comes from an impressive range of methods, including direct observation (e.g., Charles-Dominique et al., 1981; Rasmussen, 1990), trapping studies (Malcolm, 1991; Leite et al., 1996; Lambert et al., 2005), spool-and-line tracking (Miles et al., 1981; Cunha and Vieira, 2002), and artificial nests (Delciellos et al., 2006). Fortunately, all of these approaches yield broadly congruent results (Table 8.2).

By far the most commonly reported pattern of differential microhabitat use by sympatric forest opossums is vertical stratification, whereby syntopic species use different levels of the forest. For practical reasons (traps are hard to place in treetops and unspooled threads are difficult to recover much above head height), most studies of vertical stratification recognize only three strata: ground level (including

*Table 8.2*    Microhabitat Use by Rainforest Opossums

| | Principal microhabitat(s) | References[a] |
|---|---|---|
| *Caluromys* | canopy & subcanopy | Charles-Dominique et al. (1981), Rasmussen (1990), Leite et al. (1996) |
| *Caluromysiops* | canopy & subcanopy | Janson & Emmons (1990), Emmons (2008) |
| *Chironectes* | streams | Voss et al. (2001), Galliez et al. (2009) |
| *Didelphis* | ground & understory[b] | Miles et al. (1981), Rasmussen (1990), Lambert et al. (2005) |
| *Gironia* | canopy & subcanopy | Emmons (1997), Silveira et al. (2014) |
| *Gracilinanus* | understory & subcanopy | Vieira & Monteiro-Filho (2003), Delciellos et al. (2006) |
| *Marmosa* | understory & subcanopy | Charles-Dominique et al. (1981), Leite et al. (1996), Lambert et al. (2005) |
| *Marmosops* | ground & understory[c] | Patton et al. (2000), Loretto & Vieira (2008), Leiner et al. (2010) |
| *Metachirus* | ground | Miles et al. (1981), Cunha & Vieira (2002), Lambert et al. (2005) |
| *Monodelphis* | ground | Charles-Dominique et al. (1981), Malcolm (1991), Lambert et al. (2005) |
| *Philander* | ground & understory[c] | Miles et al. (1981), Charles-Dominique et al. (1981), Lambert et al. (2005) |

[a] See text (Chapter 4) for additional references.
[b] Occasionally climbs into the canopy, but does not travel from tree to tree (Miles et al., 1981).
[c] Apparently never climbs into the canopy or subcanopy.

such low, horizontal substrates as fallen branches and logs), the undergrowth (e.g., treelets, giant herbs, and other aboveground substrates below about 3 m), and the "canopy" (including the subcanopy, effectively including all vegetation >3 m above the ground). Although qualitative descriptions of vertical stratification (e.g., Charles-Dominique et al., 1981; Miles et al., 1981) are sometimes persuasive, the null hypothesis that species occur equiprobably at all levels in the forest has often been rejected based on statistical tests (e.g., by Malcolm, 1991; Leite et al., 1996; Cunha and Vieira, 2002; Lambert et al., 2005). Furthermore, inferences about use of different vertical strata are impressively correlated with morphological traits (e.g., Lemelin, 1999; Argot, 2001, 2002) and with laboratory assessments of loco-motor behavior (Delciellos and Vieira 2006, 2009a, 2009b), so the aggregate evidence that forest opossums use different vertically defined microhabitats, and that they are adapted to do so, is compelling.

In effect, all rainforest opossums seem to be vertical microhabitat specialists in the sense that each species uses some stratum more frequently than others. Additionally, closely related species typically resemble one another in vertical stratum use, such that genera can be broadly characterized as (for example) canopy or understory taxa. Thus, Central American *Caluromys derbianus* is primarily active in the canopy and subcanopy (Rasmussen, 1990), as is *C. philander* in northeastern Amazonia (Charles-Dominique et al., 1981) and *C. lanatus* in southwestern Amazonia (Patton et al., 2000). Likewise, the predominantly terrestrial habits of *Philander opossum* in northeastern Amazonia (Charles-Dominique et al., 1981) strikingly resemble those of *P. quica* in the Atlantic Forest (Cunha and Vieira, 2002). Different species of *Marmosops* seem to have essentially similar vertical distributions in both Amazonian and southeastern Brazilian rainforests (Patton et al., 2000; Voss et al., 2001; Leiner et al., 2010), and species of *Monodelphis* seem to be strictly terrestrial wherever they occur. Therefore, vertical microhabitat use seems to be a taxonomically conserved trait.

By contrast, sympatric congeners often occur in different kinds of vegetation within the local rainforest matrix, resulting in what might be termed horizontal microhabitat segregation. For example, trapping results in southwestern Amazonia suggest that *Philander mcilhennyi* occurs in upland forest whereas *P. "opossum"* (probably *P. pebas*; Voss et al., 2018) occurs in adjacent riparian (seasonally flooded) forest. In the same region, *Marmosa constantiae* ("*Micoureus demerarae*") occurs in upland forest but sympatric *M. rutteri* ("*Mic. regina*") occurs in seasonally flooded riparian forest (Patton et al., 2000). In French Guiana, *Marmosops parvidens* prefers well-drained primary forest, whereas sympatric *M. pinheiroi* prefers adjacent swamp forest (Voss et al., 2001). In southern Brazil, *Didelphis aurita* occurs in the forest interior, whereas sympatric *D. albiventris* occupies the forest edge (Cáceres and Machado, 2013). These (and other) examples suggest that horizontal microhabitat selection is more taxonomically labile than vertical segregation, a conjecture that would be interesting to test in a phylogenetic context when detailed information about both phenomena become more widely available.

## Discussion

Tropical lowland rainforest is a uniquely congenial habitat for opossums, which decline in syntopic species richness (alpha diversity) along gradients of increasing aridity, elevation, and latitude. Dry forests, montane forests, grasslands, deserts, and temperate habitats have correspondingly depauperate opossum faunas.[2] Because similar diversity gradients have been reported for other Neotropical mammals (Voss and Emmons, 1996), a common cause seems likely. Research on South American primates—which show similar ecogeographic diversity gradients—suggests that species richness is largely determined by primary productivity, which is greatest at tropical lowland sites with good soils and abundant rainfall (Kay et al., 1997a); invariably, the climax vegetation in such situations is rainforest. Although the ecological mechanisms by which higher primary productivity sustains more diverse animal communities are not entirely clear (Cusens et al., 2012), high opossum diversity in lowland rainforest surely has something to do with the prolonged annual abundance of insect prey and fruit by comparison with sparser and more seasonal trophic resources in other habitats.

In addition to high productivity, the structural complexity of rainforest vegetation may also be conducive to opossum syntopy. Because locomotion on different substrates requires disparate morphological and behavioral phenotypes (Chapters 5 and 7), structurally complex vegetation probably makes it difficult for locomotor generalists to compete with species that are specialized to move about more effectively on the ground, on vertical trunks, or on fine terminal branches. By contrast, there may be fewer opportunities for spatial partitioning and species coexistence in grasslands, deserts, and other habitats with structurally simpler vegetation.

Habitat partitioning by vertical stratification seems to be a ubiquitous property of opossum communities in tropical forests, and congeneric species seem to occupy similar vertical niches throughout Central America, Amazonia, and the Atlantic Forest. Vertical-niche conservatism is probably explained by morphological and behavioral adaptations for locomotion, but it is possibly also reinforced by evolved behaviors to escape predation by snakes, raptors, and carnivores that are also vertically distributed (Chapter 11). Additionally, animal prey and fruit are unevenly distributed from the forest floor to the canopy, so vertical stratification may be a major driver of trophic divergence (Chapter 9).

Horizontal microhabitat segregation is another correlate of high opossum diversity in rainforest communities. An apparently labile trait, it may be an important mechanism for coexistence of recently diverged species following secondary contact, and an essential prelude to evolutionary divergence in other niche dimensions. Horizontal microhabitat segregation in the complex mosaics of riverine successional vegetation, palm swamps, flooded forests, and upland forests in western Amazonia might also explain the remarkably high sympatric diversity predicted by geographic range overlap, which could exceed 20 opossum species in some parts of that region.[3]

NOTES

1. According to Murray et al. (2000), about 81% of the tree species at Monteverde (ca. 1500 m), a lower-montane cloud forest in Costa Rica, produce vertebrate-dispersed fruits versus about 91% of the tree species at nearby La Selva (30 m), a lowland rainforest research station.

2. Note however, that forest-grassland mosaics, especially where the Cerrado interdigitates with tropical or subtropical evergreen forest (e.g., at Mbaracayú, in eastern Paraguay; Owen et al., 2018), can sometimes have high sympatric diversity due to an added component of beta diversity.

3. For example, between the Yavarí and the Ucayali Rivers in northeastern Peru, where Voss et al. (2019) recorded 19 opossum species and noted that another four might also occur.

# 9

# Diets

Diet is a fundamental aspect of a species' ecological niche, and it is often correlated with other important natural history traits, such as habitat occupancy, diel activity, resource competition, and parasitism. Because trophic adaptations can include a wide range of morphological, physiological, and behavioral characteristics, dietary information is often crucial for understanding phenotypic evolution. There are, however, several methodological and conceptual issues associated with ecological and evolutionary interpretations of dietary data, and these have sometimes been ignored or oversimplified in the literature on didelphid diets.

## Sources of Dietary Information

Information about opossum diets typically comes from examination of feces or stomach contents, from observations of free-ranging animals in their native habitats, and from feeding experiments with captives. Each method has limitations, but each also contributes unique insights. Unfortunately, dietary data are entirely lacking for many species.

Frequency data based on analyses of feces obtained from trapped animals during mark-recapture studies (Table 9.1) provide unequivocal evidence that many opossums routinely eat invertebrates and fruit. Without exception, the invertebrate phylum most abundantly represented by macroscopic fragments in didelphid feces is Arthropoda, and by far the most commonly eaten arthropod class is Insecta. Based on these data, most opossums could be described as fruit-eating insectivores (or as insect-eating frugivores), but several methodological caveats are in order.

As many authors (e.g., Dickman and Huang, 1988) have acknowledged, frequency data from scat-based studies tend to overestimate the dietary importance of invertebrates with hard exoskeletons and to underestimate the importance of soft-bodied prey. In particular, the dietary contribution of invertebrates lacking

*Table 9.1*    Exemplar Didelphid Diets Based on Fecal Analyses

| Species | Habitat[a] | N[b] | Frequencies[c] In | SV | Fr | Other | Source |
|---------|-----------|------|-------------------|----|----|-------|--------|
| *Caluromys philander* | RF | 34 | 26 | — | 94 | — | Leite et al. (1996) |
| *Marmosa paraguayana*[d] | RF | 98 | 100 | — | 57 | — | Pinheiro et al. (2002) |
| *Monodelphis domestica* | CR | 70 | 100 | 14 | 31 | 7[g] | Carvalho et al. (2019) |
| *Metachirus myosuros*[e] | RF | 44 | 100 | 9 | 7 | — | Cáceres (2004) |
| *Philander quica*[f] | An | 74 | 94 | 64 | 82 | — | Ceotto et al. (2009) |
| *Didelphis albiventris* | Ar | 71 | 100 | 58 | 76 | — | Cáceres (2002) |
| *Didelphis aurita* | RF | 157 | 100 | 59 | 78 | — | Cáceres and Monteiro-Filho (2001) |
| *Marmosops incanus* | DF | 58 | 100 | — | 40 | 8[g] | Lessa and da Costa (2010) |
| *Marmosops paulensis* | MF | 61 | 79 | 5 | 54 | 39[g] | Leiner and Silva (2007b) |
| *Gracilinanus agilis* | DF | 422 | 100 | 4 | 86 | — | Camargo et al. (2014) |
| *Thylamys pallidior* | De | 123 | 100 | — | 19 | 90[h] | Albanese et al. (2012) |

[a] An = anthropogenic (human-modified landscape), Ar = Araucaria forest, CR = campo rupestre (rocky grassland), De = desert, DF = dry forest, MF = montane forest, RF = rainforest.
[b] Number of fecal samples examined.
[c] Frequencies of tabulated food categories in fecal samples (note that these do not sum to 100% because feces can simultaneously contain several items): In = invertebrates, SV = small vertebrates, Fr = fruit.
[d] Identified as *Micoureus demerarae* by Pinheiro et al. (2002).
[e] Identified as *Metachirus nudicaudatus* by Cáceres et al. (2004).
[f] Identified as *Philander frenatus* by Ceotto et al. (2009).
[g] Flowers.
[h] Leaves.

exoskeletons altogether, such as earthworms (Lumbricidae) and slugs (shell-less pulmonate gastropods) might be missed completely unless a special effort is made to detect microscopic hard parts (e.g., setae and radulas; Pernetta, 1976; Wroot, 1985). Similarly, the dietary importance of large-volume, easily digested prey tissues—such as vertebrate muscle and viscera—might be underestimated by fecal frequency data, even if some indigestible bits (fur, feathers, scales, bones, etc.) are also eaten. Additionally, feces provide no clear indication as to whether ingested vertebrate prey were killed or encountered as carrion, although the presence of dipteran larvae is sometimes cited as evidence for scavenging (e.g., by Carvalho et al., 2005).

Fecal analysis is also problematic for inferences about frugivory and other feeding behaviors. Seeds in feces provide the primary evidence for fruit consumption, but large seeds might not be swallowed. As a result, the dietary contribution of large-seeded fruits could be underestimated if only pulp is ingested, and the importance of small-seeded fruits might be correspondingly overestimated. Fungi, a potentially important food resource for many small mammals (Fogel and Trappe,

1978), entirely lack macroscopic hard parts and are also unlikely to be detected in feces that are not examined under high magnification for spores (Mangan and Adler, 2000). Other potentially important foodstuffs such as gum and nectar leave few macroscopic traces in the lower digestive tract, so gummivory and nectarivory—feeding syndromes known to be important for other Neotropical mammals—are difficult to detect by fecal analysis. In fact, the common practice of washing opossum feces on coarse screens prior to examination (Santori et al., 1995; Cáceres, 2002, 2004; Cáceres et al., 2002; Ceotto et al., 2009; Zapata et al., 2013) virtually guarantees that much potentially important microscopic dietary evidence (earthworm setae, fungal spores, pollen, etc.) is lost.

Many disadvantages of fecal analysis could be avoided by examining stomach contents, but stomachs can only be obtained from dead specimens, so this method is not an option for nondestructive ecological studies. Nevertheless, stomach contents provide uniquely important information about the relative volume of ingested items (Pineda-Munoz and Alroy, 2014), and such information can suggest different dietary inferences than frequency data. Cordero and Nicolas (1987), for example, found insects in 49% of examined stomachs of *Didelphis marsupialis*, but estimated that they comprised only 15% of the total volume, whereas birds were found in only 13% of examined stomachs but accounted for 21% of the total volume. Undigested stomach contents can also provide evidence for consumption of soft-bodied prey and other material that is hard to detect in feces. For example, stomach-based dietary studies (e.g., Hamilton, 1958; Charles-Dominique et al., 1981) suggest that didelphids do eat earthworms, sometimes in considerable amounts.

Direct observation of free-living individuals is the method of choice in dietary studies of diurnal mammals such as primates (Terborgh, 1983; Nickle and Heymann, 1996), and it has also been used successfully with radio-collared nocturnal species (e.g., kinkajous and porcupines; Kays, 1999; Lima et al., 2010). Only a few researchers have directly observed radio-collared opossums feeding in the wild, but their results confirm that fecal analyses provide a biased and incomplete picture of didelphid diets. In particular, Charles-Dominique et al. (1981) and Atramentowicz (1988) observed that rainforest opossums do not ingest the seeds of most of the fruit species they commonly eat, swallowing just the pulp; therefore, only the smallest seeds are found in feces. Field observations have also shown that some opossums eat gum and nectar (Charles-Dominique et al., 1981; Gribel, 1988; Aléssio et al., 2005; Ibarra-Cerdeña et al., 2007; Martins and Gribel, 2007), dietary habits that have yet to be reported from any study based on analyses of scat. Lastly, direct observations in the wild clearly indicate that some opossums are active predators of small vertebrates and are not simply consumers of carrion (Kluge, 1981; Tuttle et al., 1981; Rodrigues, 2005; Gómez-Martínez et al., 2008; Acosta-Chaves et al., 2018).

Feeding experiments with captive animals under controlled conditions can be used to test hypotheses about food preference, to elucidate predatory interactions, and to gain insight into food-handling behaviors that affect the interpretation of scat analyses. Experiments with these objectives have seldom been used to study opossum food habits, but Astúa de Moraes et al. (2003) summarized the results of cafeteria trials that provided compelling evidence for taxonomic differences in macronutrient consumption (Fig. 9.1). Among other unique insights about didelphid diets gained from observing captive animals, staged encounters of *Didelphis* with live pitvipers confirm anecdotal accounts that these large opossums actually kill and eat venomous snakes with impunity (Oliveira and Santori, 1999; Almeida-Santos et al., 2000). Similarly, the behavior of *Lestodelphys halli* presented with live mice makes it clear that this ferocious small predator competently attacks,

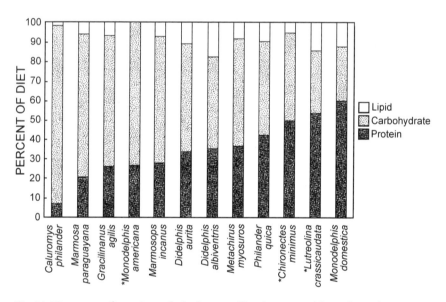

*Fig. 9.1.* Histogram of protein, carbohydrate, and lipid content of diets chosen by captive opossums in cafeteria experiments. Asterisks indicate species data from single experiments (all other species are represented by averaged data from multiple experiments; redrawn from Astúa de Moraes et al., 2003). The lack of obvious discontinuities in protein/carbohydrate ratios in these data suggest an interspecific continuum from predominant frugivory to predominant faunivory rather than discrete dietary categories.

kills, and eats sympatric rodents (Birney et al., 1996a; Martin and Udrizar Sauthier, 2011), despite the fact that examined feces collected in summer months contain mostly arthropods (Zapata et al., 2013).

In summary, fecal analysis is a biased source of information about didelphid
diets and should be supplemented with other techniques to gain a more complete
picture of what opossums really eat. On the other hand, observations of field behavior unsupported by analyses of scat or stomach contents are often insufficient
to positively identify ingested food (Emmons, 2000). Therefore, the most compelling marsupial dietary studies (e.g., Charles-Dominique et al., 1981; Dennis, 2002)
have used multiple methods.

## Diversity, Nutritional Quality, and Availability of Food Resources

Foods known to be eaten by opossums include invertebrates, small vertebrates,
fruit, nectar, gum, and fungi. Most of these are taxonomically diverse categories, some have distinct nutritional qualities, and each may be available on different schedules and in different places in didelphid habitats. Because the diversity,
nutritional value, and spatiotemporal availability of food resources are all important aspects of a species' trophic niche, these topics merit attention.

### Invertebrates

Dietary studies based on large samples of feces or stomachs suggest that most didelphids do not specialize on a single kind of invertebrate prey. Although Coleoptera (beetles) and Formicidae (ants) are the arthropod taxa most frequently
found in didelphid feces, Blattaria (roaches) and Orthoptera (katydids, crickets,
etc.) are also commonly identified in scat, and several other arthropod taxa are
sometimes consumed in large numbers. Diplopoda (millipedes) and Opiliones
(harvestmen), in particular, are often eaten by litter-foraging opossums (e.g.,
*Metachirus myosuros*; Cáceres, 2004), whereas Hemiptera and Lepidoptera seem
to be eaten more frequently by arboreal species (e.g., *Marmosa paraguayana*; Pinheiro et al., 2002; Pires et al., 2012). Some arthropod taxa only appear frequently in
opossum scat from particular habitats. Decapod crustaceans, for example, are
commonly eaten by terrestrial opossums only in coastal forests (e.g., Cáceres et al.,
2002), whereas Isoptera (termites) seem to be eaten more frequently in the Cerrado
than elsewhere (e.g., Martins et al., 2006a), and scorpions might be an important
dietary resource only in arid habitats (Zapata et al., 2013). Spiders, isopods, and
centipedes are also eaten occasionally, as are most other higher taxa of large

(>5 mm) terrestrial arthropods, with the interesting exception of Dermaptera (ear-wigs).[1] Gastropods appear to be minor elements of some opossum diets (Martins et al., 2006a; Leiner and Silva, 2007b), although their frequency is perhaps underestimated for the methodological reasons discussed above. As previously remarked, earthworms are only known to be eaten by opossums based on stomach content analyses, which are available for just few species.

Authors of many dietary studies have claimed that didelphids prey opportunistically on invertebrates, but few investigators have actually compared frequencies of prey taxa in feces or stomachs with independent estimates of invertebrate abundance in local habitats. Among those who have done so, Thielen et al. (1997b) found that *Marmosa xerophila* prefers Orthoptera, despite the fact that Coleoptera was more frequently found in feces. Camargo et al. (2014) detected a seasonal preference of *Gracilinanus agilis* for Hemiptera and Isoptera, and their results also suggested that *G. agilis* avoids eating ants (although ants were, paradoxically, more frequently identified in feces than other insect orders). Such results are too few to generalize about, but it would be consistent with the better-documented behavior of other mammals (such as tamarins; Nickle and Heymann, 1996) if didelphids were to prefer large phytophagous insects lacking mechanical or chemical defenses (e.g., cicadas and katydids; Acosta-Chaves et al., 2018) over smaller arthropods or those protected by spines or defensive toxins. Indeed, it seems likely that ants—perhaps overrepresented in fecal frequency data because they are swallowed whole and are easily identified in scat—may often be ingested accidentally with fruit, trap bait, or other food items.

By contrast with fruit, most invertebrates are high in protein but low in carbohydrates and lipids (Redford and Dorea, 1984). Although adult arthropods are protected by indigestible chitinous exoskeletons, most are easily crushed and cut into fragments by tribosphenic molars (Chapter 5), so handling costs are probably minimal. Direct observations of *Marmosa* killing and dismembering large insects, however, suggests that the hardest parts are often discarded (Enders, 1935), so heavily armored arthropods might be avoided by very small opossums.

With some noteworthy exceptions—ants, termites, aposematic species, and blood-feeding dipterans—insects are hard to find in the tropical forest understory by day (Elton, 1973). Instead, most large, biochemically unprotected tropical forest insects (e.g., katydids; Nickle and Castner, 1995) are nocturnal, concealing themselves in the daytime by hiding in the litter layer, inside hollow logs and tree cavities, under loose bark, in hanging clusters of dead leaves, in the rolled new leaves of large monocots, and so forth. By night, however, large invertebrates occur on a wide variety of forest substrates, including tree trunks (important avenues of insect migration between the litter layer and the canopy), green stems (favored by phloem-feeding Hemiptera), understory and canopy foliage (grazed by larval Lepidoptera and many orthopterans), and the forest floor (the principal habitat of detritivores and their arachnid predators; Janzen, 1973a; Penny and Arias, 1982; Adis, 1988; Basset, 2001).

At high latitudes most invertebrates are dormant in the winter, but invertebrates can fluctuate in abundance even in tropical habitats. No single sampling method is sufficient to census the entire invertebrate fauna of interest, but a variety of methods suggest that, in tropical habitats with distinctly seasonal rainfall, insect biomass increases early in the rainy season (more or less synchronously with, or just preceding, peak leaf flush; Janzen, 1973a; Boinski and Fowler, 1989; Janson and Emmons, 1990; Levings and Windsor, 1996; Smythe, 1996). Earthworms—which, in the tropics, are most abundant in forests with annual precipitation between 2000 and 4000 mm—remain deep underground during the dry season but move closer to the soil surface during the rainy season (Fragoso and Lavelle, 1992), and some large species are active in the litter layer at night after heavy rains (Charles-Dominique et al., 1981).

### Vertebrates

Dietary studies based on feces or stomach contents suggest that small mammals, birds, squamates, and anurans (in that order) are the vertebrate prey most frequently consumed by didelphids. Unfortunately, vertebrate bone fragments in didelphid scat are seldom identified even to class, so this ranking may reflect preservational artifacts. In particular, taxa lacking indigestible integumental structures such as hair, feathers, and scales may be systematically under-represented in published frequency data obtained from fecal analyses.[2] Rodents are more frequently identified in opossum scat and stomachs than other mammals (e.g., Busch and Kravetz, 1991), although small didelphids are sometimes eaten by large opossums (e.g., *Monodelphis* by *Philander*; Macedo et al., 2010), and *Didelphis virginiana* often eats shrews, moles, and rabbits in addition to rodents (Hamilton, 1958). Birds in opossum scat are so infrequently identified that no taxonomic generalizations are possible; most, perhaps, are nestlings.[3] Lizards and snakes are about equally represented among the squamates identified in dietary analyses, but anuran remains found in scat are seldom identified even to family.

Vertebrate soft tissues—muscle and viscera—are rich in protein and lipids, and they are highly digestible (Hume, 1999). However, there may be costs associated with the capture and consumption of vertebrate prey. Rodents, for example, can deliver painful and potentially crippling wounds with their sharp incisor teeth so, unless they are much smaller than their attacker, they must be dispatched with care by opossums that are dentally equipped to deliver killing bites. Vertebrates that are too large to be swallowed whole must be reduced to smaller chunks by biting through mechanically resistant skin, bones, and ligaments. Additionally, some biochemically defended vertebrate prey—such as toads (Garrett and Boyer, 1993; Laurance and Laurance, 2007), newts (Hart et al., 2019), and venomous snakes (Voss and Jansa, 2012)—are unlikely to be eaten except by opossums that have evolved molecular mechanisms for neutralizing ingested or injected toxins (Chapter 6).

Small vertebrates are longer-lived than most invertebrates, and (with the exception of ectothermic taxa at high latitudes and amphibians in seasonally dry habitats) they are active throughout the year. Vertebrate prey is therefore likely to be a persistent food resource for tropical opossums. At high latitudes, where ectothermic taxa are dormant in cold weather, small mammals and birds may be key winter food resources for the few opossum species that live in such hostile environments. Small mammals, birds, lizards, snakes, and anurans are found at all levels in tropical forests—from the litter layer to the canopy—so they are potentially available to any opossum large and active enough to eat them.

*Fruit*

Opossums eat a wide variety of fruit, as documented by several exemplary studies based on direct observation of feeding behavior and examination of stomach contents in French Guiana. Fruits eaten by three sympatric rainforest opossums, for example, included 44 plant species in 21 families, representing canopy trees, understory treelets, lianas, stranglers, and hemiepiphytes (Atramentowicz, 1988). Twenty-two species of fruits eaten by opossums were biochemically assayed in the same study and were found to have pulp that was rich in sugars or lipids but low in protein. The pulpy parts of fruits actually eaten by didelphids (aril, exocarp, epicarp, mesocarp, pericarp, sarcotesta, etc.) are either exposed or weakly protected at maturity (when the fruit is ripe), so most of these items are also consumed by other sympatric vertebrates, including birds, bats, kinkajous, and primates (Charles-Dominique et al., 1981; Guillotin et al., 1994; Simmen and Sabatier, 1996). Insofar as known, opossums seem to be generalist frugivores that lack coevolved dietary adaptations to particular plant taxa or to particular kinds of fruit (Atramentowicz, 1988). Some opossums, however, exhibit marked preferences for particular fruit taxa but actively avoid others, and it would be interesting to know the basis for such behaviors. Although nutrient content would seem like an obvious basis for choice, phenological predictability might be another (Leiner and Silva, 2007b).

Fruits produced by canopy trees—which typically have large, synchronously ripening crops—may be eaten in situ by arboreal opossums, or they may be consumed on the ground by terrestrial species (Charles-Dominique et al., 1981). The latter probably benefit from the wasteful behavior of canopy-foraging diurnal birds and monkeys, which often dislodge or pick and discard uneaten fruits, leaving a substantial fraction of the daily production of ripe pulp on the forest floor for nocturnal foragers. By contrast, understory treelets have smaller fruit crops, some of which ripen over extended fruiting periods (e.g. *Piper* spp.; Fleming, 1981; Leiner and Silva, 2007b), and these spatiotemporally dispersed resources are probably harvested in situ by arboreal or scansorial species that are small and agile enough to be supported by slender terminal branches.

Of course, the availability of fruit pulp in sufficient quantities to support a frugivore community for more than a few weeks out of the year is primarily a tropical phenomenon (Jordano, 2000), so opossums that inhabit temperate latitudes (e.g., *Didelphis virginiana* and *Lestodelphys halli*) necessarily rely on vertebrate or invertebrate prey for much of the annual cycle. Even in the tropics, however, fruit is not equally available at all times. Instead, more or less predictable temporal patterns of abundance and scarcity seem to characterize almost all tropical forests, especially those with strongly seasonal rainfall (Chapter 8). During periods of fruit scarcity—usually corresponding to the local dry season—frugivorous tropical mammals rely on stored fat reserves, switch to other (less calorically dense) food resources, or migrate to edaphically wetter habitats with asynchronous fruiting phenologies (Charles-Dominique et al. 1981; Terborgh, 1986; van Schaik et al., 1993). For species that lack significant fat reserves and are unable to move long distances, alternative food resources ("fallback" foods; Constantino and Wright, 2009) may be key to survival at such times.

### Nectar and Gum

Soon after Sussman and Raven (1978) claimed that New World marsupials do not visit flowers, multiple observations of nectar-feeding didelphids were published by researchers working in western Amazonia (Janson et al., 1981), Panama (Steiner, 1981, 1983), and French Guiana (Charles-Dominique et al., 1981). Decades later, the list of plant taxa known to be visited by opossums includes a wide diversity of growth forms from both moist and dry tropical forests at geographically widespread localities (Table 9.2). By contrast, gum consumption by didelphids is not

Table 9.2  Flowers Visited by Opossums

| Plant taxon | Locality | Habitat[a] | Growth form | Visitors[b] | References |
|---|---|---|---|---|---|
| ACANTHACEAE | | | | | |
| _Trichanthera gigantea_ | Panama | RF | small tree | 6 | Steiner (1981) |
| ARECACEAE | | | | | |
| _Calyptrogyne ghiesbreghtiana_ | Costa Rica | RF | sessile palm | 10 | Sperr et al. (2008) |
| BALANOPHORACEAE | | | | | |
| _Helosis cayennensis_[c] | French Guiana | RF | small herb[d] | 9 | Charles-Dominique et al. (1981) |
| BOMBACACEAE | | | | | |
| _Pseudobombax tomentosum_ | C Brazil | DF | large tree | 3 | Gribel (1988) |
| _Quararibea cordata_ | SE Peru | RF | large tree | 1, 2, 3 | Janson et al. (1981) |
| _Q. stenopetala_ | E Ecuador | RF | large tree | 3, 4 | Janson et al. (1981) |
| CACTACEAE | | | | | |
| _Stenocereus queretaroensis_ | Mexico | DF | small tree[e] | 7 | Ibarra-Cerdeña et al. (2007) |
| CARYOCARACEAE | | | | | |
| _Caryocar glabrum_ | French Guiana | RF | large tree | 5 | Julien-Laferrière (1999) |
| _C. villosum_ | N Brazil | RF | large tree | 3, 5 | Martins and Gribel (2007) |
| CAESALPINIACEAE | | | | | |
| _Eperua falcata_ | French Guiana | RF | large tree | 5 | Julien-Laferrière (1999) |
| _Hymenaea courbaril_ | French Guiana | RF | large tree | 3 | Charles-Dominique et al. (1981) |
| CHRYSOBALANACEAE | | | | | |
| _Couepia longipendula_ | N Brazil | RF | large tree | 3 | Gribel (1988) |
| CLUSIACEAE | | | | | |
| _Symphonia globulifera_ | French Guiana | RF | large tree | 5 | Julien-Laferrière (1999) |
| EUPHORBIACEAE | | | | | |
| _Mabea fistulifera_ | SE Brazil | SG | small tree | 8 | Vieira and Carvalho-Okano (1996) |
| _M. occidentalis_ | Panama | RF | small tree | 6 | Steiner (1981) |

| | | | | | |
|---|---|---|---|---|---|
| **LECYTHIDACEAE** | | | | | |
| *Eschweilera coriacea* | French Guiana | RF | large tree | 5 | Julien-Laferrière (1999) |
| **MALVACEAE** | | | | | |
| *Ceiba pentandra* | Panama | RF | large tree | 6 | Steiner (1981) |
| *Ochroma pyramidale* | Panama | RF | large tree | 2, 6 | Steiner (1981), Kays et al. (2012) |
| **MARCGRAVIACEAE** | | | | | |
| *Marcgravia nepenthoides* | Costa Rica | RF | liana | 2, 6 | Tschapka and von Helversen (1999) |
| *Norantea guianensis* | French Guiana | RF | shrub | 5 | Julien-Laferrière (1999) |
| **MIMOSACEAE** | | | | | |
| *Inga ingoides* | French Guiana | RF | large tree | 5 | Charles-Dominique et al. (1981) |
| *I. thibaudiana* | French Guiana | RF | large tree | 5 | Charles-Dominique et al. (1981) |
| *Parkia pendula* | N Brazil | RF | large tree | 2 | Hopkins (1984) |
| **PAPILIONACEAE** | | | | | |
| *Alexa wachenheimii* | French Guiana | RF | large tree | 5 | Julien-Laferrière (1999) |
| **STRELITZIACEAE** | | | | | |
| *Phenakospermum guyannense*[f] | N Brazil | RF | giant herb | 3 | Gribel (1988) |

[a] Habitat codes: DF = dry forest, SG = secondary growth (at a formerly rainforested site), RF = rainforest.
[b] Only opossums are recorded here; additional visitors may include other nonflying mammals (kinkajous, olingos, etc.), bats, birds, and/or insects (Chapter 12). Visitor codes: (1) *Caluromysiops irrupta*, (2) *Didelphis marsupialis*, (3) *Caluromys lanatus*, (4) *Marmosa* sp., (5) *Caluromys philander*, (6) *Caluromys derbianus*, (7) *Tlacuatzin canescens*, (8) *Didelphis aurita*, (9) *Philander opossum*, (10) *Marmosa zeledoni* (*M.* "*mexicana*").
[c] Probable identification (Mori et al., 2002).
[d] Root parasite (Mori et al., 2002).
[e] Columnar cactus.
[f] Also known as *Ravenala guyanensis* and sometimes placed in the banana family (Musaceae).

well documented, nor (to our knowledge) have plants with edible exudates been inventoried at any Neotropical forest study site. However, only a few tree species seem to be routinely exploited by local populations of gummivorous callitrichines (e.g., in northeastern Peru; Garber and Porter, 2010), so it is possible that gum is not, in general, a botanically diverse food source.

By comparison with fruit pulp and animal prey, nectar and gum are low-quality food resources. Most nectars available in large amounts to mammalian flower visitors are dilute solutions of hexose sugars—easily digestible but without significant quantities of other nutrients (Nicolson and Thornburg, 2007). Gums consist primarily of complex polysaccharides that are resistant to mammalian digestive enzymes and require fermentation by gut bacteria; however, they may contain useful minerals, such as calcium (Power, 2010). Because didelphids lack dental adaptations for gouging bark (Chapter 5), it is likely that they parasitize the bark-gouging activities of diurnal gum-eating mammals such as marmosets, and this behavior has, in fact, been observed (Aléssio et al., 2005).

Without exception, plants known to be visited for nectar by opossums at sites with highly seasonal rainfall flower in the dry season (e.g., *Quararibea cordata* at Cocha Cashu in southeastern Peru; Janson et al., 1981). At sites without a severe dry season, however, opossums have been reported to visit flowers throughout the year (Julien-Laferrière, 1999; Tschapka and von Helversen, 1999). Apparently, edible exudates are available throughout the year to primate consumers (e.g., in northeastern Peru; Garber and Porter, 2010) and, presumably, also to opportunistically gum-feeding opossums.

## Fungi

Mycophagy is not commonly reported for New World marsupials (Fogel and Trappe, 1978), but one dietary study of Panamanian cloud-forest mammals (Mangan and Adler, 2000) found spores of arbuscular mycorrhizal fungi in the feces of *Marmosa mexicana* and *Marmosops invictus*. Because conventional methods of fecal analysis are unlikely to detect mycophagy for the reasons already explained, and because fungi are abundant in most opossum habitats, this resource might be a cryptic component of many didelphid diets. Wallis et al. (2012) reviewed information about the macronutrient content of fungi and noted that some fungal species could be a rich source of amino acids and digestible protein. We are not aware of any phenological studies of fungal species known to be eaten by Neotropical mammals, but two studies of mycophagy by Neotropical rodents suggest that sporocarp consumption varies seasonally (Janos et al., 1995; Mangan and Adler,

2002). By contrast, fungi are a year-round dietary staple for some species of Neo-tropical rainforest primates (Hanson et al., 2006; Hilário and Ferrari, 2010). Future studies of opossum diets should include methodology to detect and quantify consumption of this potentially important food resource.

## Discussion

Didelphids have often been described as omnivores, but this description is simultaneously uninformative and potentially misleading. Although most opossums do, in fact, eat both animal and plant tissues, they do not eat everything. In particular, didelphids do not eat seed endosperm, the mainstay of many rodent diets, for which they lack appropriate dental adaptations. Likewise, most opossums do not normally eat leaves, the principal fodder of some sympatric placental mammals (e.g., sloths) and of many Australian marsupials. Contrary to some fanciful accounts of metatherian biogeography (Leigh et al., 2014), neither didelphids nor any other New World marsupial is known to eat pollen. If fungi, which are commonly eaten by some Australian marsupials, are actually consumed in significant amounts by opossums, compelling evidence for this is not yet available.

Instead, a mixed diet of invertebrates (primarily insects) and fruit seems to characterize most didelphids, which—in this respect—are notably similar to some sympatric primates (e.g., tamarins; Peres, 1993), bats (e.g., phyllostomids; Giannini and Kalko, 2004), and Paleotropical mammals (e.g., treeshrews; Emmons, 2000). These trophic resemblances are not surprising because insects and fruit are permissive foods, requiring no special anatomical or physiological adaptations to obtain, mechanically reduce, or digest. In fact, mammals that exploit a mixed diet of insects and fruit appear to share no morphological or physiological attributes other than those inherited from their common therian ancestor—which might well have had a similar diet.

There are, to be sure, many exceptions (Chapter 4). Small vertebrates, which supplement the insectivorous-frugivorous diets of most opossums, are important (if not predominant) in the diets of several large species (e.g., of *Didelphis*, *Lutreolina*, and *Philander*), and even small opossums that live at high latitudes (e.g., *Lestodelphys halli*) might be almost exclusively carnivorous for parts of the year when fruit is unavailable and insects are inactive. Species of *Monodelphis* appear to be mostly insectivorous and to seldom eat fruit, whereas *Caluromys* appears (by most accounts) to be primarily frugivorous and to consume only small numbers of insects. Insofar as known, *Chironectes* eats mainly fish and crustaceans and does not eat fruit at all.

Although some authors have proposed trophic-niche classifications for opossums, others have argued that discrete dietary categories are misleading because many species have broadly overlapping diets along a continuum from frugivory to animalivory (Astúa de Moraes et al., 2003). Indeed, many accounts cited above suggest that didelphids are opportunistic, generalist consumers that simply eat whatever is locally and temporally available to satisfy their metabolic requirements for carbohydrates, lipids, proteins, and other macronutrients. The lack of morphological evidence for special trophic adaptations tends to support this view, as does seasonal and geographic variation in reported diets. Implicit in many accounts of opossum trophic ecology (and explicit in our own) is the notion that food availability is largely determined by habitat or microhabitat occupancy: fruit, for example, is maximally abundant in the rainforest canopy, invertebrate prey on the ground, and crustaceans and fishes in streams. By this logic, opossum diets are secondary consequences of habitat occupancy or microhabitat segregation.[4]

This seems like a perfectly valid interpretation of the results from cafeteria experiments and fecal analyses discussed earlier. However, information about food choice in captivity and about routinely consumed diets in the wild can be uninformative about evolutionarily and ecologically important trophic adaptations. Optimal foraging theory tells us that preferred foods should be those that are easy to harvest, process, and digest; inevitably, such foods will tend to be widely used by many species, all of which will then seem like opportunistic dietary generalists. But dietary staples can be in seasonally short supply, and when they are, some species may fall back on different foods that require special phenotypes to consume effectively. Because such fallback foods (Marshall and Wrangham, 2007; Constantino and Wright, 2009) may be key to survival in hard times, these may be the principal drivers of dietary specialization.

Phenotypic specialists can function as ecological generalists because some resources are intrinsically easy to use, even by consumers that have evolved specialized traits to exploit less-favored resources. . . . As paradoxical as it may sound, consumers can be elaborately designed by natural selection to exploit resources that account for only a small portion of their diet, are utilized relatively poorly, and are frequently ignored in favor of other resources. (Robinson and Wilson, 1998: 231–232)

The concept of fallback foods was initially proposed to explain "Liem's Paradox," the observation that some species with unusual trophic morphologies often seem to have rather generalized diets (Liem, 1980). As we have seen, didelphids exhibit a limited range of trophic morphologies—for example, modest taxonomic differences in dental traits and gastrointestinal proportions—which provide some evidence of dietary divergence but do not suggest any extraordinary specialization. Not all trophic adaptations are morphological, however, and other aspects of didelphid phenotypes tell a different story.

Snake-venom resistance is perhaps the most compelling example of opossum dietary specialization and trophic adaptation. As described in Chapter 6, several members of the tribe Didelphini (species of *Didelphis, Lutreolina,* and *Philander*) are known to eat pitvipers and to be highly resistant to pitviper venom. For those opossums that can eat them, venomous snakes may be an important fallback food at times when preferred resources are scarce. Indeed, no other explanation seems plausible, because venom resistance is a genetically complex trait—almost certainly involving amino-acid substitutions in dozens of metabolically important proteins (Voss and Jansa, 2012)—that is unlikely to have evolved without sustained natural selection for ophiophagy. Because ophiophagous mammals (venom-resistant opossums, mongooses, hedgehogs, etc.) do not share any special morphological traits in common (Voss and Jansa, 2012), ophiophagy could be thought of as a cryptic dietary specialization, and if ophiophagy is only important when other foods are scarce, then snakes will not often appear in fecal samples, even of species that are physiologically adapted to eat them.

Other opossum trophic adaptations might also be morphologically inapparent and difficult to detect by routinely analyzed dietary data. As discussed earlier in this chapter, fecal samples are often rinsed on coarse screens before macroscopic sorting of seeds and insect fragments, so diets rich in soft-bodied invertebrates and fungi (for example) cannot be recognized by such methods. Because soft-bodied invertebrates and fungal fruiting structures are often biochemically protected, these items might also be important fallback foods for toxin-resistant species. Toxic amphibians (toads and newts) might be another such resource. Future researchers should be alert to the possibility that opossum dietary specializations are more likely to be behavioral or physiological than morphological, and they should pay more attention to diets during seasons of food scarcity than to what is eaten when food is abundant.

NOTES

1. Earwigs are present in most didelphid habitats, adult earwigs are well within the size range of invertebrate prey commonly eaten by opossums, and earwig cercal forceps would be unmistakable

were they often found in didelphid feces. However, in our review of dietary studies (based on a cumulative total of >2600 fecal and stomach samples from 16 opossum species), we encountered only two reports of dermapteran consumption, both consisting of single individuals (Santori et al., 1995, 1997). Earwigs are known to deploy defensive chemicals (Gasch et al., 2013), which would appear to effectively deter mammalian predation based on this circumstantial evidence.

2. See Hamilton (1958), who remarked that fecal analyses of *Didelphis virginiana* probably underestimate the dietary importance of toads (Bufonidae), which comprised a substantial fraction of undigested stomach contents in his study.

3. The Virginia opossum is often implicated as a bird-nest predator in North American habitats (e.g., by Henry, 1969; Wilcove, 1985; Crooks and Soulé, 1999), and other opossum species are apparently responsible for nest predation in the Neotropics (Brightsmith, 2005; Menezes and Marini, 2017).

4. This possibility was recently discussed by Finotti et al. (2018), who reported statistically significant correlations between data from Astúa et al.'s (2003) dietary experiments and a numerical index of vertical microhabitat use. However, some species in Astúa et al.'s study were omitted without explanation, so it is not clear whether or not Finotti et al.'s results were biased by data selection.

# 10

# Parasites

Like most other mammals, didelphids are hosts to numerous arthropod, helminth, and protozoan symbionts. Most arthropod symbionts are ectoparasites, whereas symbiotic helminths ("worms") and protozoans are endoparasites, but life cycles, effects on host fitness, and other characteristics vary greatly within each of these groups. In the absence of any published review of didelphid parasitology, we provide brief synopses of relevant information in the accounts that follow. In describing parasitic arthropods, we distinguish temporary parasites (which spend only the feeding stage of their life cycle on opossum hosts) from permanent parasites (which spend their entire life cycle on the host). In discussing parasitic helminths, the distinctions among definitive hosts (in which reproduction occurs), intermediate hosts (in which larval or nymphal stages are passed), and paratenic hosts (used only for dispersal; Chubb et al., 2009) are likewise important.

## Arthropods

Arthropod parasites of opossums include mites, ticks, and chiggers (Acari); flies (Diptera); lice (Phthiraptera); and fleas (Siphonaptera). At least 17 families belonging to these higher taxa may routinely parasitize didelphids (Table 10.1). By contrast, some mites that are occasionally found in didelphid fur (e.g., macrochelids, pygmephorids; Whitaker and Wilson, 1974; Whitaker et al., 2007) are phoretic rather than parasitic—using opossum hosts solely for transportation—and are not considered further here. Similarly, rove beetles (Staphylinidae: Amblyopinini) are sometimes found in didelphid fur and were once thought to be parasitic blood feeders, but amblyopinines are now thought to be phoretic predators that benefit their hosts by eating fleas and mites (Ashe and Timm, 1987). Lastly, several groups of hematophagous insects, including mosquitos (Diptera: Culicidae), sand flies (Diptera: Psychodidae), and kissing bugs (Hemiptera: Reduviidae) have only transient contact with opossums and are not usually considered to be truly parasitic.

*Table 10.1*    Some Arthropod Parasites of Didelphids[a]

| Taxon | Infestation site |
| --- | --- |
| ACARI | |
| Ixodida (ticks) | |
| Argasidae | skin surface |
| Ixodidae | skin surface |
| Mesostigmata (mesostigmatan mites) | |
| Laelapidae | fur |
| Macronyssidae | skin surface |
| Trombidiformes (chiggers and prostigmatid mites) | |
| Demodicidae | cutaneous glands |
| Myobiidae | fur |
| Trombiculidae | skin surface |
| Sarcoptiformes (including astigmatid mites) | |
| Atopomelidae | fur |
| Echimyopodidae | hair follicles |
| Glycyphagidae | hair follicles |
| Listropsoralgidae | fur |
| Myocoptidae (Dromiciocoptinae) | fur |
| INSECTA | |
| Diptera (flies) | |
| Oestridae (Cuterebrinae) | subcutaneous tissues |
| Phthiraptera (lice) | |
| Trimenoponidae | fur and skin surface |
| Siphonaptera (fleas) | |
| Hystrichopsyllidae | fur |
| Pulicidae | fur |
| Rhopalopsyllidae | fur |

[a] Based on references cited in the text. Some taxa known from only a few, possibly incidental, records are excluded (e.g., Cheyletidae, Lobalgidae, and Speleognathidae).

Ticks (Ixodida) are temporary blood-feeding ectoparasites of amphibians, reptiles, birds, and mammals (Hoogstraal and Kim, 1985; Durden, 2006). Most soft ticks (Argasidae) parasitize reptiles and birds, but *Ornithodoros chironectes* was described from larvae infesting the water opossum (Jones and Clifford, 1972), larvae of *O. marmosae* commonly infest *Marmosa robinsoni* (Jones et al., 1972), and adults of *Argas miniatus* have been found on *Thylamys macrurus* (Cáceres et al., 2007). Although many species of hard ticks (Ixodidae) have been reported from didelphids, Neotropical opossums are important hosts of just four species: *Ixodes amarali, I. loricatus, I. luciae,* and *I. venezuelensis* (Guglielmone et al., 2011). Adults of these species are primarily found on large opossums—especially *Didelphis* and *Philander*—but their larvae and nymphs usually feed on cricetid rodents (Díaz et al., 2009; Labruna et al., 2009). Because cricetids and didelphids have only been sympatric for a few million years (Chapter 3), such host-parasite relationships must have evolved relatively recently (Guglielmone et al., 2011). The North American opossum *Didelphis virginiana* is often parasitized by species of the ixodid genera *Amblyomma, Dermacentor,* and *Ixodes* (Durden and Wilson, 1990; Pung et al., 1994), which include known vectors of several rickettsial diseases (Azad and Beard, 1998). Indeed, ticks in general (and ixodids in particular) are important vectors of many microbial parasites and pathogens (Durden, 2006).

Laelapids are large, active mites that are often found in opossum fur, but it is not known whether or not they are true parasites. Some Old World laelapids feed on vertebrate blood or lymph (Radovsky, 1985, 1994), but blood has not been found in the digestive tracts of Neotropical laelapids, and some acarologists (e.g., Martins-Hatano et al., 2002) believe that they might be nest-inhabiting scavengers that use mammalian hosts only for transportation (phoresis). All of the laelapid species commonly recovered from didelphids (e.g., by Furman, 1972; Martins-Hatano et al., 2002; Gettinger et al., 2005) belong to the genus *Androlaelaps.* Typically, only adult female *Androlaelaps* are found in opossum fur, the immature stages and males presumably remaining in the nest (Martins-Hatano et al., 2002).

The only macronyssid mites that commonly parasitize opossums belong to the genus *Ornithonyssus.* Species of *Ornithonyssus* are blood-feeding ectoparasites, two of which—*O. bacoti* and *O. wernecki*—are often found on didelphids (e.g., in Venezuela; Saunders, 1975). *Ornithonyssus bacoti* (commonly known as the tropical rat mite) is now cosmopolitan, although it is thought to have evolved in South America (Radovsky, 1985, 1994). This species also infests many sympatric species of rodents and is known to transmit filarial nematodes, trypanosomes, and other microbial parasites and pathogens (Valiente-Moro et al., 2005).

Little is known about the biology of trombidiform mites known to parasitize didelphids. Demodicids are permanent endoparasites that inhabit mammalian hair follicles, skin glands, lymph nodes, subdermal fascia, and other internal tissues (Nutting, 1985); the only two demodicids known from opossums were both found in the cerumen (earwax) glands of the external auditory canal of *Didelphis* (Nutting et al., 1980). The myobiid mites found on didelphids are permanent ectoparasites that attach themselves to individual hairs with specialized clasping organs; they may have coevolved closely with their mammalian hosts (species of *Archemyobia* only occur on didelphids, whereas related genera are found on microbiotheriids, caenolestids, and Australian marsupials; Fain, 1994). Chiggers (Trombiculidae) are temporary skin-feeding ectoparasites; the six-legged larvae are parasitic, but the eight-legged post-larval stages are free-living predators of soil micro-arthropods. Neotropical didelphids are commonly infested with larval chiggers, especially species of *Eutrombicula* and *Odontocarus* (Brennan and Reed, 1975).

Sarcoptiform mites that parasitize opossums are even more obscure. Atopomelids, dromiciocoptines, and listropsoralgids are permanent ectoparasites that live in mammalian fur and seem to be quite host-specific; only a few atopomelids and dromiciocoptines are known to parasitize didelphids, but Neotropical listropsoralgids infest at least 13 didelphid species in six genera and seem appropriately characterized as opossum specialists (Bochkov and OConnor, 2008; Bochkov and Wauthy, 2009; Bochkov, 2011; Bochkov et al., 2013). By contrast, echimyopodids and glycyphagids are temporary dermal parasites—only the follicle-inhabiting deutonymphs are parasitic (OConnor, 1982; Fain et al., 1972)—and echimyopodids, at least, do not appear to be very host-specific (OConnor et al., 1982). Other sarcoptiform clades that occasionally parasitize didelphids are discussed by Bochkov (2011).

The larvae (maggots) of New World botflies (Diptera: Oestridae: Cuterebrinae) are temporary subcutaneous endoparasites of a wide variety of mammals (Catts, 1982). They are infrequently found on marsupials, but a few studies have reported high local infestation rates (Bossi and Bergallo, 1992; Cruz et al., 2009). Although botfly maggots are large and repulsive, their infestation is brief and the wounds they cause usually heal quickly (in part because antimicrobial substances are secreted by the feeding larva; Colwell et al., 2006). In the only field study to date of the impact of botfly parasitism on an opossum, infested individuals of *Gracilinanus agilis* were found to be mildly anemic (despite the fact that botfly larvae do not feed on blood), but their body condition was unaffected by comparison with botfly-free individuals in the same population (Zangrandi et al., 2019).

Lice are permanent fur-dwelling ectoparasites commonly found on many Neotropical mammals, but they seem to be infrequently reported from opossums (e.g., in Venezuela; Emerson and Price, 1975). The only lice consistently associated with didelphids belong to the trimenoponid genus *Cummingsia*, other species of which occur on caenolestid marsupials and cricetid rodents (Timm and Price, 1985). To date, the only opossums known to host these lice are species of *Gracilinanus*, *Marmosops*, and *Monodelphis* (Timm and Price, 1985; Price and Emerson, 1986).

Fleas are temporary blood-feeding ectoparasites (the larval and pupal stages occurring in nests or in the soil). Most of the fleas reported from didelphids in large-scale faunal surveys (e.g., Tipton and Machado-Allison, 1972) and regional studies (e.g., Pinto et al., 2009; Lareschi et al., 2010) are species of the hystrichopsyllid genus *Adoratopsylla*, but some pulicids (especially *Ctenocephalides felis* and

*Pulex simulans*) and rhopalopsyllids (especially *Polygenis* spp.) are also commonly found on opossums (Linardi, 2006). Stephanocircids, which are sometimes said to be marsupial fleas (Traub, 1985), are more frequently collected from cricetid rodents than from marsupials in South America (e.g., in Venezuela; Tipton and Machado-Allison, 1972). Like many other temporary blood-feeding ectoparasites, fleas are important vectors of microbial disease organisms, some of which are known to infect opossums (e.g., rickettsiae; Azad and Beard, 1998).

## Helminths

Helminth ("worm-like") parasites of didelphids include acanthocephalans (spiny-headed worms), cestodes (tapeworms), digeneans (flukes), nematodes (round-worms), and pentastomids (tongue worms). Didelphids are definitive hosts for most helminths, but they are intermediate hosts of pentastomids, the definitive hosts of which are often predatory reptiles, especially large snakes (Riley, 1986). Because pentastomids have only been reported in a single study of didelphid helminths (Tantaleán et al., 2010), this group is not considered further here.

Members of at least 35 helminth families are known to parasitize opossums (Table 10.2), but any account of didelphid helminth parasites must be qualified by recognizing several important sampling biases. First, only abdominal and thoracic viscera—stomach, intestines, liver, heart, lungs, and so forth—are routinely examined for endoparasites (e.g., by Ellis et al., 1999; Gomes et al., 2003; Jiménez et al., 2008, 2011; Torres et al., 2009), so helminths that infest other sites—for example, the cranial sinuses, middle ear, central nervous system, muscles, and lymph—are probably underreported in local and regional surveys. Second, most information about helminth diversity comes from just a few countries with active parasitological research programs, including the United States (e.g., Potkay, 1970; Alden, 1995), Mexico (Monet-Mendoza et al., 2005; García-Prieto et al., 2012), Brazil (Vicente et al., 1997; Thatcher, 2006; Pinto et al., 2011), and Argentina (Lunaschi and Drago, 2007). Therefore, scant information is available from vast areas with diverse opossum faunas but few resident parasitological researchers (e.g., the tropical Andes, the Guianas, and much of Central America). Lastly, most information about opossum helminths comes from large, easily collected host taxa—notably species of *Didelphis* and *Philander*—with the result that relatively little is known about the endoparasites of smaller and more elusive opossums (*Gracilinanus*, *Marmosops*, *Monodelphis*, *Thylamys*, etc.).

Adult acanthocephalans attach to the intestinal mucosa of their definitive vertebrate hosts with an evertable hooked proboscis and absorb nutrients through

Table 10.2    Some Helminth Parasites of Didelphids

| Taxa | Infestation site |
|---|---|
| ACANTHOCEPHALA (spiny-headed worms) | |
| Gigantorhynchida | |
| Gigantorhynchidae | large & small intestines |
| Oligocanthorhynchida | |
| Oligocanthorhynchidae | large & small intestines |
| Polymorphida | |
| Centrorhynchidae | small intestine |
| Plagiorhynchidae | small intestine |
| CESTODA (TAPEWORMS) | |
| Cyclophyllidea | |
| Anoplocephalidae | small intestine |
| Mesocestoididae | small intestine |
| Proteocephalidea | |
| Proteocephalidae | intestines |
| DIGENEA (flukes) | |
| Brachylaimidae | small intestine |
| Dicrocoeliidae | biliary ducts, gall bladder, pancreatic ducts |
| Diplostomidae | lungs, large & small intestines |
| Echinostomatidae | stomach, intestines |
| Opisthorchiidae | biliary ducts, gall bladder |
| Paragonimidae | lungs |
| Rhopaliidae | large & small intestines |
| Strigeidae | small intestine |
| NEMATODA (round worms) | |
| Ascaridida | |
| Ascarididae | stomach |
| Aspidoderidae | large intestine |
| Kathlaniidae | large & small intestine |
| Subuluridae | small intestine |
| Oxyurida | |
| Oxyuridae | small intestine, caecum |
| Rhabditida | |
| Strongyloididae | digestive tract, lungs |
| Spirurida | |
| Gnathostomatidae | stomach, intestines, liver |
| Gongylonematidae | esophagus |
| Onchocercidae | body cavities, connective tissue, heart, blood |
| Physalopteridae | stomach |
| Rictulariidae | small intestine |
| Spirocercidae | stomach |
| Spiruridae | esophagus, stomach, large & small intestines |

| Taxa | Infestation site |
|------|------------------|
| Strongylida | |
| Angiostrongylidae | lungs |
| Globocephalidae | small intestine |
| Heligmosomidae | small intestine |
| Trichostrongylidae | stomach, small intestine, lungs |
| Trichurida | |
| Capillariidae | lungs, esophagus |
| Trichuridae | mouth, lungs, stomach, large intestine |
| Viannaiidae | large & small intestines |
| PENTASTOMIDA (tongue worms) | |
| unidentified nymphs | lungs, liver, stomach, and other viscera |

[a] Based on references cited in the text. Does not include incidental records (e.g., of parasites with fish intermediate hosts recorded only once or twice from *Didelphis virginiana*).

their integument (Fig. 10.1A, 10.1B). Acanthocephalans that parasitize terrestrial mammals usually have insects as intermediates hosts, but some acanthocephalan life cycles also involve one or more paratenic hosts (used for dispersal) in which development does not proceed (Nickol, 1985). Therefore, opossums might acquire acanthocephalan parasites by eating insect intermediate hosts (commonly beetles and orthopteroids) or by eating paratenic hosts (typically frogs, lizards, and snakes; Nickol, 1985). Didelphids are known to be parasitized by members of four acanthocephalan families, of which Oligacanthorhynchidae is the most frequently reported. Species of the genus *Oligacanthorhynchus*, for example, have been reported from species of *Caluromys*, *Didelphis*, *Marmosa*, *Philander*, and *Thylamys*. Significant harm to opossum hosts may be caused by acanthocephalan parasitism (Alden, 1995), although few case studies have been reported. Mature adult females of *Gigantorhynchus ortizi*, which have been found in the intestines of several rainforest opossums, are impressively large (>200 mm; Sarmiento, 1954), and considerable numbers of this species are sometimes found in infested hosts (Tantaleán et al., 2010).

Adult cestodes, like acanthocephalans, are intestinal parasites that lack an alimentary canal and absorb nutrients through their integument. Most tapeworms infesting didelphids belong to the order Cyclophyllidea, but a new genus of proteocephalidean tapeworms was recently described from *Didelphis marsupialis* by Cañeda-Guzmán et al. (2001). Neotropical opossums are usually parasitized by anoplocephalid tapeworms of the genera *Linstowia* and *Mathevotaenia*, the intermediate hosts of which are probably insects. By contrast, the tapeworms most

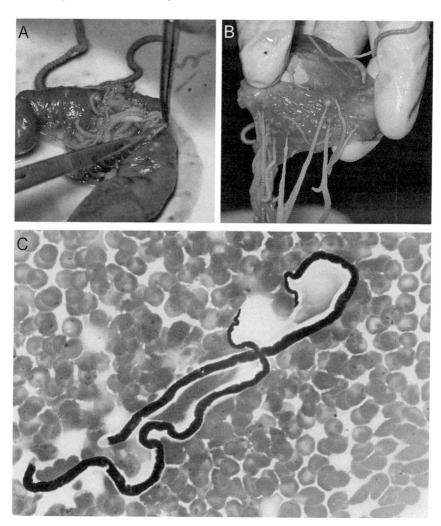

*Fig. 10.1.* Parasitic helminths: **A**, large acanthocephalans (*Gigantorhynchus ortizi*) spilling from the cut intestines of *Metachirus myosuros* (photo by Mónica Díaz); **B**, the same species clinging to the inner wall of the bowel (photo by Mónica Díaz); **C**, larval nematodes (Onchocercidae: *Skrjabinofilaria* sp.) nestled among erythrocytes in a Giemsa-stained blood smear from *Marmosa robinsoni* (photo by Kelly Kroft)

commonly found in the Virginia opossum are species of *Mesocestoides* (Mesocestoididae), the intermediate or paratenic hosts of which are small vertebrates (rodents, birds, reptiles, or amphibians). Tapeworms seem to cause little visible damage to infested opossum hosts (Alden, 1995), and no cases of morbidity or mortality attributable to cestode parasitism of didelphids seem to have been reported. Infestation rates often seem to be quite low—for example, Tantaleán et al. (2010) found no tapeworms in 40 examined specimens representing eight didelphid species—but high frequencies are sometimes reported (Ellis et al., 1999; Byles et al., 2013).

Digeneans have complex life cycles, often involving several hosts, of which the first is always a mollusk (Chubb et al., 2009). Although flukes representing at least eight families have been reported from opossum hosts, only a few genera seem to commonly parasitize didelphids, notably *Brachylaima* (Brachylaimidae; reported from species of *Didelphis* and *Lutreolina*), *Zonorchis* (Dicrocoeliidae; from *Caluromys* and *Didelphis*), *Podospathalium* (Diplostomidae; from *Didelphis, Metachirus, Monodelphis*, and *Philander*), *Paragonimus* (Paragonimidae; from *Didelphis* and *Philander*), and *Rhopalias* (Rhopaliidae; from *Chironectes, Didelphis, Lutreolina, Metachirus*, and *Philander*). The life cycles of only a few digenean species known to parasitize didelphids have been worked out, but most opossum flukes are probably acquired by ingesting intermediate or paratenic hosts (Kearn, 1998). Lung flukes are known to cause pulmonary lesions in opossums (García-Márquez et al., 2010), and intestinal flukes sometimes occur in sufficiently large numbers that they must have negative effects on host nutrition (Nettles et al., 1975). Nevertheless, digenean parasites are usually less prevalent than parasitic nematodes (e.g., in Mexican populations of *Didelphis virginiana*; Monet-Mendoza et al., 2005), so the net effect of digenean parasitism on the demography of opossum hosts is almost certainly less than that of nematodes.

Nematodes are by far the most diverse and important group of helminths that parasitize opossums. Multiple parasitic nematode species have been reported from all well-sampled didelphid populations (e.g., by Foster, 1939; Silva and Costa, 1999; Gomes et al., 2003; Lopes-Torres et al., 2009; Byles et al., 2013); nematode infestation rates are typically quite high (Alden, 1995; Ellis et al., 1999; Monet-Mendoza et al., 2005; Lopes-Torres et al., 2009); and there is clear evidence, at least from the well-studied Virginia opossum, that nematodes can cause significant harm to their hosts (Nettles et al., 1975; Gray and Anderson, 1982; Duncan et al., 1989; Snyder et al., 1991). The 14 nematode genera most commonly reported from didelphids (Table 10.3) represent a diversity of parasitic life cycles (Anderson, 2000).

Most parasitic nematodes live in the digestive tract, shedding their eggs into the gut lumen, from which they are passed in feces. In some taxa—for example, *Cruzia, Trichuris*, and (probably) *Aspidodera*—the eggs are transmitted directly from definitive host to definitive host. Alternatively, eggs passed in feces are ingested by an intermediate host (typically earthworms or insects), wherein they hatch and

*Table 10.3*     Parasitic Nematode Genera Widely Reported from Didelphid Hosts[a]

| Parasites | Hosts |
| --- | --- |
| *Aspidodera* (Aspidoderidae) | *Caluromys, Chironectes, Didelphis, Marmosa, Metachirus, Monodelphis, Marmosops, Philander* |
| *Capillaria* (Capillariidae) | *Didelphis, Philander* |
| *Cruzia* (Kathlaniidae) | *Chironectes, Didelphis, Marmosa, Metachirus, Philander* |
| *Dipetalonema* (Onchocercidae) | *Chironectes, Didelphis, Philander* |
| *Gnathostoma* (Gnathostomatidae) | *Caluromys, Didelphis, Lutreolina, Philander* |
| *Litomosoides* (Onchocercidae) | *Chironectes, Marmosa, Monodelphis* |
| *Physaloptera* (Physalopteridae) | *Gracilinanus, Metachirus, Philander* |
| *Pterygodermatites* (Rictulariidae) | *Caluromys, Didelphis, Gracilinanus, Lestodelphys, Marmosa, Thylamys* |
| *Spirura* (Spiruridae) | *Caluromys, Didelphis, Gracilinanus, Marmosa, Marmosops, Philander* |
| *Subulura* (Subuluridae) | *Caluromys, Didelphis, Gracilinanus* |
| *Travassostrongylus* (Viannaiidae) | *Didelphis, Lutreolina, Marmosa, Metachirus, Philander, Thylamys* |
| *Trichurus* (Trichuridae) | *Chironectes, Didelphis, Marmosa, Marmosops, Philander* |
| *Turgida* (Trichuridae) | *Caluromys, Chironectes, Didelphis, Philander* |
| *Viannaia* (Viannaiidae) | *Didelphis, Marmosa, Metachirus, Monodelphis, Marmosops, Philander* |

[a] Based on references cited in the text.

larval development begins; species of *Subulura, Physaloptera, Turgida, Pterygodermatites, Spirura,* and *Capillaria* exemplify this common type of life cycle, which is completed when the definitive host ingests the intermediate host or a paratenic host (often a reptile or amphibian). Less commonly, eggs hatch in the environment and free-living larvae infect an intermediate host, from which they subsequently pass up the food chain through one or more paratenic hosts to the definitive host; species of *Gnathostoma* (the first intermediate hosts of which are copepods) exemplify this type of life cycle. By contrast, many onchocercids (including species of *Dipetalonema* and *Litomosoides*) occur in the abdominal and thoracic cavities rather than digestive viscera, and their transmission does not involve the food chain; instead, microscopic larvae (microfilariae) develop in the uteri of female worms and migrate to the host's bloodstream (Fig. 10.1C), where they are accessible for transmission by blood-feeding arthropod vectors (dipterans, ticks, macronyssid mites, etc.), which also serve as intermediate hosts. Apparently, nothing definite is known about the life cycles of *Travassostrongylus* and *Viannaia*, which include many common opossum parasites.

# Protozoans

Among the single-celled eukaryotic ("protozoan") parasite genera known to infest didelphids are euglenozoans (e.g., *Leishmania, Trypanosoma*), several apicomplexans (*Babesia, Eimeria, Hepatozoon, Sarcocystis*), and metamonads (*Giardia, Tetratrichomonas*). The biology of most protozoan species that infest opossums is poorly known, and no general review is attempted here. Instead, we briefly discuss the exceptions: two relatively well-studied microbes that also cause human disease. We note in passing that species of *Plasmodium*, apicomplexan hemoparasites that cause malaria in primates, rodents, birds, and lizards, are not known to infect marsupials.

The best-studied protozoan parasites of opossums are members of the kinetoplastidean euglenozoan order Trypanosomatida. This medically important taxon includes *Trypanosoma cruzi*, the causative agent of Chagas disease, one of the most serious zoonotic illnesses in the New World tropics. Like many other trypanosomatids (Lukeš et al., 2014), the transmission cycle of *T. cruzi* typically includes both invertebrate and vertebrate hosts. The invertebrate hosts of *T. cruzi* are hematophagous bugs of the reduviid subfamily Triatominae, whereas the vertebrate hosts are mammals. In the lingo of medical epidemiologists (e.g., Brener, 1973), triatomines are said to be "vectors" and nonhuman mammals to be "reservoirs" of *T. cruzi*, but such anthropocentric terminology oversimplifies the variety of transmission cycles by which this parasite is apparently maintained in nature (Jansen et al., 2015).

Triatomine hosts acquire *Trypanosoma cruzi* when they ingest trypomastigotes (the flagellated, infective, extracellular stage of the parasite) with a blood meal from an infected mammal. Within the insect gut, trypomastigotes transform to epimastigotes (a flagellated but noninfective extracellular stage), which then undergo binary fission before transforming back into infective trypomastigotes in the rectum. Transmission to another mammalian host occurs when the insect feeds again, simultaneously defecating trypomastigotes that are probably licked up as the bitten mammal grooms the wound site. Ingested typomastigotes penetrate a wide variety of mammalian tissues, especially muscle and glia, where they transform into amastigotes (an unflagellated intracellular stage) that undergo binary fission about every 12 hours until the infected cell is filled, whereupon they transform into trypomastigotes that enter the bloodstream to infect other cells or to be taken up by the next blood-feeding triatomine (Brener, 1973).

*Trypanosoma cruzi* is a genetically diverse generalist parasite that is known to infect at least 180 species of wild, commensal, and domesticated mammals from the southern United States to central Argentina (Noireau et al., 2009). Didelphids are commonly infected with *T. cruzi* throughout the known range of the parasite, but many other sympatric mammals—including artiodactyls, bats, carnivorans, and primates—often sustain equally high infection rates. The prevalence of *T. cruzi* infections among opossums sampled by recent parasitological surveys in six Brazilian biomes—Amazonia, Atlantic Forest, Caatinga, Cerrado, Pantanal, and Pampas—ranged from 6% to 51% (Jansen et al., 2015). This parasite is most commonly reported from species of *Didelphis*, but perhaps only because these large opossums are easy to trap; other didelphid genera that *T. cruzi* is known to infect include *Caluromys*, *Gracilinanus*, *Marmosa*, *Monodelphis*, and *Philander* (Coura and Junqeira, 2015; Jansen et al., 2015), so it would be reasonable to expect *T. cruzi* to infect almost any tropical or subtropical opossum species.

The special interest of opossums for researchers working on *Trypanosoma cruzi* is due to the fact that this parasite is apparently capable of invading the anal ("scent") glands of several didelphid species, where epimastigotes (otherwise found only in triatomines) then proliferate in the extracellular lumen and produce infective trypomastigotes (Deane et al., 1984; Lenzi et al., 1984). Because the glandular contents are expelled with feces—either during normal defecation or as part of the defensive behavior previously described for *Didelphis* (Chapter 7)—it seems plausible that opossums might directly transmit the parasite to other mammals (e.g., to attacking carnivorans), thus bypassing the usual triatomine intermediary host (Carreira et al., 2001). To some researchers, this unique ability of didelphids to sustain the otherwise "invertebrate" cycle of *T. cruzi* in the anal glands is evidence for an ancient coevolutionary relationship between host and parasite.

Other medically important opossum parasites belong to the trypanosomatid genus *Leishmania*, species of which cause cutaneous, mucocutaneous, and visceral leishmaniases in human populations throughout the tropical lowlands of Central and South America. Twenty-one Neotropical species of *Leishmania* are currently recognized, of which at least seven are known to occur in opossums (Lainson, 2010; Quaresma et al., 2011). Insofar as known, all of the Neotropical species of *Leishmania* are transmitted from mammalian host to mammalian host by phlebotomine sand flies (Diptera: Psychodidae). Species of *Leishmania* are intracellular parasites of mammals; sand flies ingest mammalian blood or epithelial cells containing amastigotes, from which flagellum-bearing promastigotes develop in the insect's gut; the promastigotes then secrete a gut-blocking gel, forcing the fly to regurgitate before it ingests its next blood meal, thereby depositing the gel and the parasites it contains on mammalian skin to complete the normal transmission cycle (Bates, 2007). The species of *Leishmania* most commonly associated with didelphid hosts is *L. amazonensis*, which is known to infect *Caluromys, Didelphis, Marmosa, Marmosops, Metachirus*, and *Philander* in addition to sympatric carnivorans and rodents (Rotureau, 2006; Lainson, 2010). Marsupial infections with *Leishmania* are said to be externally inapparent (Rotureau, 2006; Lima et al., 2013), suggesting that didelphids may be coevolved natural hosts of this parasite. Infection prevalence of *Leishmania* spp. in opossums sampled by parasitological surveys in some areas of high zoonotic leishmaniasis can be as high as 33%, but infection rates in sympatric rodents can be even higher (Lima et al., 2013).

## Parasite Diversity

Parasite diversity can be quantified in several different ways (Poulin, 2007): (1) at the faunal level, by counting all the parasite species known to infest a single host species anywhere in its geographic range; (2) at the level of the so-called component community, by counting all the parasite species that infest a given local population of a host species; and (3) at the level of the so-called infracommunity, by estimating the mean number of parasite species that infest individual hosts. Obviously, infracommunities must be subsets of component communities, which must in turn be subsets of the larger parasite fauna that infests the host species throughout its range.

The total parasite fauna known to infest a host species might be large if the ecogeographic range of the host is extensive. Widespread host species that occur in several different biomes might support several distinct component communities of temporary ectoparasites if the nonfeeding stages require different soils, climates, or other abiotic attributes for off-host survival. Widespread host species might also support numerous distinct component communities of helminth parasites acquired by eating habitat- or biome-specific intermediate or paratenic hosts. These seem like reasonable explanations for the very large number of macroparasites known to infest *Didelphis virginiana*, for example, which ranges across several biomes from Central America to Canada. A total endoparasite fauna of 70 helminth species (including 7 acanthocephalans, 8 cestodes, 21 digeneans, and 34 nematodes) have been recorded from *D. virginiana* in the United States and Mexico (Alden, 1995; García-Prieto et al., 2012), whereas component communities of helminths from local parasite surveys of this host are consistently much smaller (e.g., 12 species in southern Illinois; Alden, 1995). Therefore, parasite faunas (in the comprehensive, species-wide sense) are of little, if any, ecological or physiological significance.

Of greater potential ecological interest is the number of parasite species that infest local host populations, the component-community richness. However, this number—like other measures of biological extent—is sample-size dependent, because component-community richness monotonically increases as more hosts are examined (up to some asymptotic value, presumably). Also, if many local parasite species each infest only a small number of host individuals, then component-community richness is a poor predictor of infracommunity richness. The latter is perhaps the most biologically relevant measure of parasite diversity, and because it is a measure of central tendency rather than extent, it is not sample-size dependent.

Although infracommunity richness values have seldom been reported in published studies of didelphid parasites, an estimate of the average number of parasite species per host individual can be computed from infestation frequency data (usually reported as "prevalence" and scaled as percentages) on the assumption that parasite infestations are independent.[1] Comparisons of such estimates with component-community richness from nine studies of didelphid gastrointestinal helminths (Table 10.4) suggest that individual opossums are usually infested by fewer than half the species in the corresponding component community. As reported by the single study known to us of opossum ectoparasites that provides relevant frequency data (Durden and Wilson, 1990), the component community that infests *Didelphis virginiana* in central Tennessee includes 12 species (of fleas, ticks, and mites), but the estimated mean infracommunity richness is only about two species. Based on this and other studies, it would seem reasonable to expect that opossum macroparasite component communities often include 10 or more species of helminths and perhaps an equivalent number of arthropod ectoparasites; of these, however, individual opossums might often be infested, on average, by no more than three or four helminth species and perhaps by only one or two species of ectoparasites.

Such numbers are difficult to put into comparative perspective because there are only a few quantitative results from parasitological studies of sympatric placental mammals of equivalent body size and trophic habits. Perhaps the best comparative data are available from several surveys of the endo- and ectoparasites of raccoons (*Procyon lotor*) in the southeastern United States. In the aggregate, these studies suggest that raccoons are about equally parasitized as Virginia opossums. For example, the mean infracommunity richness of gastrointestinal helminths

Table 10.4  Ten Opossum Gastrointestinal Helminth Communities[a]

| Host | N[b] | Locality | Method | CCR[c] | ICR[d] | References |
|---|---|---|---|---|---|---|
| *Didelphis albiventris* | 22 | Brazil (Minas Gerais) | necropsies | 10 | 3.27 | Silva & Costa (1999) |
| *Didelphis virginiana* | 30 | United States (Georgia) | necropsies | 12 | 6.95 | Ellis et al. (1999) |
| *Didelphis virginiana* | 46 | United States (Illinois) | necropsies | 10 | 4.00 | Alden (1995) |
| *Gracilinanus agilis* | 102 | Brazil (São Paulo) | fecal analysis[e] | 11 | 1.07 | Meyer-Lucht et al. (2010) |
| *Marmosa demerarae* | 21 | French Guiana | necropsies | 12 | 3.67 | Byles et al. (2013) |
| *Marmosa murina* | 18 | French Guiana | necropsies | 14 | 2.95[f] | Byles et al. (2013) |
| *Marmosops incanus* | 123 | Brazil (São Paulo) | fecal analysis[e] | 13 | 1.42 | Meyer-Lucht et al. (2010) |
| *Metachirus myosuros* | 11 | Peru (Loreto) | necropsies | 6 | 0.81 | Tantaleán et al. (2010) |
| *Philander opossum* | 26 | French Guiana | necropsies | 10 | 4.95 | Jiménez et al. (2011) |
| *Philander vossi* | 49 | Mexico | necropsies | 12 | 2.97 | Ramírez-Cañas et al. (2019) |

[a] From studies with >10 examined host individuals; tabulated data do not include parasite species found in nondigestive viscera (lungs, coelomic adipose tissue, etc.), which were not consistently examined in all studies.
[b] Number of necropsied host individuals or analyzed fecal samples.
[c] Total parasite component community species richness.
[d] Average parasite infracommunity species richness (mean number of parasite species per necropsied specimen, estimated from prevalence data; see text).
[e] Microscopic examination of feces for helminth eggs; parasite "species" are egg morphotypes.
[f] Differs from infracommunity richness value provided by Byles et al. (2013), which they computed only from infested host individuals (28% of necropsied *Marmosa murina* in their study had no gastrointestinal helminths).

(estimated from infestation frequencies) in a sample of 30 raccoons from Arkansas (Richardson et al., 1992) is about 5.5 species per host individual. In the only comparative study of ectoparasite faunas from sympatric opossums and raccoons (Pung et al., 1994), estimates based on infestation frequency data yield sim-· ilar infracommunity richness estimates (about 1.8 ectoparasite species per host individual). Few as they are, these data suggest that opossums do not suffer substantially higher (or lower) rates of parasitism than co-occurring placental mammals of similar size and habits.

## Discussion

Although opossums are infested by a wide range of metazoan parasites, all the arthropod and helminth higher taxa (families, orders, etc.) that commonly parasitize opossums also parasitize other small mammals (Morand et al., 2006), and research on medically important trypanosomatids suggests that the same is true of parasitic protozoans. Additionally, the scant literature on opossum parasite communities suggest that didelphids are neither much more nor much less heavily parasitized than sympatric placental species. In effect, didelphids seem, parasitologically speaking, to be rather ordinary mammalian hosts.

Parasitism is, nevertheless, a potentially important aspect of opossum natural history. As reviewed above, several groups of blood-feeding ectoparasites (fleas, ticks, and mites) are known to transmit microbial disease agents, and some helminth parasites cause visible lesions in infected tissues or are so large and/or so numerous that they must often have adverse effects on host fitness. The physiological consequences of protozoan infestations for opossum hosts have never been assessed, but it seems likely that intracellular trypanosomatid parasitism must eventually result in tissue damage and suboptimal somatic performance. Therefore, selection might be expected to favor anatomical, behavioral, and physiological traits that minimize the probability that individuals will become infested, or that minimize the severity of established infestations.

The obvious first line of defense against ectoparasitism is grooming, a very frequent mammalian behavior (Hart, 1992). Opossum grooming behavior involves licking, hind-foot scratching, and biting (McManus, 1970; González and Claramunt, 2000), and it appears to be remarkably effective in parasite removal. In an experimental study, captive *Didelphis virginiana* removed over 96% of the larval ticks with which they were inoculated, presumably by grooming (Keesing et al., 2009). Morphological traits that might facilitate ectoparasite removal during grooming include the spoon-shaped claw on digit II of the hind foot (Fig. 5.5), which has no plausible locomotory function, and the tweezer-like morphology of the first upper incisor (Fig. 5.6), which has no apparent role in feeding.

Adaptations that minimize endoparasitism might be behavioral (e.g., avoidance of prey infested by larval helminths; careful choice of nesting sites) or physiological (e.g., immune responses to helminth and protozooan infestations), but no relevant observations on these topics have been published for opossums. However, it is noteworthy that olfactory genes (which might encode odorant receptor molecules to detect parasite-infested prey) and genes associated with immune response are among the most abundant loci in the *Monodelphis domestica* genome (Goodstadt et al., 2007), as they are in many other mammals for which entire genome assemblies are currently available. Whether or not opossums (or any other mammal) are capable of olfactory discrimination between parasite-free and parasite-infested prey is apparently unknown, although the necessary experiments would seem to be relatively straightforward. Heritable genetic variation for ecto- and endoparasite resistance, presumably mediated at least in part by the immune system, is known to occur in a wide range of mammalian species (Charbonnel et al., 2006), but relevant observations are not yet available for didelphids.

The potentially important role of parasitism in other aspects of opossum biology also remains uninvestigated. For example, research on placental mammals suggests that parasite infestations can affect phenomena as diverse as mate choice (Kavaliers and Colwell, 1995), population regulation (Pedersen and Greives, 2008), and predator avoidance (Berdoy et al., 2000). Thus, there is much to learn in the years ahead as opossum parasitology moves beyond taxonomic descriptions, faunal inventories, and biomedical studies to test evolutionary hypotheses in the field and laboratory.

NOTE

1. This seems to often be true as a first approximation because the sum of infestation frequencies is very close to mean infracommunity richness in every published opossum macroparasite survey that provides relevant data.

# 11

# Predators

Predation is thought to be both a major cause of mortality and a potent agent of natural selection in most animal populations, but its evolutionary importance for marsupials has long been unappreciated (Coulson, 1996). Unfortunately, most evidence of predation on didelphids is anecdotal, so its role in shaping relevant morphological traits (e.g., concealing coloration; Chapter 5), physiological phenomena (senescence; Chapter 6), behaviors (lunar phobia; Chapter 7), and demographic variables (survivorship rates; Chapter 13) is still largely conjectural. Below we briefly consider the major vertebrate taxa known to prey on opossums and discuss the evidence for their probable impact on didelphid evolution.[1]

## Predator Taxa
### Snakes

Boas (family Boidae) are constricting snakes that locate their prey by active searching and by waiting in ambush; visual, olfactory, and infrared-sensory cues are probably used for prey capture. Neotropical boas are mostly nocturnal and include arboreal, terrestrial, and aquatic taxa. Several treeboas (genus *Corallus*) are known to eat didelphids, including species of *Caluromys* (Fig. 11.1), *Didelphis*, *Gracilinanus*, *Marmosa*, *Metachirus*, and *Philander* (Pizzatto et al., 2009; Henderson and Pauers, 2012). The apparent paradox of arboreal snakes eating nonarboreal opossums (such as *Metachirus*) is explained by the characteristic posture of some of these boas, which are said to coil around the trunks of small trees close to the ground with the head oriented downward, presumably to ambush terrestrial prey (Henderson and Pauers, 2012). The formidable terrestrial species *Boa constrictor* (to about 5 m; Murphy and Henderson, 1997) has been reported to eat *Didelphis* (Sawaya et al., 2008; Pizzatto et al., 2009) and probably takes other large didelphids as well. The genus *Epicrates* includes several smaller terrestrial boas that might eat didelphids, but only *E. crassus* is apparently known to do so (Pizzatto et al., 2009).

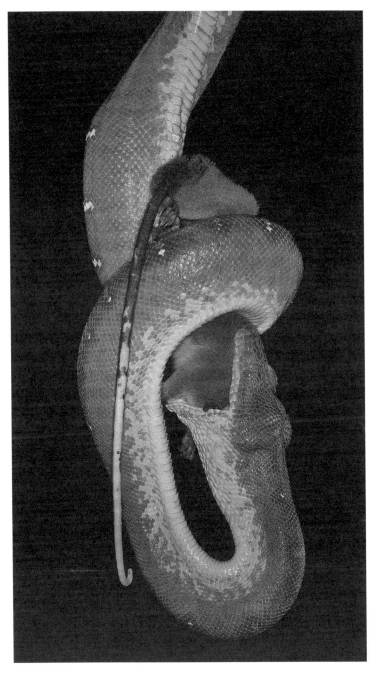

*Fig. 11.1.* Emerald treeboa (*Corallus caninus*) swallowing a woolly opossum (*Caluromys philander*) in lowland rainforest of Pará state, Brazil (photo by Jany Sauvanet)

The huge aquatic boa *Eunectes murinus* (to at least 7 m; Murphy and Henderson, 1997) lives in the same rainforest streams as the water opossum (Voss et al., 2001), but it is only definitely known to eat *Didelphis* (Pizzatto et al., 2009).

Neotropical pitvipers are mostly nocturnal ambush predators that use infrared-sensitive pit organs, olfaction, and visual cues to locate prey; both arboreal and terrestrial pitvipers are known to eat opossums (Voss, 2013). As discussed elsewhere (Chapter 9), *Didelphis*, *Lutreolina*, and *Philander* are known to attack and eat venomous snakes, but even venom-resistant opossums are apparently eaten by some large pitvipers. Other didelphids are not known to be venom resistant, and it is likely that most are eaten, at least occasionally, by sympatric pitvipers. Pitvipers that are definitely known to eat opossums include species of *Bothriechis*, *Bothriopsis*, *Bothrocophias*, *Bothrops*, *Crotalus*, and *Lachesis*.

Only a few other snakes ("colubrids" sensu lato) are definitely known to eat opossums (Pinto and Lema, 2002; Mesquita et al., 2011; Patrón et al., 2011), but most Neotropical colubrids that prey on small mammals probably take opossums on occasion. Such species include, inter alia, active diurnal predators in the genera *Drymarchon*, *Mastigodryas*, *Masticophis*, *Philodryas*, and *Spilotes*, as well as predominantly nocturnal species of *Clelia* and *Pseudoboa* (Duellman, 1990; H.W. Greene, personal commun.). Because most small mammals are nocturnal, diurnal colubrids probably attack sleeping prey in their daytime refugia, which are probably located by olfaction and visual cues.

## Birds

Opossums are preyed upon by many raptors, including hawk-like birds (Accipitridae) and owls (Strigidae and Tytonidae). Neotropical accipitrids definitely known to eat opossums include species commonly known as eagles (Izor, 1985; Klein et al., 1988; Piana, 2007), hawks (Granzinolli and Motta, 2006; Tortato, 2009), and kites (Scheibler, 2007). Because accipitrids are diurnal, most opossums eaten by these visually oriented raptors are probably taken at dawn or dusk. Numerous species of owls probably eat didelphids, but dietary data for these nocturnal raptors are usually obtained from regurgitated pellets, which are seldom found in rainforests (where they decompose quickly). Therefore, most published analyses of owl diets are from dry-forest and savanna habitats, where barn owls (*Tyto alba*), in particular, are often reported to eat small opossums (e.g., by Motta, 2006; Carmona and Rivadeneira, 2006; Teta et al., 2009; Baglan and Catzeflis, 2016). The spectacled owl (*Pulsatrix perspicillata*) is a rainforest species large enough to take

*Didelphis*, but it also preys on *Caluromys* and smaller opossums (Charles-Dominique et al., 1981; Timm et al., 1989; Gómez de Silva et al., 1997).

## Mammals

Although didelphids sometimes prey on one another, definite evidence of such behavior is rare. Among the few known examples are Wilson's (1970) report of *Didelphis marsupialis* attacking and eating an individual of *Philander melanurus*, one report of "*Marmosa* sp." in the diet of *Lutreolina crassicaudata* (see Monteiro-Filho and Dias, 1990), one report of an unidentified species of *Monodelphis* in the diet of *Philander quica* (see Macedo et al., 2010), and one report of unidentified young opossums in the diet of *Marmosa paraguayana* (see Pires et al., 2010a). By contrast, reports of placental mammals preying on didelphids are much more abundant.

Placentals that prey on opossums include cats (Felidae), dog-like carnivorans (Canidae), weasel-like carnivorans (Mustelidae), coatis and raccoons (Procyonidae), and monkeys (Cebidae). Neotropical cats, predominantly nocturnal, include active and ambush predators, and most of them are known to prey on opossums. In particular, didelphids appear to be eaten frequently by pumas (*Puma concolor*), ocelots (*Leopardus pardalis*), margays (*Leopardus wiedii*), and jaguarundis (*Puma yagouaroundi*) (Konecny, 1989; Chinchilla, 1997; Meza et al., 2002; Bianchi and Mendes, 2007; Rocha-Mendes et al., 2010). Interestingly, opossums are absent or rare in reported jaguar (*Panthera onca*) diets, except in the Chaco (Taber et al., 1997). Elsewhere, common opossums (*Didelphis*) are seldom eaten by big cats (Pratas-Santiago et al., 2017), and there is statistical evidence from at least one study that jaguars actively avoid eating *D. marsupialis* (see Weckel et al., 2006).

Canids are active diurnal predators, and most species are likely to prey on sympatric opossums. However, published reports of didelphids in Neotropical canid diets are less common than published records of felid predation, and all are from "wolves" or "foxes" inhabiting semiarid habitats rather than rainforest (Simonetti et al., 1984; Motta et al., 1996; Gatti et al., 2006; Varela et al., 2008; Castro and Emmons, 2012). Dietary knowledge of the widespread rainforest canids *Atelocynus microtis* and *Speothos venaticus* is very limited and neither is certainly known to eat opossums, but both probably do so when the opportunity offers. In temperate North America, *Didelphis virginiana* is preyed upon by coyotes (*Canis latrans*), foxes (*Vulpes vulpes*), and feral dogs (Gardner, 1982).

Records of other carnivorans preying on didelphids are uncommon, but diurnal tayras (*Eira barbara*), grisons (*Galictis vittata*), coatis (*Nasua* spp.), and the nocturnal

Neotropical raccoon (*Procyon cancrivorus*) are tireless foragers that probably overpower and eat most of the smaller vertebrates they encounter, and all are known to eat opossums occasionally (Bisbal, 1986; Konecny, 1989; Gatti et al., 2006; Rocha-Mendes et al., 2010). Monkeys may seem like unlikely predators, but species of *Cebus* are known to hunt and kill a variety of arboreal small mammals, including opossums (Resende et al., 2003; Milano and Monteiro-Filho, 2009), and even marmosets (*Callithrix*) are known to prey opportunistically on small didelphids (Camargo et al., 2017).

## Discussion

Snakes, raptors, and carnivorans are important vertebrate predators in most terrestrial ecosystems (e.g., Jaksic et al., 1981; Greene, 1988), and all have been reported to prey on opossums (Table 11.1). Unfortunately, the relevant literature is clearly biased by abundance, because more published records of predation are available for common than for rare prey taxa. In fact, no published records of predation are available for the rare (or elusive) genera *Caluromysiops*, *Glironia*, and

Table 11.1    Recorded Predators of Opossums[a]

|  | Snakes | Raptors | Carnivorans |
| --- | --- | --- | --- |
| *Caluromys* | X | X[b] | X |
| *Marmosa* | X | X | X |
| *Monodelphis* | X | X | X |
| *Tlacuatzin* | X | X[c] | X |
| *Metachirus* | X | X | X |
| *Chironectes* |  | X |  |
| *Didelphis* | X | X | X |
| *Lutreolina* |  | X |  |
| *Philander* | X | X[d] | X |
| *Chacodelphys* |  | X[e] |  |
| *Cryptonanus* |  | X |  |
| *Gracilinanus* | X | X | X |
| *Lestodelphys* |  | X[f] | X[g] |
| *Marmosops* | X |  | X |
| *Thylamys* | X[h] | X | X |

[a] Based on literature cited in the text except as noted.
[b] L.H. Emmons (personal commun.).
[c] Zarza et al. (2003).
[d] Curay et al. (2019)
[e] Teta and Pardiñas (2007).
[f] Udrizar Sauthier et al. (2007).
[g] Martínez et al. (2012).
[h] H. W. Greene (personal commun.).

*Hyladelphys*. Additionally, owl and hawk vomitus is more easily found in semiarid habitats (deserts, savannas, and scrubland) than in rainforest, plausibly accounting for the lack of avian predation records for some common rainforest opossums (e.g., species of *Marmosops*).

Because of these (and other) biases, few generalizations emerge from the literature cited above. However, it is noteworthy that no predator seems to specialize on didelphids. Instead, all of the snakes, raptors, and carnivorans known to eat opossums are also known to prey on a wide range of other small vertebrates, especially rodents. Although opossums sometimes comprise a large fraction of reported predator diets (e.g., of ocelots in Belize; Konecny, 1989), no published comparisons of observed dietary frequencies with estimated encounter rates (e.g., Emmons, 1987; Granzinolli et al., 2006) suggest that didelphids are preferred over other mammalian prey.

Predation pressure on opossum populations has seldom been quantified, but two field studies of radio-collared *Didelphis virginiana* suggest that it can have a major impact. Predation accounted for more than half (53%) of 30 naturally occurring opossum deaths observed in a protected area in Florida (Ryser, 1990), whereas coyotes alone accounted for 76% of opossum mortality in a protected area in Kansas (Kamler and Gipson, 2004). Although comparable information is lacking for tropical opossum populations, the cumulative demographic impact of predation by sympatric snakes, raptors, and placental mammals is probably substantial. At one well-studied Amazonian rainforest site, for example, at least 40 vertebrate species could prey on opossums, including 13 snakes, 15 raptors, and 12 placentals (Table 11.2). Almost equivalent numbers of potential predators are present in Central American rainforests (Greene, 1988), in Brazilian savannas (Rodrigues et al., 2002; Valdujo et al., 2009), and perhaps in other tropical biomes. Of course, not every opossum species is vulnerable to every sympatric predator, but most tropical opossums probably have multiple predators with diverse hunting behaviors and sensory abilities. Thus, *Didelphis marsupialis* is known to be preyed upon by at least 10 taxa that probably co-occur throughout most of its geographic range, including two boas (*Boa constrictor* and *Corallus* spp.), two large accipitrids (*Harpia harpyja* and *Spizaetus ornatus*), spectacled owls, pumas, ocelots, jaguarundis, grisons, and tayras (Bisbal, 1986; Klein et al., 1988; Konecny, 1989; Chinchilla, 1997; Piana et al., 2007; Pizzatto et al., 2009; Henderson and Pauers, 2012). Predator communities are less diverse at south-temperate latitudes than in the tropics, but at least 17 species (2 snakes, 12 raptors, and 3 placental mammals) are potential predators of small mammals in the semiarid shrublands of north-central

*Table 11.2*    The Río Manu Predator Community[a]

| Predator taxa | Species | Activity[b] | Microhabitat[c] | Behavior[d] | Sensory cues[e] |
|---|---|---|---|---|---|
| SERPENTES | | | | | |
| Boidae | 3 | N | T, Arb | active, ambush | V, O, IR |
| Colubridae | 6 | D, N | T | active | V, O |
| Viperidae | 4 | N | T, Arb | ambush | V, O, IR |
| AVES | | | | | |
| Accipitridae | 11 | D | Aer | active, ambush | V |
| Strigidae | 4 | N | Aer | ambush | V, A |
| MAMMALIA | | | | | |
| Cebidae | 2 | D | Arb | active | V |
| Canidae | 1 | D | T | active | V, A, O |
| Felidae | 5 | N | T, Arb | active, ambush | V, A |
| Mustelidae | 2 | D | T | active | V, A, O |
| Procyonidae | 2 | D, N | T | active | V, A, O |

[a] Nonaquatic tetrapods known or expected to prey on didelphids based on published faunal lists from Cocha Cashu and Pakitza, Madre de Dios, Peru (Rodriguez and Cadle, 1990; Terborgh et al., 1990; Robinson, 1994; Morales and McDiarmid, 1996; Voss and Emmons, 1996) and other literature cited in the text.
[b] Predominantly diurnal (D) or nocturnal (N).
[c] Predominantly terrestrial (T), arboreal (Arb), or aerial (Aer).
[d] Actively searches for prey or waits in ambush.
[e] Primarily detects prey by vision (V), auditory cues (A), olfaction (O), or by sensing infrared radiation (IR).

Chile, and most of them probably eat the single species of local opossum (*Thylamys elegans*), although only three owls and the fox *Pseudalopex culpaeus* are definitely known to do so (Jaksic et al., 1993).

Predation pressure is not necessarily correlated with numbers of sympatric predator species because ecological interactions can sometimes reduce the net impact of predation on prey populations (Sih et al., 1998). Additionally, the presence of refugia (concealing vegetation, burrows, etc.) can significantly decrease the vulnerability of prey, so species with only a few sympatric predators in open habitats (such as the Patagonian steppe inhabited by *Lestodelphys halli*) might well suffer higher predation than an otherwise similar species with more predators in densely vegetated habitats. Some didelphids with unusual lifestyles might escape attack from many of the taxa that persecute other opossums, but habitat specialists may be vulnerable to habitat-specific risks; water opossums, for example, perilously share their aquatic habitats with mammal-eating caimans and anacondas (Magnusson et al., 1987; Voss et al., 2001).

In summary, it seems likely that no opossum species is immune from predation, and that most (if not all) didelphids are routinely eaten by multiple sympatric predator species. Especially in the tropical rainforests where most didelphids

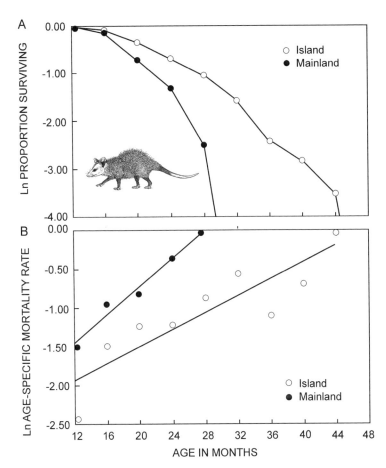

*Fig. 11.2.* Survivorship curves (*top panel*) and mortality rates (*bottom panel*) for adult *Didelphis virginiana* at a mainland study site with carnivoran predators and on an island without carnivorans (redrawn from Austad, 1993)

live today, and where most of didelphid evolutionary history probably transpired (Chapter 14), the diversity of sympatric predators using disparate tactics and sensory systems to encounter mammalian prey would seem to guarantee that opossums have been subject to relentless selection for behavioral and morphological crypsis, sensory vigilance, and other phenotypic traits that tend to prevent, deter, or deflect predatory attacks (Coulson, 1996; Ruxton et al., 2004; Caro, 2005).

Predation also seems likely to be responsible for the high extrinsic mortality that has been hypothesized to explain the remarkably short reproductive lifespans and early senescence of many—if not all—opossums (Chapter 13). Evidence for the probable role of predation in shaping relevant demographic rates was reported by Austad (1993), who documented higher survivorship and lower age-specific mortality in a population of *Didelphis virginiana* on an island lacking mammalian apex predators (bobcats and feral canids) by comparison with survivorship and mortality in an adjacent mainland population with a full complement of predators (Fig. 11.2). Consistent with theory, opossums from the island population exhibited retarded senescence relative to mainland individuals.

The potential role of predation in shaping other aspects of opossum autecology is unclear. However, it is possible that sufficiently numerous or effective predators could keep opossum population densities well below the carrying capacity of local habitats, which might account for the lack of evidence for density dependence and the absence of any demonstrable effects of interspecific competition in some demographic studies (e.g., Troyer et al., 2014). To address such issues, however, we need more than anecdotes and assumptions. As in so many other aspects of ecological research, quantification is key, and future research on opossum natural history would benefit from effective efforts to numerically assess the impact of predators on their marsupial prey.

NOTE

1. To our knowledge, only a single example of invertebrate predation on opossums—a large mygalomorph spider preying on *Marmosops noctivagus* (von May et al., 2019)—has been reported in the literature, but large mygalomorphs ("tarantulas") are abundant in many tropical habitats and might often prey on small opossums despite the lack of additional evidence for such interactions.

# 12

# Competitors and Mutualists

In addition to parasites and predators, animals have to contend with species that compete with them for essential resources. Competition—by definition, any interspecific interaction that harms both participants—is usually a less conspicuous phenomenon than either parasitism or predation, but it is often thought to be important in structuring ecological communities, and it has sometimes been invoked as an explanation for extinctions in biogeographic scenarios (e.g., the Great American Biotic Interchange; Chapter 3). To keep this account within manageable limits, we restrict our consideration of taxa that potentially compete with opossums to placental mammals, although we cannot rule out the possibility that resource competition with birds and other vertebrates might also be important.

By contrast with competition, mutualisms are interspecific interactions that yield reciprocal benefits, and they are believed to be crucial for maintaining ecosystem functionality (Bascompte and Jordano, 2007). Seed dispersal and pollination, in particular, are important mutualisms sustained between plants on the one hand and frugivorous and nectarivorous animals on the other, but plants do not necessarily benefit from all of the frugivores and nectarivores that visit them. There is, furthermore, much taxonomic variation in the quality of seed-dispersal and pollination services provided by animals that mutualistically eat fruits and visit flowers. In the absence of much relevant detail, our treatment of these topics is necessarily superficial, but the data at hand seem adequate to suggest several interpretations of opossum-plant interactions.

## Competition

Everywhere that opossums live, they coexist with, and are usually outnumbered by, placental mammals. In the deciduous woodlands of temperate North America, the Virginia opossum (*Didelphis virginiana*) co-occurs with several dozen placental species, whereas Neotropical opossums can co-occur with almost 200 placental species at some Amazonian localities (Voss and Emmons, 1996). Many sympatric placental taxa seem likely to compete with opossums for food and other resources, but there is not much compelling evidence to support this expectation.

Demonstrating interspecific competition in the wild is notoriously difficult, and only experimental results are widely regarded as conclusive. The only relevant field experiment known to us compared Virginia opossum diets and population characteristics between control plots and plots from which raccoons (*Procyon lotor*) had been removed by live-trapping (Kasparian et al., 2002, 2004). Raccoons and Virginia opossums live in the same habitats and eat much the same food, so it is reasonable to expect that Virginia opossum diets and/or demography would be affected by raccoon removal, but the results of Kasparian et al.'s experiment were inconclusive.

In the absence of other experimental studies, we assess the potential for niche overlap as the basis for inferring competition. In particular, we focus on trophic overlap to identify potential competitors, because trophic resources are thought to be limiting for didelphid populations (e.g., by Charles-Dominique et al., 1981), and because taxonomic differences in other niche dimensions (e.g., circadian activity, locomotor behavior) do not preclude competition for food. Fruit eaten by diurnal mammals, for example, is not subsequently available for consumption by nocturnal species, nor are insects gleaned from foliage by bats available for consumption by nonflying mammals that forage in the same vegetation. Such exploitative competition, which results simply from the depletion of shared resources without direct interspecific contact, is probably much more common than interference competition, whereby individuals of one species physically displace individuals of another while foraging. Interference competition between marsupials and placentals is known to occur—for example, at flowering or fruiting trees (Kays et al., 2012; Breviglieri and Kuhnen, 2016)—but it is seldom reported. By contrast, published dietary data suggest that a wide range of sympatric placentals might compete exploitatively with opossums for food.

### Potentially Competing Taxa

Dasypodid armadillos (*Dasypus* spp.) consume small ectothermic vertebrates, litter invertebrates, and fallen fruit (Szeplaki et al., 1988; Emmons, 1997; Loughry and McDonough, 2013; Voss and Fleck, 2017), including many of the same invertebrate taxa (e.g., Diplopoda, Coleoptera, Oligochaeta) also eaten by terrestrial opossums that live in the same habitats. In lowland tropical rainforests, dasypodid armadillos are nocturnal, increasing the likelihood that they encounter the same prey as sympatric opossums. Chlamyphorid armadillos, on the other hand, are either fossorial (e.g., *Cabassous* spp.) or so much larger and more divergent in their foraging behavior (e.g., *Priodontes maximus*, which tears apart rotten logs) as to seem unlikely to overlap substantially in diet with sympatric opossums.

Members of the bat family Phyllostomidae are adaptively diverse, including frugivorous, nectarivorous, omnivorous, and animalivorous species (Giannini and Kalko, 2004). Among other resources, phyllostomids exploit many of the same fruits (e.g., *Cecropia, Ficus, Piper*; Charles-Dominique, 1986; Fleming, 1986), nectar sources (e.g., *Mabea, Ochroma*; Steiner, 1983; Kays et al., 2012), insects (e.g., katydids, caterpillars; Kalka and Kalko, 2006), and small vertebrates (e.g., frogs; Ryan et al., 1981) that sympatric opossums are known to eat. With the exception of vampires, all phyllostomids can be considered potential competitors of at least some sympatric didelphids.

New World monkeys trophically resemble opossums in having diets that usually include both ripe fruit pulp and insects, although some atelids and pitheciids also eat foliage and seeds (Rosenberger, 1992). Large monkeys primarily forage in the canopy and subcanopy (Mittermeier and van Roosmalen, 1981; Terborgh, 1983), where they eat many of the same fruits consumed in situ by arboreal opossums; however, the same fruits, once fallen, are also eaten on the forest floor by terrestrial species. By contrast, marmosets (*Callithrix* spp.), tamarins (*Saguinus* spp.), and other callitrichines forage for insects and small fruits in the primary forest understory, along forest edges, and in secondary growth (Rylands, 1993), where they might compete trophically with small opossums (e.g., *Marmosa* spp.) that frequent such habitats and eat similar items. Primates have also been observed to compete with opossums for nectar (Kays et al., 2012) and exudates (Aléssio et al., 2005).

Living procyonids include several taxa that overlap broadly in diet, substrate use, and/or diel activity with opossums and might compete directly with them for essential resources. Kinkajous (*Potos flavus*) and olingos (*Bassaricyon* spp.), in particular, are nocturnal-arboreal frugivores/nectarivores that consume many of the same plant resources as syntopic arboreal and scansorial opossums (Charles-Dominique et al., 1981; Julien-Laferrière, 1999; Kays, 1999, 2000), and kinkajous have been observed to physically displace opossums feeding in the same canopy trees (Kays et al., 2012). Coatis (*Nasua* spp.), although diurnal, have a mixed diet of fallen fruits, litter arthropods, and small vertebrates (Kaufmann, 1962; Alves-Costa et al., 2004; Hirsch, 2009) that is strikingly similar to the diets of large syntopic terrestrial opossums (e.g., *Didelphis* spp.). As noted above, the North American raccoon (*Procyon lotor*) forages nocturnally in temperate woodland habitats for items that are also consumed by the Virginia opossum (Kissell and

Kennedy, 1992), whereas the Neotropical raccoon (*P. cancrivorus*) is a riparian species that might compete with syntopic opossums such as *Philander* and *Chironectes* for frogs, crustaceans, and other semiaquatic or aquatic prey.

The North American striped skunk (*Mephitis mephitis*) and spotted skunks
(*Spilogale* spp.) are terrestrial omnivores with diets that include many of the
same items consumed by syntopic Virginia opossums (e.g., Hamilton, 1936;
Cantú-Salazar et al., 2005). Neotropical hog-nosed skunks (*Conepatus*) are poorly
characterized ecologically, but they are nocturnal, terrestrial, and appear to favor
open habitats. Available dietary data are all from high latitudes (e.g., Travaini et al.,
1998; Donadio et al., 2004), where *C. chinga* is known to eat a wide variety of invertebrates and small vertebrates, many taxa of which are also consumed by sympatric opossums (e.g., *Lestodelphys* and *Lutreolina*).

Mustelids include several taxa that are potential competitors with opossums.
In North America, weasels (e.g., *Mustela frenata*) take many of the same species of
rodents and other small mammals eaten by syntopic Virginia opossums. In the
Neotropics, a case for potential competition could be made for the river otter
(*Lontra longicaudis*), which occurs syntopically with the water opossum (*Chironectes minimus*) and, although diurnal, feeds on essentially similar items (fish and
crustaceans; Pardini, 1998; Quadros and Monteiro-Filho, 2001; Galliez et al., 2009).
Grisons (*Galictis* spp.) might also overlap trophically with some of the larger sympatric opossums, but their food habits, especially in rainforest, are not well documented, and their primary interaction with opossums might be predatory rather
than competitive.

Although most ungulates (artiodactyls and perissodactyls) are folivorous grazers or browsers, Neotropical rainforest deer, peccaries, and lowland tapirs eat substantial quantities of fruit (Bodmer, 1989; Henry et al., 2000; Gayot et al., 2004),
including many of the same families, genera, and species of fruits eaten by terrestrial opossums.

Many terrestrial caviomorph rodents—including nocturnal spiny rats (*Proechimys* spp.,), diurnal agoutis (*Dasyprocta* spp.), diurnal acouchies (*Myoprocta* spp.),
and nocturnal pacas *Cuniculus paca*)—are at least partially frugivorous (Smythe,
1978; Dubost, 1988; Dubost and Henry, 2006). The few available lists of fruits consumed by terrestrial caviomorphs (e.g., Adler, 1995; Beck-King et al., 1999) contain many taxa eaten by opossums. In addition, some arboreal caviomorphs are
partially frugivorous, including the widespread porcupine *Coendou prehensilis*
(Charles-Dominique et al., 1981) and some smaller rat-like species (*Mesomys* spp.
and *Makalata didelphoides*; Emmons, 1997).

Neotropical cricetid rodents (subfamily Sigmodontinae) are an adaptively diverse group that includes many open-country granivores or grazers that are unlikely to compete with opossums. Forest-dwelling sigmodontines, however, are all at least partially insectivorous, and many also eat fruits that are eaten by sympatric opossums (Guillotin, 1982; Casella and Cáceres, 2006). Some semiaquatic taxa (e.g., *Ichthyomys*) occur syntopically with water opossums (*Chironectes minimus*) and eat many of the same prey items (aquatic insects, crustaceans, and sometimes fish; Voss, 1988).

## Diffuse Competition

Multiple sympatric species in most of these placental clades are present in many Neotropical habitats, where the aggregate potential for trophic-resource competition seems correspondingly high (Table 12.1). Of course, not every opossum species is likely to compete with every sympatric placental, but for every opossum species there are likely to be at least several placentals that forage for similar kinds of food in similar places. Water opossums (*Chironectes*), for example, share rainforest streams with otters and semiaquatic cricetids; woolly opossums share tree crowns with primates and arboreal procyonids; small didelphines (e.g., species of *Gracilinanus*, *Marmosa*, and *Marmosops*) search for insects and small fruits in the same understory vegetation patrolled by insectivorous and frugivorous phyllostomids; and terrestrial opossums must subsist on what is missed or rejected by foraging armadillos, coatis, ungulates, and rodents. Therefore, most opossums are probably exposed to diffuse competition, the summed effect of resource depletion by multiple competing species (MacArthur, 1972; Pianka, 1974).

Although diffuse competition is impossible to quantify in species-rich tropical communities, there are reasons to expect that it might have a substantial effect on opossum populations. First, the numbers of potentially competing placental species at several well-sampled Neotropical localities are large, far exceeding the number of opossum species in most faunas known to us.[1] Second, in any particular microhabitat, opossums are often sandwiched between potential competitors that are larger (e.g., primates, carnivores, ungulates) and others that are smaller (e.g., bats), an awkward position that (at least in theory; MacArthur, 1972) could make them especially vulnerable to resource depletion. Lastly, as small to medium-sized nonvolant mammals, opossums are less mobile than many sympatric placentals— e.g., bats, ungulates, and the larger primates—some of which are known to migrate seasonally from local habitats with low resource availability to neighboring sites where resources are more abundant. In other words, most opossums probably

Table 12.1 Numbers of Placental Mammals that Might Compete Trophically with Opossums at Four South American Localities

| Locality | Habitat[b] | Opossum species | Potentially competing species[a] | | | | | | |
|---|---|---|---|---|---|---|---|---|---|
| | | | Armadillos | Bats | Primates | Carnivores | Ungulates | Rodents | Totals |
| Paracou, French Guiana[c] | RF | 14 | 2 | 47 | 6 | 2 | 5 | 19 | 81 |
| Manaus, Brazil[d] | RF | 9 | 2 | 46 | 6 | 3 | 5 | 14 | 76 |
| Balta, Peru[e] | RF | 12 | 2 | 39 | 10 | 5 | 4 | 19 | 79 |
| Emas National Park, Brazil[f] | DF | 10 | 2 | 15 | 2 | 4 | 4 | 16 | 43 |

[a] "Armadillos" include only *Dasypus* spp.; "bats" include only phyllostomids (except vampires); "primates" include all species known to occur at each site; "carnivores" include procyonids, the river otter, and skunks; "ungulates" include cervids, tayassuids, and tapirs; "rodents" include cricetids and frugivorous caviomorphs.

[b] RF = rainforest, DF = dry forest (including savanna).

[c] Simmons and Voss (1998), Voss et al. (2001).

[d] Sampaio et al. (2003), Voss and Emmons (1996: Appendix 7).

[e] Voss and Emmons (1996: Appendix 9), Voss et al. (2018).

[f] Rodrigues et al. (2002).

stay put and make do with what is left when more mobile competitors depart for richer pickings elsewhere.

To quantify the potential for resource overlap between opossums and sympatric placentals, we tabulated the fruit and prey taxa consumed by potentially competing mammalian species in two rainforested regions. Data on fruit consumption was compiled from several field studies in French Guiana, where research on seed dispersal has resulted in numerous published dietary studies of frugivorous mammals. The most extensive available data on prey consumption, by contrast, is from the Atlantic Forest of southeastern Brazil, from which many dietary studies of secondary consumers have been published over the last several decades. Both datasets are incomplete in the sense that dietary data are not available from all potentially competing local taxa, and both are coarsely resolved taxonomically (fruits identified usually only to genus, prey commonly identified only to order), but for now this is the best information available.

In French Guiana, three species of opossums are known to consume fruits belonging to 45 genera of plants, of which sympatric placentals (bats, primates, procyonids, and ungulates) consume fruits of 40 genera, or about 89% (Appendix 2). Of the five plant genera whose fruits were found to be eaten only by opossums in these studies, four—*Xylopia* (Annonaceae), *Monstera* (Araceae), *Pera* (Euphorbiaceae), and *Loreya* (Melastomataceae)—are known from other Neotropical studies to be consumed by bats, primates, and/or procyonids. Just one genus—*Dichapetalum* (Dichapetalaceae)—produces fruits for which opossums are the only known mammalian consumers; however, this genus includes only two local species of lianas (Mori et al., 2002) and seems unlikely to constitute an important trophic resource for any mammal.

In southeastern Brazil, 11 species of opossums are known to prey on 25 animal higher taxa (mostly orders in the Linnaean hierarchy), of which sympatric placental mammals (primates, procyonids, and cricetids) also prey on 24, or about 96% (Appendix 3). The sole higher taxon exclusively preyed upon by opossums in this tabulation is Psocoptera, a group of tiny insects (commonly known as barklice) that are unlikely to be an important food source.

## Mutualistic Interactions
### Seed Dispersal

As reviewed earlier in this volume (Chapter 9), many opossums eat fruit, and as we have just seen, a large number of plants have fruit that is at least occasionally eaten by opossums. However, although the purpose of fleshy fruits is to provide nutritional rewards for seed dispersal, not all frugivores disperse seeds. Many species of mammals and birds are known to spit out large seeds, consuming only the pulp of large-seeded fruits. In such cases, the animal-plant interaction is not mutualistic, because only the animal benefits.

Opossums eat many species of fruits without dispersing the seeds. Of 44 species of fruits eaten by three species of radio-tracked didelphids in French Guiana, the seeds of only 19 species (43%) were ingested (Charles-Dominique et al., 1981; Atramentowicz, 1988). These opossum-dispersed plant species (Table 12.2) represent a diversity of growth forms (trees, stranglers, lianas, epiphytes, shrubs, etc.), but all have small seeds (<10 mm long). Many are from pioneer plant taxa that usually occur in canopy gaps and secondary vegetation, but other seeds dispersed by opossums in this study are from plants that occur in primary forest.

Studies of opossum-mediated seed dispersal have also been carried out in the Atlantic Forest of southeastern Brazil. As in French Guiana, seeds dispersed by opossums in the Atlantic Forest are invariably small, and they are usually from pioneer species (Cáceres et al., 1999; Pinheiro et al., 2002; Carvalho et al., 2005; Leiner and Silva, 2007b). Some of the same plant families that are opossum-dispersed in French Guiana are also opossum-dispersed in southeastern Brazil (e.g., Araceae, Cecropiaceae, Melastomataceae, Moraceae, and Piperaceae), but Solanaceae (a family that is primarily dispersed by birds and bats in the Guianas) is notably well represented in lists of opossum-dispersed seeds from the Atlantic Forest.

Only a few relevant studies have been carried out in the Cerrado of central Brazil, where seeds of Cecropiaceae, Moraceae, and Piperaceae are conspicuously absent from didelphid feces (Lessa and da Costa, 2010; Camargo et al., 2011; Lessa et al., 2013). Instead, species of Melastomataceae (especially *Miconia*), Myrtaceae, and Rubiaceae seem to predominate among the seeds ingested by opossums in this dry-forested biome. The seed taxa dispersed by opossums in other tropical biomes are poorly documented, but *Marmosa xerophila* disperses the seeds of at least four species of cacti in the desert thornscrub of northern Venezuela (Thielen et al., 1997b). Because fruit is only a minor component of didelphid diets at temperate latitudes (Chapter 9), seed dispersal by opossums seems unlikely to be ecologically important there.

Of course, seed dispersal is only effective if seeds are still viable after they have passed through the gut of the animal that swallowed them, and several experimental studies have assessed the viability of seeds ingested by opossums (e.g., Cáceres et al., 1999; Cáceres and Monteiro-Filho, 2000; Lessa et al., 2013). The results

Table 12.2   Plant with Seeds Dispersed by Opossums (*Caluromys philander*, *Didelphis marsupialis*, *Philander opossum*) in French Guiana

| Taxon[a] | Growth form[b] | Seed size[b,c] | Other seed dispersers[b] |
|---|---|---|---|
| **ANNONACEAE** | | | |
| *Rollinia exsucca* | tree | 6 mm | bats, primates |
| *Xylopia cayennensis* | tree | 5 mm | primates, birds[d] |
| *X. nitida* | tree | ≤5 mm[e] | primates, birds |
| **ARACEAE** | | | |
| *Monstera* sp. | epiphyte | 7 mm[f] | primates |
| **CECROPIACEAE** | | | |
| *Cecropia obtusa* | tree | 3 mm | bats, kinkajous, primates, birds |
| *C. palmata* | tree | ≤2 mm[e] | bats, kinkajous, birds |
| *C. sciadophylla* | tree | 3 mm | bats, kinkajous, primates, birds |
| **CLUSIACEAE** | | | |
| *Clusia scrobiculata* | strangler | ? | primates[g] |
| **EUPHORBIACEAE** | | | |
| *Pera bicolor* | tree[h] | 6–8 mm[h] | birds |
| **MELASTOMATACEAE** | | | |
| *Bellucia grossularioides* | tree | 1 mm | bats, primates, birds |
| *Henrietta succosa* | tree | 1 mm | bats, birds |
| *Loreya mespiloides* | tree | ≤5 mm[e] | birds[i] |
| **MORACEAE** | | | |
| *Ficus broadwayi* | strangler | ? | ? |
| *F. catappaefolia* | strangler | ? | ? |
| *F. nymphaefolia* | tree | 1 mm | bats, kinkajous, primates |
| **PASSIFLORACEAE** | | | |
| *Passiflora coccinea* | liana | 6 mm | ? |
| *P. glandulosa* | liana | 8 mm | primates |
| **PIPERACEAE** | | | |
| *Piper* spp. | shrubs & treelets | ≤3 mm | bats, birds |
| **RUBIACEAE** | | | |
| *Guettarda macrantha* | tree | ? | ? |

[a] After Atramentowicz (1988), but nomenclature follows Mori et al. (1997, 2002).
[b] After Roosmalen (1985a), Mori et al. (2002), and Lobova et al. (2009) except as noted.
[c] Length, rounded to the nearest millimeter.
[d] Cummings and Read (2016).
[e] Prevost (1983).
[f] In *Monstera adansonii*, the only species for which seed measurements are provided (Roosmalen, 1985a).
[g] Roosmalen (1985b).
[h] Bigio and Secco (2012).
[i] Charles-Dominique (1986)

suggest that many seeds recovered from opossum feces germinate at about the same rate as "control" seeds (seeds extracted directly from ripe fruits), although some species of opossum-ingested seeds fare better (have higher germination rates) and a few fare worse than conspecific "controls." Seed latency—the average time required for opossum-ingested seeds to germinate in these experiments— varied considerably according to plant species. In Lessa et al.'s (2013) study, latency ranged from $12 \pm 3$ days for seeds of *Cipocereus minensis* (Cactaceae) passed by *Marmosops incanus* to $171 \pm 8$ days for seeds of *Cordiera sessilis* (Rubiaceae) passed by *Didelphis albiventris*; as suggested by the authors, such disparate latencies suggest different germination strategies in habitats with seasonally variable conditions for seedling establishment.

Plants with seeds dispersed by opossums are often dispersed by other vertebrates as well. In French Guiana, opossum-dispersed seeds are usually also dispersed by bats, primates, kinkajous, and/or birds (Table 12.2); the few apparent exceptions are uncommon and poorly researched species. Bats, monkeys, and birds are also implicated as dispersal agents for many Atlantic Forest plants with opossum-dispersed seeds (Cáceres et al., 1999; Cáceres and Monteiro-Filho, 2000; Vieira and Izar, 1999). Relevant data from other biomes are sparse, but *Stenocereus griseus*, the cactus whose seeds are most commonly recovered from the feces of *Marmosa xerophila* in Venezuelan coastal deserts, is also known to be dispersed by bats and birds (Naranjo et al., 2003).

### Pollination

Similar to the distinction we have just discussed between frugivory and seed dispersal, nectarivory is not synonymous with pollination. Many animals that visit flowers are nectar robbers, not pollinators, so their flower-visiting habits do not benefit the plant. Pollen transport is obviously a minimal qualification for a flower-visiting animal to act as a pollinator, but to be effective in this role—at least for self-incompatible, dioecious, or dichogamous plants—animals must carry pollen from the anthers of one individual to the stigmas of another conspecific.[2] Pollination is hard to observe directly, so plant-pollinator relationships are usually inferred from indirect evidence, typically animal phenotypes and floral traits. Only animals with color vision, for example, are likely to be important pollinators of brightly pigmented flowers, only flying animals are likely to pollinate flowers suspended in midair on slender peduncles, and only nocturnal animals are likely to pollinate flowers that are only open at night. Thus, floral attractants (e.g., color, scent, and nectar chemistry), floral morphology, and the timing of anthesis (flower opening) are typically associated with different pollinator taxa, with the result that various "syndromes" of correlated traits are widely recognized as evidence for pollination by butterflies, moths, bees, birds, bats, and so forth (Fenster et al., 2004).

Many flowers are visited by multiple species of animals, so careful observations are often necessary to judge which animals are the primary pollinators, which are secondary, and which are exploiting floral resources without any reciprocal benefit to the plant. Inflorescences of the night-flowering understory palm *Calyptrogyne ghiesbreghtiana*, for example, are visited by opossums, bats, katydids, and beetles; field studies of this system suggest that bats are the primary pollinators, although visiting opossums may also result in some seed set, whereas katydids and beetles are floral herbivores whose visits are mainly destructive (Cunningham, 1995; Sperr et al., 2008). Blossoms of the night-flowering canopy tree *Ochroma pyramidale* are visited by opossums, carnivorans, primates, birds, and bees, but exclusion experiments suggest that bees are not pollinators, and whereas some opossums lap nectar from the cup-shaped flowers while contacting the anthers (Fig. 12.1), others do not. Apparently, only bats and large nonflying mammals—kinkajous, olingos, and night monkeys—routinely transport pollen of this tree species (Wilkinson and Boughman, 1998; Kays et al., 2012).

Similar observations and experiments suggest that the primary pollinators of most opossum-visited flowers are bats or other nocturnal mammals, but two species appear to be pollinated primarily by birds, and another two by insects (Table 12.3). Opossums are probably nectar robbers for some of these species—such as *Phenakospermum guyannense*—but they may be secondary pollinators for others (e.g., *Calyptrogyne ghiesbreghtiana*). Only three species remain as candidates for effective pollination by opossums.

Opossums might be important pollinators of the rainforest tree *Quararibea cordata*, although flowers of this species are also visited by other animals (primates, procyonids, bats, birds, and insects), and no exclusion experiments have been carried out to narrow the range of possible mutualists; however, the floral morphology of *Q. cordata* does not conform to any of the syndromes typically associated with bat, bird, or insect pollination, so pollination by some sort of non-flying mammal seems plausible. Inflorescences of the rainforest liana *Marcgravia nepenthoides* are visited by both bats and opossums, but Tschapka and von Helversen (1999) interpreted the unusual floral morphology of this plant, which has enormous nectaries that are widely separated from the flowers, as evidence for opossum pollination. Stronger evidence for opossums as primary pollinators of the night-blooming tree *Pseudobombax tomentosum* was summarized by Gribel

*Fig. 12.1. Caluromys derbianus* visiting balsa flowers (Malvaceae: *Ochroma pyramidale*) on Barro Colorado Island, Panama (photos by Christian Ziegler). Note that the opossum's head in the lower photo is in contact with the anthers as it laps nectar from the bottom of the upright corolla.

*Table 12.3*   Primary Pollinators of Plants with Flowers Visited by Opossums

| Plant taxon | Primary pollinators | References |
|---|---|---|
| ARECACEAE | | |
| *Calyptrogyne ghiesbreghtiana* | bats | Cunningham (1995) |
| BOMBACACEAE[a] | | |
| *Ceiba pentandra* | bats | Gribel et al. (1999) |
| *Ochroma pyramidale* | bats and nonflying mammals[b] | Wilkinson and Boughman (1998), Kays et al. (2012) |
| *Pseudobombax tomentosum* | opossums | Gribel (1988) |
| *Quararibea cordata* | nonflying mammals[c] | Janson et al. (1981) |
| CACTACEAE | | |
| *Stenocereus queretaroensis* | bats | Ibarra-Cerdeña et al. (2007) |
| CARYOCARACEAE | | |
| *Caryocar glabrum* | bats | Fleming et al. (2009) |
| *C. villosum* | bats | Martins and Gribel (2007) |
| CAESALPINIACEAE | | |
| *Eperua falcata* | bats | Fleming et al. (2009) |
| *Hymenaea courbaril* | bats | Fleming et al. (2009) |
| CHRYSOBALANACEAE | | |
| *Couepia longipendula* | bats | Fleming et al. (2009) |
| CLUSIACEAE | | |
| *Symphonia globulifera* | birds | Gill et al. (1998) |
| EUPHORBIACEAE | | |
| *Mabea fistulifera* | bats | Vieira and Carvalho-Okano (1996) |
| *M. occidentalis* | bats | Steiner (1983) |
| LECYTHIDACEAE | | |
| *Eschweilera coriacea* | bees | Dulmen (2001) |
| MARCGRAVIACEAE | | |
| *Marcgravia nepenthoides* | opossums? | Tschapka and von Helversen (1999) |
| *Norantea guianensis* | birds | Dulmen (2001) |
| MIMOSACEAE | | |
| *Inga ingoides* | moths | Kimmel et al. (2010) |
| *Parkia pendula* | bats | Hopkins (1984) |
| STRELITZIACEAE | | |
| *Phenakospermum guyannense* | bats | Kress and Stone (1993) |

[a] Sometimes referred to Malvaceae.
[b] Kinkajous, olingos, and night monkeys.
[c] Opossums, monkeys (*Cebus, Aotus*), and arboreal procyonids.

(1988), who recorded 169 flower visits by *Caluromys lanatus* but only six visits by bats; moreover, the faces of *C. lanatus* visitors were in close contact with the stamens while the animals were drinking nectar from the upright, cup-shaped flowers, and individual opossums were observed to move among conspecific flowering individuals on the same night, thus potentially effecting cross pollination.

## Discussion

The data available to assess trophic-resource overlap between opossums and sympatric placental mammals are (to say the least) far from what could be desired. Among other difficulties, methods for quantifying diets have not been consistently applied in field studies of different taxa. Primate diets, for example, are usually studied by direct observation of feeding behavior (summarized as time budgets), whereas dietary studies of nocturnal species are typically based on stomach contents (summarized volumetrically) or fecal analysis (summarized as occurrence frequencies). Therefore, the only common currency among relevant dietary studies for our purposes is presence/absence data, which—due to uncertain taxonomy and identification difficulties—could only be tabulated for fruits at the generic level and for prey at the ordinal level.

Such crude procedures are problematic for inferring competition. Coarse taxonomic resolution of dietary data always inflates estimates of trophic-resource overlap because sympatric predators (or frugivores) can eat many or all of the same higher taxa but do not necessarily eat the same species (Greene and Jaksic, 1983). For example, two sympatric species of rainforest tamarins (*Saguinus*) each consume large numbers of katydids (Orthoptera: Tettigoniidae), but of the 39 katydid species identified in their combined diets, only three were eaten by both (Nickle and Heymann, 1996). Presence/absence scoring of data can also inflate estimates of resource overlap if different taxa are eaten in different proportions. For example, sympatric species of short-tailed fruit bats (*Carollia*) eat most of the same species of *Piper*, but each bat species prefers different fruit species, which they eat in different proportions (Andrade et al., 2013).

Given such inadequacies, the data compiled for this chapter simply suggest that there are no obvious differences in the kinds of foods consumed by opossums on the one hand and by sympatric placental mammals on the other, and we doubt that any additional insights can be wrested from these crude dietary comparisons. In effect, the potential for diffuse trophic competition surely exists, but our data are not informative about whether such competition is ecologically important. Instead, progress in understanding potentially competitive relationships between opossums and sympatric placental mammals is more likely to come from detailed field studies of trophic behavior accompanied by fully resolved taxonomic determinations of diet. The only such study known to us suggests that two species with superficially similar autecologies—*Caluromys philander* and *Potos flavus*, both nocturnal-arboreal frugivores—are not close competitors, largely as a consequence of their body size difference (Julien-Laferrière, 1999). More studies of this kind would contribute to a better assessment of opossum-placental trophic competition than we are currently able to provide.

If the existing evidence for trophic competition between opossums and placental mammals is weak, the evidence that opossums are significant participants in mutualistic relationships with plants is weaker still. Opossums are sometimes said to be important seed dispersers (Cáceres, 2006), but the importance of opossum-mediated seed dispersal relative to seed dispersal by other vertebrate taxa that consume the same fruit species has not been convincingly assessed in any of the studies we reviewed. Given that most opossum-dispersed plants are also serviced by other frugivores, just how ecologically important is opossum seed dispersal?

Two important factors that affect the contribution of seed dispersers to plant reproduction are the numbers of seeds dispersed and the quality of dispersal (Howe and Smallwood, 1982; Schupp, 1993), neither of which have been quantified for participants in any plant-frugivore system for which opossums are candidate mutualists. True, opossum scats sometimes contain impressive quantities of seeds—scats of *Didelphis marsupialis* each had an average of about 500 seeds of *Cecropia obtusifolia* in one Mexican rainforest study (Medellín, 1994). However, the proportion of seeds dispersed by opossums versus those dispersed by other sympatric frugivores was not estimated, and the short distances that seeds were moved from the parent trees by *D. marsupialis* (about 13 m, on average) suggest that more mobile taxa such as bats and birds (which also disperse *C. obtusifolia* seeds; Estrada et al., 1984) might make greater contributions to the fitness of this gap-colonizing plant.

More generally, if ideal sites for seedling establishment and growth are unpredictable in time and space, as seems likely for pioneer species of plants, then flying vertebrates are probably more important seed dispersers than opossums.[3] By contrast, opossums might be important mutualists for plants with seeds that must be deposited in particular places where opossums predictably go. Conceivably, opossums might be important dispersers of seeds of epiphytes and hemiepiphytes (including stranglers) that need to germinate on branches or in tree crotches that are often used for travel or as resting places by arboreal didelphids. There are hints of such possible mutualisms in the literature cited above, but none have been the primary focus of any study. In fact, many studies of didelphid seed dispersal have been carried out in human-modified landscapes that are not ideal places to detect coevolved interactions with primary-forest plants.

Similar considerations suggest that opossums are not important mutualists for most of the plant-pollinator systems in which they occasionally participate. More mobile species—flying or nonflying—seem likely to be more effective pollinators, especially in tropical rainforests where conspecific plants are often very widely dispersed. The unique exception, a tree for which opossums may, in fact, be important pollinators, grows along the edges of gallery forest, where conspecific individuals occur close enough together to be visited by individual opossums in a single night (Gribel, 1988).

We suspect that opossums are more frequently exploiters of mutualisms sustained by other species than they are important sustainers of mutualisms on their own or in combination with other taxa. Fortunately, not all exploiters have significant negative impacts on mutualisms (Bronstein, 2001). Seed-spitting, for example, is far less destructive than seed predation, and it is possible that pulp removal actually enhances seedling recruitment even in the absence of seed dispersal (Fedriani et al., 2012). Nectar-robbing also seems benign by comparison with flower destruction (e.g., by insects; Cunningham, 1995). Therefore, even if opossums seldom provide reciprocal benefits to the plants whose fruit and nectar they consume, it seems unlikely that their exploitative activities have a net destabilizing influence on the webs of mutualism that sustain the diversity of Neotropical ecosystems.

NOTES

1. The only exception seems to be in the desert thornscrub of northwestern Venezuela (Fig. 8.5), where *Marmosa xerophila* is said to be the only resident species of nonvolant small mammal (Thielen et al., 1997a).

2. The breeding systems of most tropical plants with growth forms that produce opossum-visited flowers—trees, shrubs, and lianas—seem to require cross-pollination (Bawa, 1974; Zapata and Arroyo, 1978).

3. The average distance that seeds are moved obviously depends on many factors, but mobility seems likely to be important. By comparison with the modest nightly movements of opossums (e.g., 423 m traveled per night, on average, for *Marmosa paraguayana*; Moraes and Chiarello, 2005a), bats are far more mobile (e.g., 4.7 km/night for *Carollia perspicillata*; Heithaus and Fleming, 1978).

# 13

# Population Biology

Population-level phenomena are the emergent results of interactions between individual phenotypes and the environment over time, and they ultimately determine the geographic range and evolutionary fate of species. Demographic information is important for a wide range of research purposes, but populations of behaviorally cryptic animals are difficult to study, and for most species of opossums, relevant data are entirely lacking. The best we can do at this point is to summarize current knowledge about the use of space by conspecific individuals, reproductive seasonality, population turnover, and a few other related topics. Fortunately, considerable progress in our understanding of tropical opossum demography has been made in recent years, notably by teams of Brazilian researchers committed to multiyear studies of populations in the Cerrado and the Atlantic Forest. This account is richer and more substantive as a consequence of their ongoing efforts.

## Spatial Distribution
### Home Range and Territoriality

Although older studies of didelphid ecology based on capture-mark-recapture methods often described opossums as "nomadic" (e.g., Reynolds, 1945; Fleming, 1972; O'Connell, 1979), subsequent research using radiotelemetry has shown that most nondispersing adult opossums occupy well-defined home ranges qualitatively resembling those of other small mammals (Gillette, 1980; Sunquist et al., 1987; Julien-Laferrière, 1995; Ryser, 1995; Moraes and Chiarello, 2005a; Leite et al., 2016). Estimating the home range of an organism—the area normally traversed by individuals in the course of their subsistence and reproductive activities—is the topic of a formidably large technical literature (Laver and Kelly, 2008). We will not attempt a methodological critique here, except to note that home-range estimates from radiotelemetry are probably more accurate than those obtained by capture-mark-recapture studies, unless the trapping effort is very intensive and employs multiple grids (Lira and Fernandez, 2009). However, estimates from capture-mark-recapture data are all that are available for some species.

The geometry of the home range varies with microhabitat use and introduces a source of potential bias that is easily understood but difficult to correct for in taxonomic comparisons. Whereas home ranges of terrestrial species are approximately two-dimensional and can appropriately be quantified as areas, arboreal species live in three dimensions. However, home ranges of arboreal opossums are never quantified as volumes, and the bias introduced by representing three-dimensional living spaces as areas is unknown. Uniquely, water opossums (which occur only along streams) have linear home ranges, which are appropriately quantified as lengths (Galliez et al., 2009; Leite et al., 2016).

Available measurements of home-range size for a handful of didelphid species (Table 13.1) suggest that males have larger home ranges than females, although the observed difference is thought to be insignificant for some species (e.g., *Caluromys philander*; Julien-Laferrière, 1995), and sample sizes are often too small to test for sex differences. However, sexual dimorphism in home-range size in Ryser's (1995) study of *Didelphis virginiana* and Shibuya et al.'s (2018) study of *Gracilinanus agilis* are highly significant, the former documenting an order-of-magnitude difference between the sexes and the latter a twofold difference. The only other inference suggested by these scant data is that larger species may have larger home ranges than smaller species, but it should be noted that the largest reported home ranges are both from studies of *Didelphis* (which might have large home ranges for some other reason), and that the two smallest home-range estimates are from capture-mark-recapture studies (which might be prone to underestimation; Moraes and Chiarello, 2005a).

Opossums are sometimes described as nonterritorial, in the sense that individuals do not seem to defend any exclusive space (Charles-Dominique, 1983). This does, in fact, appear to be true of *Caluromys philander* (Fig. 13.1), in which male and female home ranges overlap with one another in all combinations: male-male, male-female, and female-female (Atramentowicz, 1982; Julien-Laferrière, 1995). Apparently, male home ranges of other didelphids typically also overlap those of multiple individuals of both sexes, but in several species—including *Chironectes minimus*, *Didelphis marsupialis*, *Gracilinanus agilis*, *Marmosa paraguayana*, *M. robinsoni*, and *Marmosops paulensis*—female home ranges are reported to not overlap with one another, or to exhibit only narrow overlap, suggesting intrasex-

Table 13.1  Estimates of Home-Range Size for Seven Opossum Species

| Species | Method[a] | Location | Habitat | Mean home range[b] | References |
|---|---|---|---|---|---|
| Caluromys philander | RT | Fr. Guiana | tropical rainforest | 3.1 ha (males, $N=7$) / 2.4 ha (females, $N=8$) | Julien-Laferrière (1995) |
| Chironectes minimus | RT | Brazil | tropical rainforest | 5780 m (males, $N=6$)[c] / 1690 m (females, $N=4$)[c] | Leite et al. (2016) |
| Didelphis marsupialis | RT | Venezuela | tropical dry forest | 123 ha (males, $N=3$) / 11 ha (females, $N=5$) | Sunquist et al. (1987) |
| Didelphis virginiana | RT | USA | temperate woodland | 142 ha (males, $N=23$) / 64 ha (females, $N=64$) | Ryser (1995) |
| Gracilinanus agilis | CMR | Brazil | savanna woodland | 0.5 ha (males, $N=14$) / 0.2 ha (females, $N=64$) | Shibuya et al. (2018) |
| Marmosa paraguayana[d] | RT | Brazil | tropical rainforest | 12.7 ha (males, $N=3$)[e] / 5.3 ha (females, $N=3$)[e] | Moraes & Chiarello (2005a) |
| Marmosops paulensis | CMR | Brazil | tropical rainforest | 1.5 ha (males, $N=1$)[e] / 0.3 ha (females, $N=2$)[e] | Leiner & Silva (2009) |

[a] CMR = capture-mark-recapture, RT = radiotelemetry.
[b] Average areas (in hectares, ha) except as noted. Sample sizes ($N$) are numbers of males or females from which estimates were derived)
[c] Linear home ranges (in meters, m), measured along stream courses.
[d] Originally identified as Micoureus demerarae.
[e] Data from individuals with ≥10 captures.

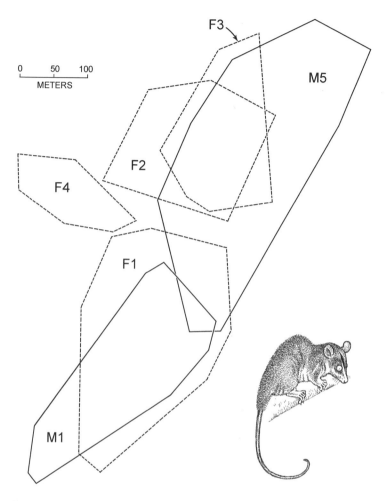

Fig. 13.1. Home ranges (approximated by minimum convex polygons) of two male and four female individuals of *Caluromys philander* mapped by radiotelemetry in primary forest in French Guiana (redrawn from Julien-Laferrière, 1995)

ual territoriality (Fleming, 1972; Sunquist et al., 1987; Pires and Fernandez, 1999; Leiner and Silva, 2009; Leite et al., 2016; Shibuya et al., 2018).

Sex differences in territoriality are thought to result from sex differences in reproductive strategies and resource defense (Ostfeld, 1990). In taxa without paternal care, males maximize reproductive success by mating with multiple females and, when females are widely dispersed, must have large home ranges to monitor

the sexual receptivity of as many neighbors as possible. Consistent with this expectation, male home ranges of *Didelphis virginiana* are twice as large in the breeding season (when they overlap the home ranges of five to seven neighboring females) than they are in the nonbreeding season (Ryser, 1992). By contrast, female
reproductive success could be enhanced by territoriality if resources are relatively
sparse, patchily distributed, and slowly renewed. During the reproductive season,
for example, *Marmosops paulensis* is said to consume the fruit of understory treelets (*Piper* spp.) that occur in patches and produce only a few ripe infructescences
per night over an extended fruiting period; thus, female territoriality in *M. paulensis* might have evolved to secure exclusive access to such a resource (Leiner and
Silva, 2009). An alternative explanation for female territoriality is that it deters infanticide (Wolff, 1993). In species with promiscuous mating systems—such as the
Virginia opossum (Ryser, 1992; Beasley et al., 2010)—infanticide by males might be
prevented by multiple matings, which effectively confuse paternity (Wolff and Macdonald, 2004); therefore, the main threat of infanticide is from conspecific females.
Because the danger of infanticide is greatest for mammals that leave offspring in the
nest (Wolff and Peterson, 1998), young didelphids are probably most vulnerable to
infanticide during the weeks-long interval between teat detachment and weaning
(Chapter 7). Unfortunately, available information is insufficient to assess these alternative explanations of female territoriality in didelphids, although there are tanta-
lizing hints that infanticide might be a relevant factor (e.g., Pires et al., 2010a).

## Population Density

Estimating animal abundance is another methodologically challenging research
topic, especially when the organisms of interest are behaviorally cryptic. Mammalian population density estimates are usually based either on sight transects or
capture-mark-recapture trapping programs, both of which have been applied in
studies of opossum ecology. However, whereas sight transects are the method of
choice for large diurnal mammals, trapping studies seem to provide more reliable
data for estimating the abundance of small nocturnal taxa (Malcolm, 1990). Unfortunately, some species are more easily captured than others using standard
trapping equipment (Voss and Emmons, 1996; Voss et al., 2001), so no single trapping protocol is likely to yield optimal results for all species. Additionally, opossum population densities are known to fluctuate dramatically from one year to the
next—a local population of *Didelphis virginiana*, for example, was observed to
vary fivefold in abundance over a three-year study (Kasparian et al., 2004), so results
from short-term projects may be unreliable.

Estimates of didelphid population density based on multiyear capture-mark-recapture trapping programs vary widely among the few species for which data are available (Table 13.2). Although theory predicts that, all else being equal, larger species should be less abundant than smaller species (Peters and Raelson, 1984), no consistent relationship between size and population density is apparent in these results. To be sure, the lowest density estimates (6–7 individuals per square kilometer) are both from studies of large species (*Didelphis marsupialis* and *D. virginiana*) and the highest (1130 individuals per square kilometer) is from the smallest species studied (*Gracilinanus microtarsus*), but there is considerable variation among estimated densities of conspecific populations (e.g., a 21-fold difference between the highest and lowest estimates for *D. marsupialis*), and some large species (e.g., *Metachirus nudicaudatus*) have estimated densities equivalent to those of much smaller species (e.g., *Marmosa demerarae*).

Regressions of population density on body size always fail to account for a substantial residual variance that can sometimes be correlated with trophic level and biogeography (at constant body size, for example, herbivores tend to be more abundant than carnivores, and temperate-zone mammals are said to be more abundant than tropical mammals; Peters and Raelson, 1984), so it is instructive to compare these empirically estimated opossum densities with predictions from regression models based on Neotropical forest mammals sorted by trophic categories (Robinson and Redford, 1986). Such comparisons (Table 13.3) suggest remarkably poor agreement between model predictions and empirically estimated densities for most species. Apparently, size is an inadequate predictor of opossum population density, even when trophic level is also taken into account.

Site-to-site variation in densities of conspecific populations could have several nonexclusive explanations, including estimation error (Pollock et al., 1990), geographic variation in habitat productivity (Emmons, 1984), and other circumstances that might cause local populations to seem anomalously sparse or unusually dense. Some of these issues have been discussed in the didelphid literature. Julien-Laferrière (1991), for example, attributed large differences in estimated abundances of three species between two study sites in French Guiana to habitat productivity: much higher densities of *Caluromys philander*, *Didelphis marsupialis*, and *Philander opossum* were reported from a high-productivity secondary forest than in a lower-productivity primary forest. Similarly, Malcolm (1990) attributed the low densities of *D. marsupialis* and other mammals at his central Amazonian study site—by comparison with densities previously reported by other investigators for the same taxa in western Amazonia—to the decreased productivity of local forests

Table 13.2  Estimates of Opossum Population Density[a]

| Species | Location | Habitat | Months[b] | Density[c] | References |
|---|---|---|---|---|---|
| *Caluromys philander* | Brazil | tropical rainforest | 25 | 69 | Malcolm (1990) |
| *Caluromys philander* | Fr. Guiana | tropical rainforest | 26 | 143 | Atramentowicz (1986b) |
| *Caluromys philander* | Fr. Guiana | tropical rainforest | 27 | 51 | Julien-Laferrière (1991) |
| *Didelphis marsupialis* | Brazil | tropical rainforest | 25 | 7 | Malcolm (1990) |
| *Didelphis marsupialis* | Fr. Guiana | tropical rainforest | 26 | 45 | Atramentowicz (1986b) |
| *Didelphis marsupialis* | Fr. Guiana | tropical rainforest | 27 | 22 | Julien-Laferrière (1991) |
| *Didelphis marsupialis* | Venezuela | tropical dry forest | 26 | 150 | O'Connell (1989) |
| *Didelphis virginiana* | USA | temperate woodland | 33 | 6 | Kasparian et al. (2004) |
| *Gracilinanus microtarsus* | Brazil | tropical dry forest | 31 | 1130 | Martins et al. (2006b) |
| *Marmosa demerarae*[d] | Brazil | tropical rainforest | 25 | 26 | Malcolm (1990) |
| *Marmosa robinsoni* | Venezuela | tropical dry forest | 26 | 180 | O'Connell (1989) |
| *Metachirus nudicaudatus* | Brazil | tropical rainforest | 25 | 26 | Malcolm (1990) |
| *Philander opossum* | Fr. Guiana | tropical rainforest | 26 | 137 | Atramentowicz (1986b) |
| *Philander opossum* | Fr. Guiana | tropical rainforest | 27 | 17 | Julien-Laferrière (1991) |

[a] Based on multiyear capture-mark-recapture studies. We omit density estimates for species that, in our experience, are not effectively captured using methods employed by these studies.

[b] Duration of trapping program.

[c] Individuals per square kilometer (extrapolated from per-hectare estimates in cited references when necessary and rounded to the nearest integer), averaged over the duration of the study. From Malcolm's (1990) study, which calculated densities separately based on arboreal and terrestrial trapping, we used estimates based on terrestrial trapping for terrestrial species and estimates based on arboreal trapping for arboreal species.

[d] Originally identified as *Marmosa cinerea*.

*Table 13.3*    Predicted and Observed Population Densities of Six Neotropical Opossum Species

| Species | Weight[a] | Trophic category[b] | Density Predicted[c] | Density Observed[d] |
|---|---|---|---|---|
| *Caluromys philander* | 315[e] | Frugivore-omnivore | 42 | 51–143 |
| *Didelphis marsupialis* | 1060[e] | Frugivore-omnivore | 20 | 7–150 |
| *Gracilinanus microtarsus* | 33[f] | Insectivore-omnivore | 113 | 1130 |
| *Marmosa demerarae* | 103[g] | Insectivore-omnivore | 53 | 26 |
| *Marmosa robinsoni* | 60[h] | Insectivore-omnivore | 76 | 180 |
| *Metachirus nudicaudatus* | 385[i] | Insectivore-omnivore | 22 | 26 |
| *Philander opossum* | 455[e] | Insectivore-omnivore | 19 | 17–137 |

[a] Mean values (in grams) from cited studies.
[b] After Robinson and Redford (1986).
[c] Individuals per square kilometer (rounded to the nearest integer) predicted by regressions of population density on body weight by trophic category (Robinson and Redford, 1986: Table 2).
[d] From Table 13.2.
[e] From Julien-Laferrière (1991: Table 1).
[f] From Martins et al. (2006b).
[g] Midpoint of male and female means from Voss et al. (2001: Table 12).
[h] From O'Connell (1989: Table 1).
[i] Midpoint of male and female means from Voss et al. (2001: Table 11).

growing on poor soils. Habitat productivity, however, seems unlikely to explain why *D. marsupialis* appears to be at least three times more abundant at a Venezuelan dry-forest site than it is at any rainforest study site, nor why *Gracilinanus microtarsus* is 10 times more abundant in a Brazilian dry forest than could be expected based on its size and diet.

In effect, the ecological determinants of opossum population densities remain to be understood. Based on the best information at hand, it would seem reasonable to think that most opossums occur at densities of more than 10 but fewer than 200 individuals per square kilometer in most tropical habitats, and perhaps no greater precision can be expected for inference over such a broad range of taxa and environments.

## Reproduction and Demography
### Reproductive Seasonality

Although some opossum species appear to reproduce throughout the annual cycle at equatorial rainforest sites that lack a well-defined dry season (e.g., in the Guianas; Atramentowicz, 1986a; Julien-Laferrière and Atramentowicz, 1990; Catzeflis et al., 1997, 2019), most didelphids in other circumstances are distinctly seasonal breeders, even where sympatric placental mammals are not (O'Connell,

1989; Gentile et al., 2000).[1] In general, opossum reproductive seasonality increases with latitude and with wet/dry seasonality. The effect of latitude on reproductive seasonality is best illustrated by Rademaker and Cerqueira's (2006) reproductive studies of *Didelphis*, which suggest that length of the breeding season (first mating to last weaning in the annual cycle) tends to decrease with increasing distance from the equator in both hemispheres. Even at tropical latitudes, however, annual breeding activity can be constrained by precipitation schedules, and this climatic factor probably accounts for much of the scatter about the central tendency in Rademaker and Cerqueira's (2006) data.

Reproductive seasonality in didelphids is widely believed to result from the timing of breeding activity such that late lactation and weaning coincide with maximum availability of local food resources (Fleming, 1973; O'Connell, 1989; Julien-Laferrière and Atramentowicz, 1990; Gentile et al., 2000; Martins et al., 2006b; Rademaker and Cerqueira, 2006; Kajin et al., 2008; Barros et al., 2015). As described previously (Chapter 6), late lactation is when the energy requirements of reproductive females are highest, and newly weaned juveniles are unpracticed foragers that presumably benefit from abundant resources in the environment. For most didelphids, insects and fruit are important dietary items, and both of these tend to be most abundant in the rainy season at tropical latitudes (Chapter 8). In the temperate zone, insects and fruit are most abundant in the late spring and summer. Therefore, in order that late lactation and weaning coincide with peak resource abundance, tropical didelphids in habitats with pronounced wet and dry seasons usually begin to breed in the dry season, whereas didelphids at high latitudes begin to breed in the winter.

In *Didelphis virginiana*, for example, the interval from conception to weaning (gestation plus lactation; Table 6.1) is about 113 days, so the first matings of the year in central Missouri (ca. 38° N) occur in early February, and the resulting cohort of young is weaned in May (Reynolds, 1945). Another round of mating (following the post-lactational estrus) results in a second cohort that is mostly weaned by late August. In this more or less typical North American population, females are nonreproductive from mid-September through January, about five and a half months.[2]

At tropical latitudes, reproductive episodes can also be tightly synchronized within a single breeding season. In the Venezuelan Llanos, where there is a severe six-month dry season, *Didelphis marsupialis* produces discrete cohorts of young in two well-defined annual peaks of littering (Sunquist and Eisenberg, 1993). The first cohort is born near the middle of the dry season and is weaned

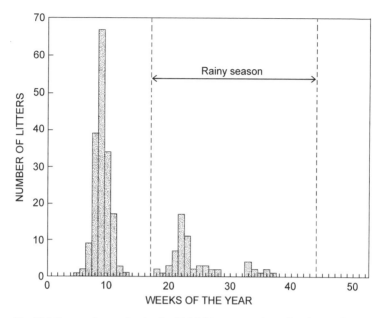

*Fig. 13.2.* Seasonal reproduction by *Didelphis marsupialis* in the Llanos of northern Venezuela (redrawn from Sunquist and Eisenberg, 1993). Three distinct cohorts are born in each annual breeding season, of which the first is weaned just at the start of the rainy season and the last just at the end.

just as the rains begin, whereas the second (the product of post-lactational mating) is born several weeks into the rainy season (Fig. 13.2). The young of a sparse third cohort (born near the end of the rainy season) seldom survive to weaning.

Most didelphid populations with conspicuously seasonal reproduction seem to produce two cohorts of young per year in successive reproductive episodes within the same breeding season, but some populations of small opossums are thought to produce only one annual cohort in a single synchronized breeding episode. The best-documented examples of populations thought to produce only one annual cohort include *Gracilinanus microtarsus* in seasonally dry savanna woodlands of the Brazilian Cerrado (Martins et al., 2006b, 2006c) and *Marmosops paulensis* at a wet premontane site in the Atlantic Forest (Leiner et al., 2008). Although superficially dissimilar in climate and vegetation, both populations are subject to seasonally limited food resources (Martins et al., 2006a; Leiner and Silva, 2007b), and both live at relatively high latitudes (22° and 24° S, respectively). We have not seen compelling evidence that any equatorial opossum population reproduces only once per breeding season.[3]

## Mating System

Information about opossum mating systems is only available for *Didelphis virginiana*. In this species, which was studied by radiotelemetry and direct observation in Florida (Ryser, 1992), adult males patrol the home ranges of five to seven neighboring adult females, visiting each one starting about 10 days before her estrus begins. Approximately six male home ranges overlap the home range of each female, so several males may visit the same female simultaneously, and on the night of estrus they compete directly with one another for mating; associated male behaviors include aggressive vocalizations, chasing, and fighting. Male mating success is nonrandom with respect to body size, because successful males are significantly heavier than unsuccessful males. Observed mating success of individual males per breeding season ranged from zero to three matings, with most (30 out of 50) males failing to mate even once. Although individual females were not observed to mate more than once per breeding episode in this Florida study, genetic analyses of 64 females and their pouch young in Indiana suggest than many litters of *D. virginiana* have two or more sires (Beasley et al., 2010).

As described in Ryser's (1992) study, the mating system of the Virginia opossum corresponds to "overlap promiscuity," a system believed to be the direct consequence of solitary habits and at least temporarily overlapping male and female home ranges (Wittenberger, 1979). Given that solitary habits and intersexually overlapping home ranges appear to be common opossum attributes, it seems reasonable to expect that most didelphids have similar mating systems. However, the apparent lack of sexual size dimorphism in some species (Chapter 5) suggests that taxonomic variation exists in the traits that determine male mating success.

## Dispersal

Overlap promiscuity would seem to favor male-biased dispersal—in which juvenile males disperse farther than juvenile females—to avoid kin competition and to prevent inbreeding (Lawson Handley and Perrin, 2007), but it is not clear that this is really the case. Although some behavioral studies (cited by Wright et al., 1995) suggest that female *Didelphis virginiana* are philopatric and that males disperse, a landscape-scale study of microsatellite variation in this species found no evidence for male-biased dispersal (Beatty et al., 2012). Thus, even in this well-studied taxon, the topic of dispersal is obviously one that merits additional research. Unfortunately, no more than anecdotal observations of dispersal are available from field studies of other didelphid species.

*Generation Time, Fecundity, and Survivorship*

For continuously breeding populations in wet tropical habitats, generation time is perhaps equivalent to the age of female sexual maturity, possibly anywhere from four to nine months depending on the species (Table 6.1). However, very few (if any) females in seasonally reproducing populations seem to breed in the season of their birth (Sunquist and Eisenberg, 1993; Regidor and Gorostiague, 1996; Martins et al., 2006c; Kajin et al., 2008; Leiner et al., 2008; Lopes and Leiner, 2015), so generation times even for small species can be almost as long as a year. In the only study to have computed this key demographic parameter by life-table analysis, the mean generation time in a seasonally breeding rainforest population of *Didelphis aurita* was estimated to be 331 days (Kajin et al., 2008).

Based on a handful of multiyear studies of wild opossum populations, it appears that most individuals of most species reproduce for just one breeding season. This phenomenon, which is accompanied by high post-reproductive mortality in both sexes, has been reported for several small marmosines and thylamyines (Pine et al., 1985; Lorini et al., 1994; Thielen et al., 1997a; Martins et al., 2006c; Leiner et al., 2008; Baladrón et al., 2012; Barros et al., 2015; Lopes and Leiner, 2015). Some of these are thought to be truly semelparous (with a single reproductive episode, followed by complete adult mortality), but, to date, true semelparity has only been convincingly documented for *Marmosops paulensis* in the Atlantic Forest (Leiner et al., 2008).[4] Possibly more widespread is complete adult mortality following two reproductive episodes in the same breeding season (as documented for *Gracilinanus agilis*; Lopes and Leiner, 2015) or high but incomplete adult mortality following a single reproductive episode (as in *G. microtarsus*; Martins et al., 2006c). However, high post-reproductive adult mortality is also characteristic of several large species (e.g., *Didelphis aurita*, *D. albiventris*, and *D. virginiana*), most individuals of which are also thought to breed for a single season only (Sunquist and Eisenberg, 1993; Regidor and Gorostiague, 1996; Woods and Hellgren, 2003; Kajin et al., 2008).

Fecundity can be defined in various ways, but per-female annual reproductive output is the most readily computed statistic from data on opossum reproductive activity. Obtained by multiplying the average number of litters birthed per year by average litter size, this quantity ranges from a little less than 9 (for *Didelphis marsupialis*) to almost 20 (for *Marmosa robinsoni*), but most estimates seem to cluster in the range of 10 to 12 young per year (Table 13.4). On the assumption that roughly half are females, the potential for rapid population growth is obvious.

Table 13.4  Per-Female Annual Reproductive Output (Fecundity) for Seven Opossum Species

| Species | Habitat | Litters[a] | Litter size[b] | Fecundity[c] | References |
|---|---|---|---|---|---|
| *Didelphis marsupialis* | tropical dry forest | 1.5 | 5.9 (N=15) | 8.8 | O'Connell (1989) |
| *Didelphis marsupialis* | tropical rainforest | 1.4 | 7.5 (N=11) | 10.5 | O'Connell (1989) |
| *Didelphis virginiana* | temperate woodland | 1.8 | 6.6 (N=265) | 11.9 | Sunquist and Eisenberg (1993) |
| *Gracilinanus microtarsus* | tropical dry forest | 1.0[d] | 10.9 (N=15) | 10.9 | Martins et al. (2006a, 2006b) |
| *Marmosa robinsoni* | tropical dry forest | 1.4 | 14.0 (N=13) | 19.6 | O'Connell (1989) |
| *Marmosa xerophila* | tropical dry forest | 1.3 | 7.9 (N=55) | 10.3 | Thielen et al. (2009) |
| *Marmosops carri*[e] | tropical rainforest | 1.6 | 6.0 (N=10) | 9.6 | O'Connell (1989) |
| *Monodelphis palliolata*[f] | tropical rainforest | 1.5 | 7.5 (N=5) | 11.2 | O'Connell (1989) |

[a] Average number of litters birthed per female per year.
[b] Average number of attached or suckled young per litter (estimated from samples shown in parentheses).
[c] Average number of young nurtured per female per year.
[d] Inferred from the presence of a single annual breeding episode per year, during which all adult females are said to be pregnant or lactating.
[e] Originally identified as *Marmosa fuscata*.
[f] Originally identified as *Monodelphis brevicaudata*

In habitats that are not chronically overrun with opossums, such high fecundity must be counterbalanced by substantial pre-reproductive mortality. Not surprisingly, survivorship curves reconstructed from capture-mark-recapture data (O'Connell, 1989; Kajin et al., 2008) are concave (type III), but demographic inference from such results is not entirely straightforward. In general, mortality cannot be distinguished from emigration in capture-mark-recapture studies, so survivorship curves reconstructed from study-grid residency could reflect either or both processes. Nevertheless, concave survivorship functions are what we would expect of species that produce large numbers of offspring in habitats with periodically deficient food resources (Chapter 8) and with numerous resident predators of various kinds (Chapter 11).

## Population Dynamics

The obvious consequence of high post-reproductive adult mortality is rapid population turnover. In "semelparous" (or "semi-semelparous") opossums, the generations are almost non-overlapping and, for several months of the year, populations of such species consist only (or primarily) of immature individuals (Leiner et al., 2008; Baladrón et al., 2012; Lopes and Leiner, 2015). However, population turnover is rapid even for unambiguously iteroparous species. For continuously reproducing populations of rainforest opossums in French Guiana, for example, annual turnover was estimated to be 84% for *Caluromys philander*, 97% for *Philander opossum*, and 100% for *Didelphis marsupialis* (Atramentowicz, 1986b).

Sensitivity analyses of matrix models based on data from a long-term study of *Didelphis aurita* in the Atlantic Forest suggest that changes in pouch-young and weanling survival have larger effects on population growth than changes in other vital rates (Ferreira et al., 2013), confirming the intuition of many researchers that these vulnerable early stages are demographically crucial. Because pouch young are known to be "aborted" in times of food scarcity (Atramentowicz, 1986a), when weanlings might also be expected to starve, environmental factors that determine resource availability during lactation and weaning are probably the main causes of year-to-year fluctuations in opossum population densities.

Environmental effects on juvenile survivorship and other demographic parameters can be density independent (or "exogenous," such as precipitation) or density dependent (or "endogenous," such as intraspecific competition). The relative importance of such effects on opossum populations has been analyzed in several recent analyses of data from long-term field projects (Lima et al., 2001; Ferreira et al., 2013, 2016; Troyer et al., 2014). Although generalizations from this handful of studies might be premature, it seems that rainfall (with or without a lag time) has generally positive effects on opossum population growth rates, and that density-dependent effects (thought to be mediated by intraspecific competition) might also be important. By contrast, two studies that looked for effects of interspecific competition on population growth rates both failed to find any (Troyer et al., 2014; Ferreira et al., 2016).

## Discussion

In an early generalization about opossum population biology, Atramentowicz (1986b) described the three species of large didelphids that she studied in French Guiana as having high population densities, high reproductive rates, and rapid population turnover. As we have seen, however, opossum population densities exhibit substantial taxonomic and site-to-site variability, as one might well expect of any ecologically diversified and widely distributed group of mammals. Although high fecundity and rapid population turnover do appear to be common themes of opossum demography among the taxa studied to date, there is a dearth of relevant information about species with reproductive morphologies that suggest different life history traits and that may have correspondingly disparate population dynamics.

Some opossums have only a few nipples—*Hyladelphys*, for example, has just four (Voss et al., 2001)—and must, therefore, have small litters (Chapter 6). To sustain fecundity equivalent to that suggested above (10–12 offspring annually per female), female *Hyladelphys* must routinely have several litters per year. Alternatively, such fecundity might be unnecessary to sustain population replacement if immature and subadult survivorship is high, or if females survive to reproduce over multiple years. The demography of small-littered opossums (which include *Glironia*, also with four nipples) would be interesting to investigate, as would the population dynamics of species with unusual lifestyles (such as *Chironectes*) or that live in extreme habitats (such as *Lestodelphys*).

The topic of reproductive synchrony has not, perhaps, received the attention it deserves from opossum demographers. It is not at all clear why female didelphids should produce their litters almost simultaneously within breeding seasons. Two possibilities merit consideration in this context. One is female infanticide, the risk of which could be minimized by individuals timing their reproduction to coincide with that of adjacent conspecifics, such that neighbors are simultaneously attempting to raise their own litters and protect their own young rather than marauding for juvenile victims (Agrell et al., 1998). Another possible explanation is that female reproductive synchronization tends to increase sperm competition, which (under some circumstances) might enhance female fitness (Fisher et al., 2006, 2013; Shuster et al., 2013). To date, however, compelling evidence to support either (or both) of these nonexclusive alternative hypotheses is essentially nonexistent.

Whatever terminology is adopted for the opossum syndrome of brief (typically uniseasonal) reproductive activity and early senescence, didelphids clearly occupy the "fast" end of the fast-slow continuum of marsupial life histories, especially by comparison with folivorous Australasian taxa that have only one or two young per year over long reproductive lifespans (Fisher et al., 2001). Small dasyurids, however, are also "fast" reproducers, an apparently convergent resemblance for which various explanations have been proposed (Cockburn, 1997; Kraaijeveld et al., 2003; Fisher et al., 2013). A detailed critique of this literature is beyond the scope of our volume, but several aspects of dasyurid "semelparity," including strong sexual dimorphism in post-reproductive mortality, seem to have no exact analogue among opossums. Future research carefully designed to quantify survival functions and test relevant causal hypotheses (e.g., about the role of sexual selection versus high extrinsic mortality) should help explain why both clades have evolved similar demographic strategies, or whether their demographic similarities are more apparent than real.

NOTES

1. According to Streilein (1982a, 1982b), *Monodelphis domestica* reproduces throughout the year in the semiarid Caatinga of northeastern Brazil, but an analysis of a large series of Caatinga specimens suggests that *M. domestica* is, in fact, a seasonal breeder (Bergallo and Cerqueira, 1994).

2. Similar reproductive scheduling has been reported for other populations of *Didelphis virginiana*, including those in much warmer climates (e.g., Florida; Sunquist and Eisenberg, 1993).

3. Streilein (1982a) claimed that *Didelphis albiventris* produces only a single litter annually in the Caatinga (ca. 7° S), but his data (Streilein, 1982c) suggest two reproductive episodes per breeding season—the same pattern typically observed elsewhere for this species and for other congeners.

4. "Semelparity" means reproducing only once (Cole, 1954). For female mammals this means having a single litter, but the term is nonsensical when applied to males, which (by definition) are never parous, and which are not known, in any mammalian species, to die after a single copulation. Unfortunately, the term has been corrupted in current usage to refer to dasyurid species in which males die after one breeding season but females can survive to reproduce in another (Braithwaite and Lee, 1979), and it has recently been used for opossum species in which females produce two litters in a single breeding season (Lopes and Leiner, 2015). The semantic damage is probably irreversible, so more precise language should be used to distinguish how reproductive effort is distributed within or between breeding seasons. In Kirkendall and Stenseth's (1985) suggested terminology, females of *Marmosops paulensis* in the population studied by Leiner et al. (2008) seem to be uniseasonal-uniparous, whereas females of *Gracilinanus agilis* studied by Lopes and Leiner (2015) are uniseasonal-iteroparous.

# V

# SYNTHESIS

# 14

# Adaptive Radiation

As we have now seen, opossums are deeply embedded in the web of trophic relationships and symbioses that sustain the Neotropical ecosystems they inhabit, and for which they are adapted morphologically, physiologically, and behaviorally. What we still lack is a plausible narrative of when and how members of this remarkable marsupial clade came to occupy the geographic regions and ecological niches they currently occupy. What was their ancestral habitat, and when did they begin to enter others? How quickly did they diversify and evolve novel phenotypes? Was their evolutionary trajectory a smooth one, or was it punctuated by unusual opportunities or traumas?

Such questions are commonly bundled together under the rubric of adaptive radiation, a central concept in evolutionary biology, but a somewhat controversial one. Whereas some researchers reserve the term for the evolution of ecological diversity within a rapidly speciating lineage, others apply it more generally to the evolutionary divergence of a monophyletic group into a variety of adaptive forms regardless of rate (Glor, 2010; Givnish, 2015). We understand adaptive radiation in the latter sense, and we use it in contradistinction to nonadaptive radiation, which results when lineage diversification is not accompanied by ecological divergence (Gittenberger, 1991; Czekanski-Moir and Rundell, 2019). In fact, both phenomena can occur at different hierarchical levels in the evolution of a single clade (Losos et al., 2006; Rundell and Price, 2009), and both may have been important in opossum evolution.[1]

In the absence of an adequately informative fossil record, our reconstruction of historical patterns of lineage accumulation (diversification), phenotypic divergence (disparity), and biogeography is based almost entirely on information from Recent species. Unfortunately, studies of adaptive radiation based on living forms can be misleading for a variety of reasons (Hunt and Slater, 2016). We acknowledge this shortcoming, and, toward the end of this chapter, we mention several examples of vanished clades and extinct morphologies in the sparse geological record of opossum evolution.

## Diversification in Ecological and Geographic Context

Inferences about diversification phenomena based on molecular phylogenies can be biased by incomplete taxon sampling and inconsistent taxonomic resolution (Nee et al., 1994; Heath et al., 2008; Faurby et al., 2016). Although various fixes have been proposed for incomplete taxon sampling, none are needed for our analyses of opossum diversification, because sequence data are now available from almost all currently recognized species. Fortunately too, most didelphid genera have been revised within the last decade using comparable methods and data (e.g., by Giarla et al., 2010; Gutiérrez et al., 2010; Rossi et al., 2010; Pavan et al., 2014; Díaz-Nieto and Voss, 2016; Voss et al., 2018), such that more or less uniform criteria for species delimitation have been applied throughout. Phylogenetic error and inaccurate fossil calibrations are other relevant issues for diversification analyses, but highly congruent opossum phylogenies have been obtained from analyses of different datasets (e.g., by Voss and Jansa, 2009; Mitchell et al., 2014; Amador and Giannini, 2016), and similar timetrees have been recovered by recently published relaxed-clock analyses using different sets of fossil calibrations and assumptions (Jansa et al., 2014, unpublished; Beck and Taglioretti, 2019). Therefore, there is reason for modest confidence that the diversification results reported below are unlikely to be grossly misleading, except insofar as the absence of fossil information might bias inference.

Taxon-dense opossum phylogenies obtained from analyses of molecular sequence data (Fig. 14.1) include the usual mix of depauperons (species-poor lineages; Donoghue and Sanderson, 2015) and species-rich clades. Several ancient opossum lineages (e.g., caluromyines, *Glironia*, and *Hyladelphys*) are species-poor, but younger depauperons (e.g., *Chacodelphys, Chironectes, Lestodelphys, Metachirus, Tlacuatzin*) are scattered among more speciose lineages throughout the tree. Although most genera are polytypic, just three of them (*Marmosops, Monodelphis*, and *Marmosa*) contain over 50% of extant opossum species. Some diversity contrasts between sister groups—for example, *Hyladelphys* (with just one currently recognized species) versus Didelphinae (with >100)—are striking and invite speculation about evolutionary causes, but stochastic modeling of speciation and extinction suggests that net diversification rates have been approximately constant across opossum lineages (Jansa et al., unpublished). This is not to say that causal explanations for unbranched lineages on the one hand or speciose clades on the

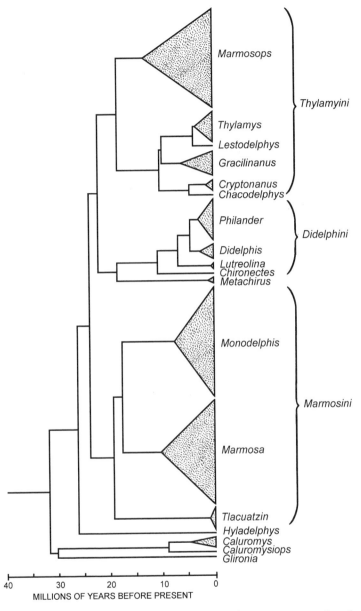

*Fig. 14.1.* A time-calibrated phylogeny (timetree) of Recent opossums based on partitioned maximum-likelihood analysis of six nuclear loci for 116 species (some of which have yet to be named; Jansa et al., unpublished). The contents of polytypic genera are cartooned as isosceles triangles with bases proportional to the number of species they contain and heights proportional to crown ages. Only the contents of polytypic tribes are indicated.

other are inappropriate, but simply that no example of either category appears to be a statistically significant outlier.

Timetrees based on relaxed-clock methods suggest that the most recent common ancestor of living opossums lived sometime in the Oligocene. Point estimates range from about 32 Ma (in the early Oligocene; Jansa et al., unpublished) to about 24 Ma (in the late Oligocene; Beck and Taglioretti, 2019), but 95% credible intervals include a wide temporal range—from the late Eocene to the early Miocene. Geographic optimizations (Jansa et al., 2014) securely locate the ancestral homeland as South America, so if the Oligocene dates are accurate, the opossum common ancestor must have shared the continent with members of numerous other metatherian groups, presumably including caroloameghiniids, sternbergiids, microbiotheriids, argyrolagids, pichipilids, polydolopids, rosendolopids, "hathliacynids," "borhyaenids," and proborhyaenids (Chapter 2).

Phylogenetic optimizations of habitat occupancy suggest that this hypothetical ancestor was a rainforest species (Jansa et al., 2014, unpublished; Mitchell et al., 2014), and that most subsequent opossum evolution leading to extant taxa occurred in South American rainforests until the late Miocene, when several lineages (including the ancestors of *Chacodelphys* + *Cryptonanus*, *Lestodelphys* + *Thylamys*, and *Lutreolina*) may have invaded dry-forest or nonforest (open) habitats. This scenario is plausible because humid tropical forests essentially similar to modern Amazonian vegetation have been a permanent feature of South America since at least the Eocene (Burnham and Johnson, 2004; Maslin et al., 2005). The Mexican endemic genus *Tlacuatzin*—which split away from other marmosines at about 20 Ma (in the early Miocene)—is an ecogeographic exception, but it is uncertain when this lineage crossed the seaway (or seaways) that still separated North and South America and began its ecological transition to dry-forest occupancy. Other North American endemic species (*Didelphis virginiana*, *Marmosa mexicana*, and *Philander vossi*) are descended from much younger immigrant lineages that may have crossed an emergent Panamanian isthmus in the Pliocene (Jansa et al., 2014).

Phylogenetic optimizations of behavioral traits suggest that the Oligocene common ancestor was arboreal or scansorial, and that most of its descendants remained arboreal or scansorial throughout the Tertiary (Jansa et al., unpublished). However, the ancestors of *Monodelphis* and *Metachirus* may have become terrestrial by the early or middle Miocene, and the ancestor of *Lutreolina* may have done so in the late Miocene; other terrestrial lineages, including the ancestors of *Chacodelphys* and *Lestodelphys*, did not diverge from their scansorial sister taxa

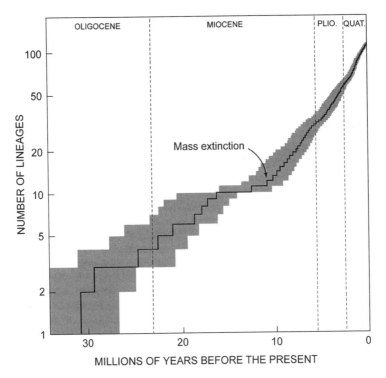

*Fig. 14.2.* A lineage-through-time graph based on the timetree in Figure 14.1. This semi-log plot illustrates an approximately constant rate of lineage accumulation from the Oligocene to the Recent, interrupted only by a hypothesized mass extinction event at about 11 Ma. Gray shading indicates 95% credibility intervals around date estimates.

until the Pliocene. The ancestor of *Chironectes* may not have begun transitioning to semiaquatic life until the late Miocene.

Lineage-through-time graphs indicate an approximately constant rate of net diversification from the Oligocene to the present, with just one statistically significant interruption (Fig. 14.2). Bayesian analyses that allow for episodic changes in speciation and extinction rates (May et al., 2016) suggest that a mass extinction event at about 11 Ma might have exterminated as much as 90% of contemporaneous opossum diversity (Jansa et al., unpublished).[2] Whereas previous estimates of opossum diversification parameters based on a uniform (constant-rate) birth-death model suggested a low rate of net diversification (ca. 0.11 species/Ma) and a high extinction fraction (ca. 0.77) throughout the course of opossum phylogeny (Jansa et al., 2014), estimates from models that allow for mass extinction suggest a somewhat higher background net-diversification rate (ca. 0.13 species/Ma) and a lower background extinction fraction (ca. 0.57).[3] Inferring extinction from molecular phylogenies is notoriously problematic, but simulations suggest that current estimators of background extinction are reasonably unbiased in the absence of among-lineage diversification-rate variation (Rabosky, 2010), whereas estimates of mass extinction are robust even in the presence of such heterogeneity (May et al., 2016). Given that there appears to be no significant rate heterogeneity among didelphid lineages, these results plausibly indicate that both background and episodic extinction have had a substantial impact on opossum diversification.

## Historical Patterns of Disparity

Optimizations of phenotypic traits on opossum phylogenies suggest that the Oligocene ancestor of the crown clade was small (<200 g), black-masked, and pouchless, with unpatterned dorsal pelage, paraxonic forefeet, unwebbed hind feet, a slender prehensile tail, and subequal P2/P3 (Jansa and Voss, 2009; Amador and Giannini, 2016; Jansa et al., unpublished). Phenotypes closely resembling this ancestral condition are conserved among arboreal Recent taxa with mixed diets of insects and fruit (e.g., *Marmosa*), whereas living taxa with markedly divergent morphologies are terrestrial or semiaquatic and predominantly or exclusively faunivorous (Chapter 5). Most of the distinct adaptive types represented among modern opossums (Table 4.1) seem likely to have been established or incipient by the end of the Miocene, when 12 lineages directly ancestral to Recent genera had already appeared. Nevertheless, a few highly divergent phenotypes, notably the suite of distinctive traits that distinguish *Didelphis* from *Philander*, *Chacodelphys* from *Cryptonanus*, and *Lestodelphys* from *Thylamys* did not evolve until the early Pliocene.

Evolutionary models fitted to morphometric data provide statistical estimates of rates of phenotypic divergence that are not obtainable from discrete character data. Size accounts for a large fraction of the variation among measured external and craniodental dimensions of adult opossums, but residual (size-independent) shape variation in external dimensions is significantly correlated with locomotor habits, and shape variation in craniodental dimensions is possibly correlated with diet (Jansa et al., unpublished). Bayesian analyses of morphometric divergence suggest that opossum size evolution proceeded at a more or less constant pace— without significant changes in rate—throughout the radiation of the crown clade. Evolution of external shape also appears to have proceeded without significant rate variation, but craniodental shape disparity increased significantly in several lineages with unusual or unknown trophic niches (*Hyladelphys, Monodelphis, Chironectes*, and *Chacodelphys*; Jansa et al., unpublished).

## Discussion

Analyses of opossum adaptive radiation provide no evidence for the early burst of rapid diversification and phenotypic evolution that theory predicts can occur when evolving lineages encounter new ecological opportunities (Schluter, 2000). New ecological opportunities can arise in a variety of circumstances (Stroud and Losos, 2016), but none seem to fit the ecogeographic context of opossum evolution. As we have seen, the opossum crown clade evolved on a continent that was already occupied by numerous other metatherian taxa (Chapter 2), any or all of which might have constrained its subsequent diversification and disparity. Additionally, our reviews of morphological, physiological, and behavioral phenotypes (Chapters 5–7) suggest that opossums are not—as a group—characterized by any unique trait or combination of traits likely to have conferred a special adaptive advantage. Therefore, if any ecological opportunity was encountered by the most recent common ancestor of living opossums, it is not readily apparent from evidence reviewed in this volume. We conjecture that didelphimorphian stem taxa (still unknown as fossils) insinuated themselves ecologically amid other contemporaneous metatherians, and that lineages ancestral to Recent opossums began to slowly diversify as archaic lineages (e.g., caroloameghiniids and sternbergiids) dwindled to extinction toward the end of the Paleogene.

The analyses reported above likewise provide no evidence for density-dependent cladogenesis, an allegedly common phenomenon that is often hypothesized to result either from niche-filling or from declining opportunities for allopatric speciation with clade age (Phillimore and Price, 2008; Pigot et al., 2010). Instead, the approximately constant background rate of opossum diversification suggests that the didelphid radiation was not constrained by either process, and that Neotropical landscapes are not yet saturated with opossums. Interestingly, analyses of several other South American vertebrate radiations of comparable crown age—ovenbirds (Derryberry et al., 2011), parrots (Schweizer et al., 2014), monkeys (Aristide et al., 2015), and bats (Rojas et al., 2016)—also suggest density-independent diversification, which might be a shared feature of continental radiations in tectonically and climatically dynamic regions (Derryberry et al., 2011).

By contrast, an episode of mass extinction at or near the transition from the middle to the late Miocene is a unique feature of opossum diversification history. Previously, we (Jansa et al., 2014) suggested two explanations for this event:

(1) flooding of the western Amazon Basin by Lake Pebas, an enormous (>$10^6$ km$^2$) wetlands complex that may have compressed rainforest faunas onto neighboring shields and mountains; or (2) the arrival of immigrant predators that might have decimated native faunas of vulnerable prey taxa. Of these alternatives, the first might be expected to have equally affected the diversification histories of other nonaquatic Amazonian vertebrate clades. However, recently published analyses of platyrrhine primates (Aristide et al., 2015) and noctilionoid bats (Rojas et al., 2016) show no indication of Miocene mass extinctions, nor does the timing of lake formation (which began at about 23 Ma; Wesselingh et al., 2002) coincide with the estimated date for opossum mass extinction. Therefore, the Lake Pebas hypothesis now seems implausible, and the predation hypothesis merits closer scrutiny.

The early evolution of the opossum crown clade did not, of course, occur in predator-free circumstances. Among other vertebrate secondary consumers in the Paleogene and early Neogene of South America were several groups of formidable aspect: large boas, "terror birds" (phorusrhacids), and sparassodonts (Argot, 2004; Degrange et al., 2010; Albino and Brizuela, 2014). Presumably, opossums and other South American mammals of the middle Miocene were adept at avoiding or surviving encounters with these long-familiar native taxa. By contrast, the sudden appearance of novel predators may have had abrupt and devastating effects on native prey populations.

The novel predators in question were crotaline viperids (pitvipers) and procyonid carnivorans, both of which arrived from North America, almost certainly by overwater dispersal, sometime in the Miocene (Chapter 3). Neotropical pitvipers are nocturnal ambush predators that locate their endothermic prey by olfaction, vision, and infrared imaging. Numerous extant pitvipers are known to eat Recent opossums (Chapter 11), and there is no reason to suppose that ancestral pitvipers, newly arrived in South America, did not prey on Miocene didelphids. Although opossums had previously been exposed to boas, which have a superficially similar array of sensory abilities, the infrared-imaging organs and supporting neural structures of pitvipers are far more sophisticated than those of boas, and it is assumed that their ability to image nocturnal endothermic prey is correspondingly superior (Goris, 2011). Also possessed of a highly effective venom delivery system (Cundall, 2002), pitvipers were formidable predators loosed on a continental fauna of naïve prey, and they radiated rapidly to fill a wide range of niches throughout tropical and temperate South America (Wüster et al., 2002). Although some didelphids subsequently evolved biochemical resistance to pitviper venom toxins (Chapter 6), such complex adaptations presumably take time to evolve and may not have done so rapidly enough to save many contemporaneous opossum lineages from extinction. The principal difficulty with the notion of pitvipers as primary agents of opossum mass extinction, however, is timing. Although the earliest recorded fossil South American pitviper is from the late Miocene (Chapter 3), viperid timetrees based on molecular phylogenies (Alencar et al., 2016, 2018) suggest that crotalines began to radiate in South America around 17 Ma, seemingly too early to account for an extinction event at around 11 Ma.

By contrast, the arrival of procyonid carnivorans cannot have occurred much before about 12 Ma, the youngest date associated with the well-sampled fossil fauna from La Venta, Colombia, which contains neither procyonids nor any other immigrant North American mammal (Kay et al., 1997b). Procyonids first appear in the South American record at about 7 Ma as an already-endemic genus (*Cyonasua*) in northwestern Argentina (Reguero and Candela, 2011), but the interval from 12 to 7 Ma remains sufficiently poorly known, especially at tropical latitudes, that an earlier arrival cannot be ruled out. Miocene South American procyonids were substantially larger and more carnivorous than Recent procyonids, some of which do eat opossums, at least occasionally (Chapter 11). As inheritors of carnivoran phenotypes honed by long interaction with Holarctic prey taxa, it would be surprising if Miocene procyonids did not possess novel predatory behaviors for which the native fauna was unprepared. Forelimb functional morphology suggests that Miocene procyonids were primarily terrestrial (Tarquini et al., 2019), so their predatory impact was perhaps most keenly felt by ground-dwelling taxa.

Ecological research on invasive species has conclusively shown that novel predators are a frequent cause of biotic extinctions (Mooney and Cleland, 2001; Salo et al., 2007). The outcome of predator invasions, however, depends crucially on innate prey vulnerabilities and complex ecological interactions (Sih et al., 2010), so not all native prey taxa are necessarily affected, or affected to the same degree. We speculate that several South American vertebrate radiations with reconstructed timetrees that show no indication of a Miocene mass extinction event (e.g., furnariids, platyrrhines, and phyllostomids; see above) may have been less vulnerable than opossums to pitviper and/or carnivoran predation, perhaps due to larger body size and greater encephalization (in the case of primates) or to flight and arboreal roosting habits (birds and bats).

Recent opossums are descended from just 10 lineages that escaped the middle-to-late Miocene mass extinction, but other lineages that survived the Miocene left no living descendants. Among these were members of a remarkable high-latitude Pliocene fauna—including species of †*Hyperdidelphys*, †*Sparassocynus*, and †*Thylophorops*—with unusual phenotypes (large size, highly carnassialized dentitions, and/or accessory auditory sinuses) that have been interpreted as adaptations for trophic niches (hypercarnivory) and sensory abilities (low-frequency hearing) that are not represented among living opossums (Goin and Pardiñas, 1996; Goin et al., 2009b; Zimicz, 2014; Beck and Taglioretti, 2019). These taxa were contemporaneous with fossil taxa of modern aspect (species of *Didelphis* and *Lutreolina*), and their disappearance by the end of the Pliocene—possibly due to biotic interactions with landbridge immigrant taxa such as canids, felids, and mustelids—is compelling evidence that extinctions may have substantially reduced the ecological scope of the didelphid radiation and so affected our perception of what it means to be an opossum.

History written by the victors is often biased by suppressed testimony, and evolutionary narratives based exclusively on molecular phylogenies and Recent phenotypes suffer from a similarly limited perspective. Living opossums are consummate survivors, victors in the struggle for existence throughout a long and eventful Cenozoic history. That their story is not the whole story of the clade to which they belong is indisputable. But, among the numerous metatherian lineages that once populated the Western Hemisphere, didelphids alone remain as vital participants in numerous modern ecosystems on both American continents. Returning at last to the question posed in our introduction: how did opossums achieve this unique evolutionary success?

The answer to any such question is necessarily speculative, but it seems relevant to note that ecological equivalents to opossum phenotypes are present in most Paleotropical faunas. Resemblances between opossums and strepsirrhine ("prosimian") primates have been discussed at some length by authors (e.g., Charles-Dominique, 1983; Rasmussen, 1990; Lemelin and Schmitt, 2007), whereas treeshrews (Emmons, 2000; Selig et al., 2019) and erinaceids (especially gymnures; Corbet, 1988) are other groups with striking morphological and ecobehavioral similarities to opossums. Such observations suggest that there is a stable niche, especially in lowland rainforest habitats, for small nonvolant mammals with mixed diets of insects, small vertebrates, and fruit. As we have seen, this can be regarded as the core opossum niche, from which other adaptive types subsequently evolved.

But even stable niches can be invaded, and if marsupials are poor competitors, why were they not eventually displaced when immigrant placental mammals with similar diets arrived in South America? Again, no definitive answers are possible, but a plausible explanation is niche incumbency (Rosenzweig and McCord, 1991): opossums were already there and intricately adapted to local environments and the native biota, so they might have been hard to displace. Indeed, the preceding chapters provide a substantial review of the numerous phenotypic adaptations and ecological interactions that might have anchored opossums to their native habitats and made them resilient incumbents. Although the role of niche incumbency is hard to assess in historical scenarios, its potential importance is supported by ecological studies of invasive species in modern faunas. Such studies suggest that competitive interactions with invaders seldom drive resident taxa to extinction (Mooney and Cleland, 2001; Davis, 2003). In fact, competition with the native opossum fauna may have constrained the adaptive radiation of immigrant placental taxa. For example, opossum incumbency may explain why the Neotropical fauna includes only a single genus of nocturnal primates (Wright, 1996), and it may explain the otherwise puzzling failure of shrews—a speciose group in African and Asian rainforest faunas—to radiate in the lowland Neotropics.

Finally, we are led to question the logic by which marsupials have come to be regarded as poor competitors. As we have seen (Chapter 1), marsupials did not become extinct in North America, because they were never there. North American Paleogene "marsupials" were stem metatherians: archaic mammals that eventually disappeared from the continental fauna along with many other ancient lineages, including stem eutherians, "condylarths," stem primates, stem carnivorans, and so forth (Rose, 2006; Halliday et al., 2017). Similarly, most of the metatherian lineages that went extinct in South America were also stem taxa rather than members of the crown group (Chapter 2). Although it is certainly true that paucituberculatan marsupials were more diverse in the South American fossil record than they are today, they reached their peak diversity after the arrival of primates and caviomorph rodents, and their decline preceded the arrival of other potentially competing placental clades in the Pliocene. The evidence that marsupials compete poorly with placentals is similarly weak in Australia, where almost all end-Pleistocene and Recent extinctions can be attributed to predation—either by humans (Saltré et al., 2016) or by introduced carnivorans (Woinarski et al., 2015, 2019)—not to competition.

To all appearances, opossums continue to thrive in modern ecosystems, including anthropogenic landscapes; clearly, they are here to stay. Nevertheless, it behooves us to learn as much as we can about their current diversity, phenotypic adaptations, and ecological relationships in natural habitats with intact faunas and floras while opportunities to do so yet remain. This volume will, we hope, provide a foundation for future advances in opossum research.

NOTES

1. Closely related opossum species (e.g., members of the subgenus *Sciophanes* of *Marmosops*; Díaz-Nieto and Voss, 2016) are usually allopatric, differ in characters that have no obvious adaptive significance, and seem to be ecologically similar, so it is possible that the initial stages of lineage diversification are nonadaptive, and that ecological divergence accompanies the establishment of sympatry following secondary contact. This scenario, which essentially resembles the hierarchical process hypothesized by Losos et al. (2006), suggests a key role for community assembly in shaping opossum adaptive radiation.

2. The 95% credible interval for the proportion of contemporaneous lineages that went extinct at about 11 Ma is 0.68–0.92, with a median value of around 0.82 (Jansa et al., unpublished).

3. Net diversification rates and extinction fractions are calculated from fitted estimates of the "birth" (speciation) rate, $\lambda$, and the "death" (extinction) rate, $\mu$. Net diversification is then computed as $\lambda - \mu$ and the extinction fraction as $\mu/\lambda$ (Nee et al., 1994). Terminology is not standardized, however, and $\mu/\lambda$ is also known as the "relative extinction rate" (e.g., by Rabosky, 2010).

# Appendix 1

## A Checklist of Living Opossums (Didelphidae)

Below we list all of the didelphid species currently recognized as valid (to January 2020). Our subfamilial, tribal, and generic classification follows the system explained in Chapter 4. Additionally, we use subgeneric nomenclature to identify monophyletic groups of species in several genera following Voss and Jansa (2009) for *Caluromys*, Giarla et al. (2010) for *Thylamys*, Voss et al. (2014) for *Marmosa*, Díaz-Nieto et al. (2016) for *Marmosops*, and Pavan and Voss (2016) for *Monodelphis*. For each species we provide an indication of their continental distribution (NA, North America [including Central America]; SA, South America) and macrohabitat(s). Macrohabitat abbreviations (De, desert [including semideserts]; DF, dry forest; Gr, grassland; MF, montane forest; RF, rainforest; TF, temperate forest) are separated by commas and listed in order of predominance for species occurring in more than one type of vegetation if predominance is known (*Caluromys philander*, for example, occurs primarily in rainforest, but is sometimes also found in dry forest); however, if it is not known which of two or more macrohabitats is predominant, these are listed in alphabetical order and separated by slashes (e.g., *Marmosa mexicana*).

Caluromyinae (4 spp.)
    *Caluromys* (*Caluromys*) *philander* (SA; RF, DF)
    *Caluromys* (*Mallodelphys*) *derbianus* (SA, NA; RF, DF, MF)
    *Caluromys* (*Mallodelphys*) *lanatus* (SA; RF)
    *Caluromysiops irrupta* (SA; RF)
Glironiinae (1 sp.)
    *Glironia venusta* (SA; RF, DF)
Hyladelphinae (1 sp.)
    *Hyladelphys kalinowskii* (SA; RF)
Didelphinae (110 spp.)
    Marmosini (50 spp.)
        *Marmosa* (*Eomarmosa*) *rubra* (SA; RF)
        *Marmosa* (*Exulomarmosa*) *isthmica* (SA, NA; RF)
        *Marmosa* (*Exulomarmosa*) *mexicana* (NA; DF/MF/RF)
        *Marmosa* (*Exulomarmosa*) *robinsoni* (SA, NA; DF, RF, MF)
        *Marmosa* (*Exulomarmosa*) *simonsi* (SA; DF)

*Marmosa (Exulomarmosa) xerophila* (SA; De)

*Marmosa (Exulomarmosa) zeledoni* (SA, NA; MF/RF)

*Marmosa (Marmosa) macrotarsus* (SA; RF)

*Marmosa (Marmosa) murina* (SA; RF)

*Marmosa (Marmosa) tyleriana* (SA; MF)

*Marmosa (Marmosa) waterhousei* (SA; RF)

*Marmosa (Micoureus) alstoni* (NA; MF)

*Marmosa (Micoureus) constantiae* (SA; RF)

*Marmosa (Micoureus) demerarae* (SA; RF)

*Marmosa (Micoureus) germana* (SA; RF)

*Marmosa (Micoureus) paraguayana* (SA; RF)

*Marmosa (Micoureus) phaea* (SA; MF)

*Marmosa (Micoureus) rapposa* (SA; DF/MF)

*Marmosa (Micoureus) rutteri* (SA; RF)

*Marmosa (Stegomarmosa) andersoni* (SA; RF)

*Marmosa (Stegomarmosa) lepida* (SA; RF)

*Monodelphis (Microdelphys) americana* (SA; RF)

*Monodelphis (Microdelphys) gardneri* (SA; MF)

*Monodelphis (Microdelphys) iheringi* (SA; RF)

*Monodelphis (Microdelphys) scalops* (SA; RF)

*Monodelphis (Monodelphiops) dimidiata* (SA; Gr, RF)

*Monodelphis (Monodelphiops) unistriatus* (SA; RF?)

*Monodelphis (Monodelphis) arlindoi* (SA; RF)

*Monodelphis (Monodelphis) brevicaudata* (SA; RF)

*Monodelphis (Monodelphis) domestica* (SA; DF)

*Monodelphis (Monodelphis) glirina* (SA; RF)

*Monodelphis (Monodelphis) palliolata* (SA; RF, DF)

*Monodelphis (Monodelphis) sanctaerosae* (SA; DF?)

*Monodelphis (Monodelphis) touan* (SA; RF)

*Monodelphis (Monodelphis) vossi* (SA; Gr?)

*Monodelphis (Mygalodelphys) adusta* (SA; RF, MF)

*Monodelphis (Mygalodelphys) handleyi* (SA; RF)

*Monodelphis (Mygalodelphys) kunsi* (SA; DF)

*Monodelphis (Mygalodelphys) osgoodi* (SA; MF)

*Monodelphis (Mygalodelphys) peruviana* (SA; RF, MF)

*Monodelphis (Mygalodelphys) pinocchio* (SA; RF)

*Monodelphis (Mygalodelphys) reigi* (SA; MF)

*Monodelphis (Mygalodelphys) ronaldi* (SA; RF)

*Monodelphis (Mygalodelphys) saci* (SA; RF)

*Monodelphis* (*Pyrodelphys*) *emiliae* (SA; RF)

*Tlacuatzin balsasensis* (NA; DF)

*Tlacuatzin canescens* (NA; DF)

*Tlacuatzin gaumeri* (NA; DF)

*Tlacuatzin insularis* (NA; DF)

*Tlacuatzin sinaloae* (NA; DF)

Metachirini (2 spp.)

*Metachirus myosuros* (SA, NA; RF)

*Metachirus nudicaudatus* (SA; RF)

Didelphini (17 spp.)

*Chironectes minimus* (SA, NA; RF, MF)

*Didelphis albiventris* (SA; DF/Gr)

*Didelphis aurita* (SA; RF)

*Didelphis imperfecta* (SA; Gr)

*Didelphis marsupialis* (SA, NA; RF)

*Didelphis pernigra* (SA; MF)

*Didelphis virginiana* (NA; TF)

*Lutreolina crassicaudata* (SA; Gr)

*Lutreolina massoia* (SA; MF)

*Philander andersoni* (SA; RF)

*Philander canus* (SA; DF/RF)

*Philander mcilhennyi* (SA; RF)

*Philander melanurus* (SA, NA; RF)

*Philander opossum* (SA; RF)

*Philander pebas* (SA; RF)

*Philander quica* (SA; RF)

*Philander vossi* (NA; DF)

Thylamyini (41 spp.)

*Chacodelphys formosa* (SA; Gr?)

*Cryptonanus agricolai* (SA; DF/Gr?)

*Cryptonanus chacoensis* (SA; DF/Gr?)

*Cryptonanus guahybae* (SA; DF)

*Cryptonanus unduaviensis* (SA; DF/Gr?)

*Gracilinanus aceramarcae* (SA; MF)

*Gracilinanus agilis* (SA; DF)

*Gracilinanus dryas* (SA; MF)

*Gracilinanus emiliae* (SA; RF)

*Gracilinanus marica* (SA; MF)

*Gracilinanus microtarsus* (SA; RF)

*Gracilinanus peruanus* (SA; RF)

*Lestodelphys halli* (SA; De)

*Marmosops* (*Marmosops*) *caucae* (SA; MF, RF)

*Marmosops* (*Marmosops*) *creightoni* (SA; MF)

*Marmosops* (*Marmosops*) *incanus* (SA; RF)

*Marmosops* (*Marmosops*) *noctivagus* (SA; RF, MF)

*Marmosops* (*Marmosops*) *ocellatus* (SA; DF)

*Marmosops* (*Marmosops*) *paulensis* (SA; MF)

*Marmosops* (*Marmosops*) *soinii* (SA; RF)

*Marmosops* (*Sciophanes*) *bishopi* (SA; RF, DF)

*Marmosops* (*Sciophanes*) *carri* (SA; MF/RF)

*Marmosops* (*Sciophanes*) *chucha* (SA; MF/RF)

*Marmosops* (*Sciophanes*) *fuscatus* (SA; MF/RF)

*Marmosops* (*Sciophanes*) *handleyi* (SA; MF)

*Marmosops* (*Sciophanes*) *invictus* (NA; MF?)

*Marmosops* (*Sciophanes*) *juninensis* (SA; MF)

*Marmosops* (*Sciophanes*) *magdalenae* (SA; MF/RF)

*Marmosops* (*Sciophanes*) *ojastii* (SA; MF/RF)

*Marmosops* (*Sciophanes*) *pakaraimae* (SA; MF)

*Marmosops* (*Sciophanes*) *parvidens* (SA; RF)

*Marmosops* (*Sciophanes*) *pinheiroi* (SA; RF)

*Thylamys* (*Thylamys*) *elegans* (SA; De)

*Thylamys* (*Thylamys*) *macrurus* (SA; DF)

*Thylamys* (*Thylamys*) *pallidior* (SA; De)

*Thylamys* (*Thylamys*) *pusillus* (SA; DF)

*Thylamys* (*Thylamys*) *sponsorius* (SA; MF)

*Thylamys* (*Thylamys*) *tatei* (SA; DF?)

*Thylamys* (*Thylamys*) *venustus* (SA; MF)

*Thylamys* (*Xerodelphys*) *karimii* (SA; DF)

*Thylamys* (*Xerodelphys*) *velutinus* (SA; DF)

# Appendix 2

## Fruit Taxa Eaten by Opossums and Sympatric Placental Mammals in French Guiana

| Fruit family | Fruit genus | Consumers | | | | |
|---|---|---|---|---|---|---|
| | | Opossums[a] | Bats[b] | Primates[c] | Procyonids[d] | Ungulates[e] |
| Anacardiaceae | *Tapirira* | X | | X | X | X |
| Annonaceae | *Duguetia* | X | | X | | X |
| | *Oxandra* | X | | | X | |
| | *Rollinia* | X | X | | | |
| | *Xylopia* | X | | | | |
| Araceae | *Monstera* | X | | | | |
| Arecaceae | *Astrocaryum* | X | | | | X |
| | *Attalea* | X | | | | X |
| Boraginaceae | *Cordia* | X | X | X | X | |
| Burseraceae | *Dacryodes* | X | | | X | |
| | *Protium* | X | X | X | X | X |
| | *Tetragastris* | X | | X | | |
| Cecropiaceae | *Cecropia* | X | X | X | X | X |
| | *Coussapoa* | X | | X | X | X |
| Chrysobalanaceae | *Couepia* | X | X | | X | |
| | *Licania* | X | X | | X | |
| | *Parinari* | X | | | X | X |
| Clusiaceae | *Clusia* | X | | | X | |
| | *Moronobea* | X | | | | X |
| | *Symphonia* | X | X | | X | X |
| | *Vismia* | X | X | | | |
| Dichapetalaceae | *Dichapetalum* | X | | | | |
| Euphorbiaceae | *Pera* | X | | | | |
| Fabaceae | *Andira* | X | | | | X |
| Humiriaceae | *Humiriastrum* | X | | | X | |
| | *Sacoglottis* | X | X | X | X | X |
| Lauraceae | *Ocotea* | X | | X | X | X |
| | *Sextonia* | X | | | X | X |
| Malpighiaceae | *Byrsonima* | X | | | | X |
| Melastomataceae | *Bellucia* | X | X | | | X |
| | *Henriettea* | X | X | | | |
| | *Loreya* | X | | | | |
| | *Mouriri* | X | | | X | X |

*continued*

| Fruit family | Fruit genus | Opossums[a] | Bats[b] | Primates[c] | Procyonids[d] | Ungulates[e] |
|---|---|---|---|---|---|---|
| | | | | Consumers | | |
| Mimosaceae | *Inga* | X | | X | X | X |
| Moraceae | *Ficus* | X | X | X | X | X |
| | *Maquira* | X | | X | | |
| Myristicaceae | *Virola* | X | | X | X | X |
| Myrtaceae | *Eugenia* | X | | X | X | |
| Passifloraceae | *Passiflora* | X | X | X | | |
| Piperaceae | *Piper* | X | X | | | X |
| Rubiaceae | *Guettarda* | X | | X | X | |
| | *Posoqueria* | X | | X | | |
| Sapotaceae | *Ecclinusa* | X | | X | X | X |
| | *Pouteria* | X | X | X | X | X |
| Simaroubaceae | *Simarouba* | X | | | X | |
| Utilization | | 100% | 33% | 42% | 56% | 51% |

[a] *Caluromys philander, Didelphis marsupialis, Philander opossum* (Charles-Dominique et al., 1981; Atramentowicz, 1988; Julien-Laferrière, 1999).

[b] Thirty-nine species of phyllostomids (Lobova et al., 2009: Table II).

[c] *Alouatta seniculus, Ateles paniscus, Cebus apella* (Guillotin et al., 1994; Simmen and Sabatier, 1996).

[d] *Potos flavus* (Julien-Laferrière, 1999, 2001).

[e] *Mazama americana, M. "gouazoubira" (= M. nemorivaga),* and *Tapirus terrestris* (Gayot et al., 2004; Henry et al., 2000; Hibert et al., 2011).

# Appendix 3

## Prey Taxa Eaten by Opossums and Sympatric Placental Mammals in Southeastern Brazil

| Prey higher taxon | Prey subordinate taxon | Consumers | | | |
|---|---|---|---|---|---|
| | | Opossums[a] | Primates[b] | Procyonids[c] | Cricetids[d] |
| ANNELIDA | Oligochaeta | X | | X | |
| ARTHROPODA/ Arachnida | Acari | X | | | X |
| | Araneae | X | | X | X |
| | Opiliones | X | | X | X |
| | Scorpiones | X | | X | |
| ARTHROPODA/ Crustacea | Decapoda | X | | X | X |
| | Isopoda | X | | X | X |
| ARTHROPODA/ Myriapoda | Chilopoda | X | | X | X |
| | Diplopoda | X | | X | |
| ARTHROPODA/ Insecta | Blattariae | X | X | X | X |
| | Coleoptera | X | X | X | X |
| | Dermaptera | X | | X | X |
| | Diptera | X | | X | X |
| | Hemiptera | X | X | X | X |
| | Hymenoptera | X | | X | X |
| | Isoptera | X | X | X | X |
| | Lepidoptera | X | X | X | X |
| | Orthoptera | X | X | X | X |
| | Phasmida | X | X | | |
| | Psocoptera | X | | | |
| MOLLUSCA | Gastropoda | X | X | X | |
| VERTEBRATA | Anura | X | X | X | |
| | Squamata | X | X | X | |
| | Aves | X | X | X | |
| | Rodentia | X | | X | |
| Utilization | | 100% | 42% | 88% | 58% |

[a] *Caluromys lanatus, C. philander, Didelphis albiventris, D. aurita, Gracilinanus microtarsus, Lutreolina crassicaudata, Marmosa paraguayana, Marmosops incanus, Metachirus myosuros, Monodelphis sorex, Philander quica* (see Santori et al., 1995, 1997; Cáceres, 2002, 2004; Martins and Bonato, 2004; Carvalho et al., 2005; Casella and Cáceres, 2006; Ceotto et al., 2009; Facure and Ramos, 2011).

[b] *Callithrix flaviceps* and *C. geoffroyi* (see Passamani and Rylands, 2000; Hilário and Ferrari, 2010).

[c] *Nasua nasua* and *Procyon cancrivorus* (see Aguiar et al., 2011; Ferreira et al., 2013; Quintela et al., 2014).

[d] *Akodon paranaensis, Brucepattersonius soricinus, Delomys sublineatus, Euryoryzomys russatus,* and *Thaptomys nigrita* (see Casella and Cáceres, 2006; Pinotti et al., 2011).

# References

Abello, M.A. 2013. Analysis of dental homologies and phylogeny of Paucituberculata (Mammalia: Marsupialia). Biological Journal of the Linnean Society 109: 441–465.

Abello, M.A., E. Ortiz-Jaureguizar, and A.M. Candela. 2012. Paleoecology of the Paucituberculata and Microbiotheria (Mammalia, Marsupialia) from the late Early Miocene of Patagonia. Pages 156–172 in S.F. Vizcaíno, R.F. Kay, and M.S. Bargo, eds. Early Miocene paleobiology in Patagonia: high-latitude paleocommunities of the Santa Cruz Formation. Cambridge, UK: Cambridge University Press.

Abello, M.A., N. Toledo, and E. Ortiz-Jaureguizar. 2018. Evolution of South American Paucituberculata (Metatheria: Marsupialia): adaptive radiation and climate change at the Eocene-Oligocene boundary. Historical Biology doi:10.1080/08912963.2018.1502286.

Abouheif, E., and D.J. Fairbairn. 1997. A comparative analysis of allometry for sexual size dimorphism: assessing Rensch's rule. American Naturalist 149: 540–562.

Acosta-Chaves, V.J., A. Sosa-Bartuano, B.H. Morera-Chacón, and J.E. Jiménez-Castro. 2018. Records of preys hunted by the Zeledon's mouse opossum Marmosa zeledoni Goldman, 1911 (Didelphimorphia: Didelphidae) in Costa Rica. Food Webs 16: e00094.

Adis, J. 1988. On the abundance and density of terrestrial arthropods in central Amazonian dryland forests. Journal of Tropical Ecology 4: 19–24.

Adler, G.H. 1995. Fruit and seed exploitation by Central American spiny rats, Proechimys semispinosus. Studies on Neotropical Fauna and Environment 30: 237–244.

Adler, G.H., A. Carvajal, S.L. Davis-Foust, and J.W. Dittel. 2012. Habitat associations of opossums and rodents in a lowland forest in French Guiana. Mammalian Biology 77: 84–89.

Agrell, J., J.O. Wolff, and H. Ylönen. 1998. Counter-strategies to infanticide in mammals: costs and consequences. Oikos 83: 507–517.

Aguiar, L.M., R.F. Moro-Rios, T. Silvestre, J.E. Silva-Pereira, D.R. Bilski, F.C. Passos, M.L. Sekiama, and V.J. Rocha. 2011. Diet of brown-nosed coatis and crab-eating raccoons from a mosaic landscape with exotic plantations in southern Brazil. Studies on Neotropical Fauna and Environment 46: 153–161.

Albanese, S., and R.A. Ojeda. 2012. Habitat use by a Neotropical desert marsupial (Thylamys pallidior): a multi-scale approach. Mammalian Biology 77: 237–243.

Albanese, S., D. Rodríguez, and R.A. Ojeda. 2011. Differential use of vertical space by small mammals in the Monte Desert, Argentina. Journal of Mammalogy 92: 1270–1277.

Albanese, S., M.A. Dacar, and R.A. Ojeda. 2012. Unvarying diet of a Neotropical desert marsupial inhabiting a variable environment: the case of Thylamys pallidior. Acta Theriologica 57: 185–188.

Albino, A.M., and S. Brizuela. 2014. An overview of the South American fossil squamates. Anatomical Record 297: 349–368.

Albino, A.M., and C.I. Montalvo. 2006. Snakes from the Cerro Azul Formation (upper Miocene), central Argentina, with a review of fossil viperids from South America. Journal of Vertebrate Paleontology 26: 581–587.

Alden, K.J. 1995. Helminths of the opossum, *Didelphis virginiana*, in southern Illinois, with a compilation of all helminths reported from this host in North America. Journal of the Helminthological Society of Washington 62: 197–208.

Alencar, L.R.V., T.B. Quental, F.G. Grazziotin, M.L. Alfaro, M. Martins, M. Venzon, and H. Zaher. 2016. Diversification in vipers: phylogenetic relationships, time of divergence, and shifts in speciation rates. Molecular Phylogenetics and Evolution 105: 50–62.

Alencar, L.R.V., M. Martins, and H.W. Greene. 2018. Evolutionary history of vipers. Pages 1–10 *in* eLS [Encyclopedia of Life Sciences]. Chichester, UK: John Wiley and Sons.

Aléssio, F.M., A.R.M. Pontes, and V.L. da Silva. 2005. Feeding by *Didelphis albiventris* on tree gum in the northeastern Atlantic Forest of Brazil. Mastozoología Neotropical 12: 53–56.

Almeida-Santos, S.M., M.M. Antoniazzi, O.A. Sant'Anna, and C. Jared. 2000. Predation by the opossum *Didelphis marsupialis* on the rattlesnake *Crotalus durissus*. Current Herpetology 19: 1–9.

Alves-Costa, C.P., G.A.B. da Fonseca, and C. Christófaro. 2004. Variation in the diet of the brown-nosed coati (*Nasua nasua*) in southeastern Brazil. Journal of Mammalogy 85: 478–482.

Amador, L.L., and N.P. Giannini. 2016. Phylogeny and evolution of body mass in didelphid marsupials (Marsupialia: Didelphimorphia: Didelphidae). Organisms, Diversity, and Evolution 16: 641–657.

Anderson, R.C. 2000. Nematode parasites of vertebrates, their development and transmission, 2nd ed. New York: CABI Publishing.

Andrade, T.Y., W. Thies, P.K. Rogeri, E.K.V. Kalko, and M.A.R. Mello. 2013. Hierarchical fruit selection by Neotropical leaf-nosed bats (Chiroptera: Phyllostomidae). Journal of Mammalogy 94: 1094–1101.

Antoine, P.-O., L. Marivaux, D.A. Croft, G. Billet, M. Ganerød, C. Jaramillo, T. Martin, et al. 2011. Middle Eocene rodents from Peruvian Amazonia reveal the pattern and timing of caviomorph origins and biogeography. Proceedings of the Royal Society B 279: 1319–1326.

Antoine, P.-O., M.A. Abello, S. Adnet, A.J.A. Sierra, P. Baby, G. Billet, M. Boivin, et al. 2016. A 60-million-year Cenozoic history of western Amazonian ecosystems in Contamana, eastern Peru. Gondwana Research 31: 30–59.

Antunes, V.Z., A.C. Delciellos, and M.V. Vieira. 2016. Postural climbing behaviours of didelphid marsupials: parallels with primates. Oecologia Australis 20: 54–69.

Aplin, K.P., K.M. Helgen, and D.P. Lunde. 2010. A review of *Peroryctes broadbenti*, the giant bandicoot of Papua New Guinea. American Museum Novitates 3696: 1–41.

Arcangeli, J., J.E. Light, and F.A. Cervantes. 2018. Molecular and morphological evidence of the diversification in the gray mouse opossum, *Tlacuatzin canescens* (Didelphimorphia), with description of a new species. Journal of Mammalogy 99: 138–158.

Ardente, N., D. Gettinger, R. Fonseca, H.G. Bergallo, and F. Martins-Hatano. 2013. Mammalia, Didelphimorphia, Didelphidae, *Glironia venusta* Thomas, 1912, and *Chironectes minimus* (Zimmermann, 1780): distribution extension for eastern Amazonia. Check List 9: 1104–1107.

Argot, C. 2001. Functional-adaptive anatomy of the forelimb in the Didelphidae, and the paleobiology of the Paleocene marsupials *Mayulestes ferox* and *Pucadelphys andinus*. Journal of Morphology 247: 51–79.

Argot, C. 2002. Functional-adaptive analysis of the hindlimb anatomy of extant marsupials, and the paleobiology of the Paleocene marsupials *Mayulestes ferox* and *Pucadelphys andinus*. Journal of Morphology 253: 76–108.

Argot, C. 2003. Functional-adaptive anatomy of the axial skeleton of some extant marsupials and the paleobiology of the Paleocene marsupials *Mayulestes ferox* and *Pucadelphys andinus*. Journal of Morphology 255: 279–300.

Argot, C. 2004. Evolution of South American mammalian predators (Borhyaenoidea): anatomical and palaeobiological implications. Zoological Journal of the Linnean Society 140: 487–521.

Arguero, A., L. Albuja, and J. Brito. 2017. Nuevos registros de *Glironia venusta* Thomas, 1912 (Mammalia, Didelphidae) en el suroriente de Ecuador. Mastozoología Neotropical 24: 219–225.

Aristide, L., A.L. Rosenberger, M.F. Tejedor, and S.I. Perez. 2015. Modeling lineage and phenotypic diversification in the New World monkey (Platyrrhini, Primates) radiation. Molecular Phylogenetics and Evolution 82: 375–385.

Ashe, J.S., and R.M. Timm. 1987. Predation by and activity patterns of "parasitic" beetles of the genus *Amblyopinus* (Coleoptera: Staphylinidae). Journal of Zoology London 212: 429–437.

Ashwell, K.W.S. 2010. Overview of marsupial brain organization and evolution. Pages 18–39 *in* K.W.S. Ashwell, ed. The neurobiology of Australian marsupials. Cambridge, UK: Cambridge University Press.

Astúa, D. 2006. Range extension and first Brazilian record of the rare *Hyladelphys kalinowskii* (Hershkovitz, 1992) (Didelphimorphia, Didelphidae). Mammalia 70: 174–176.

Astúa, D. 2010. Cranial sexual dimorphism in New World marsupials and a test of Rensch's rule in Didelphidae. Journal of Mammalogy 91: 1011–1024.

Astúa, D., R.A. Carvalho, P.F. Maia, A.R. Magalhães, and D. Loretto. 2015. First evidence of gregarious denning in opossums (Didelphimorphia, Didelphidae), with notes on their social behavior. Biology Letters 11: 20150307.

Astúa de Moraes, D., E. Hingst-Zaher, L.F. Marcus, and R. Cerqueira. 2000. A geometric morphometric analysis of cranial and mandibular shape variation of didelphid marsupials. Hystrix (n.s.) 10: 115–130.

Astúa de Moraes, D., R.T. Santori, R. Finotti, and R. Cerqueira. 2003. Nutritional and fibre content of laboratory-established diets of Neotropical opossums (Didelphidae). Pages 228–237 *in* M. Jones, C. Dickman, and M. Archer, eds. Predators with pouches, the biology of carnivorous marsupials. Collingwood, Australia: CSIRO Publishing.

Atramentowicz, M. 1982. Influence du milieu sur l'activité locomotrice et la reproduction de *Caluromys philander* (L.). Revue d'Ecologie (La Terre et la Vie) 36: 373–395.

Atramentowicz, M. 1986a. Disponibilités trophiques et rythmes de reproduction chez trois marsupiaux didelphidés de Guyane. Memoires du Muséum National d'Histoire Naturelle (nouvelle série) Série A, Zoologie 132: 123–130.

Atramentowicz, M. 1986b. Dynamique de population chez trois marsupiaux didelphidés de Guyane. Biotropica 18: 136–149.

Atramentowicz, M. 1988. La frugivorie opportuniste de trois marsupiaux didelphidés de Guyane. Revue d'Ecologie (La Terre et la Vie) 43: 47–57.

Atramentowicz, M. 1995. Growth of pouch young in the bare-tailed woolly opossum, *Caluromys philander*. Journal of Mammalogy 76: 1213–1219.

Augustiny, G. 1942. Die Schwimmanpassung von *Chironectes*. Zeitschrift für Morphologie und Ökologie der Tiere 39: 276–319.

Austad, S.N. 1993. Retarded senescence in an insular population of Virginia opossums (*Didelphis virginiana*). Journal of Zoology London 229: 695–708.

Austad, S.N., and K.E. Fischer. 1991. Mammalian aging, metabolism, and ecology: evidence from the bats and marsupials. Journal of Gerontology 46: B47–B53.

Azad, A.F., and C.B. Beard. 1998. Rickettsial pathogens and their arthropod vectors. Emerging Infectious Diseases 4: 179–186.

Bacon, C.D., D. Silvestro, C. Jaramillo, B.T. Smith, P. Chakrabarty, and A. Antonelli. 2015. Biological evidence supports an early and complex emergence of the Isthmus of Panama. Proceedings of the National Academy of Science USA: 112: 6110–6115.

Baglan, A., and F. Catzeflis. 2016. Barn owl pellets collected in coastal savannas yield two additional species of small mammals for French Guiana. Mammalia 80: 91–95.

Baker, A., and C. Dickman. 2018. Secret lives of carnivorous marsupials. Canberra: CSIRO Publishing.

Baladrón, A.V., A.I. Malizia, M.S. Bó, M.S. Liébana, and M.J. Bechard. 2012. Population dynamics of the southern short-tailed opossum (*Monodelphis dimidiata*) in the pampas of Argentina. Australian Journal of Zoology 60: 238–245.

Barchan, D., S. Kachalsky, D. Neumann, Z. Vogel, M. Ovadia, E. Kochva, and S. Fuchs. 1992. How the mongoose can fight the snake: the binding site of the mongoose acetylcholine receptor. Proceedings of the National Academy of Sciences USA 89: 7717–7721.

Barnes, B.V. 1991. Deciduous forests of North America. Pages 219–344 *in* E. Röhrig and B. Ulrich, eds. Temperate deciduous forests (Ecosystems of the World, vol. 7). Amsterdam: Elsevier.

Barnes, R.D. 1977. The special anatomy of *Marmosa robinsoni*. Pages 387–413 *in* D. Hunsaker, ed. The biology of marsupials. New York: Academic Press.

Barnes, R.D., and S.W. Barthold. 1969. Reproduction and breeding behaviour in an experimental colony of *Marmosa mitis* Bangs (Didelphidae). Journal of Reproduction and Fertility (supplement) 6: 477–482.

Barnes, R.D., and H.G. Wolf. 1971. The husbandry of *Marmosa mitis* as a laboratory animal. International Zoo Yearbook 11: 50–54.

Barros, C.S., T. Püttker, and R. Pardini. 2015. Timing and environmental cues associated with triggering of reproductive activity in Atlantic Forest marsupials. Mammalian Biology 80: 141–147.

Bascompte, J., and P. Jordano. 2007. Plant-animal mutualistic networks: the architecture of biodiversity. Annual Review of Ecology, Evolution, and Systematics 38: 567–593.

Basset, Y. 2001. Invertebrates in the canopy of tropical rainforests. How much do we really know? Plant Ecology 153: 87–107.

Bates, P.A. 2007. Transmission of *Leishmania* metacyclic promastigotes by phlebotomine sand flies. International Journal for Parasitology 37: 1097–1106.

Battistella, T., F. Cerezer, J. Bubadué, G. Melo, M. Graipel, and N. Cáceres. 2018. Litter size variation in didelphid marsupials: evidence of phylogenetic constraints and adaptation. Biological Journal of the Linnean Society 126: 40–54.

Bawa, K.S. 1974. Breeding systems of tree species of a lowland tropical community. Evolution 28: 85–92.

Beach, F.A. 1939. Maternal behavior of the pouchless marsupial *Marmosa cinerea*. Journal of Mammalogy 20: 315–322.

Beasley, J.C., W.S. Beatty, Z.H. Olson, and O.E. Rhodes, Jr. 2010. A genetic analysis of the Virginia opossum mating system: evidence of multiple paternity in a highly fragmented landscape. Journal of Heredity 101: 368–373.

Beatty, W.S., J.C. Beasley, G. Dharmarajan, and O.E. Rhodes, Jr. 2012. Genetic structure of a Virginia opossum (*Didelphis virginiana*) population inhabiting a fragmented agricultural landscape. Canadian Journal of Zoology 90: 101–109.

Beck, R.M.D. 2012. An "ameridelphian" marsupial from the early Eocene of Australia supports a complex model of southern hemisphere biogeography. Naturwissenschaften 99: 715–729.

Beck, R.M.D. 2017. The skull of *Epidolops ameghinoi* from the early Eocene Itaboraí Fauna, southeastern Brazil, and the affinities of the extinct marsupialiform order Polydolopimorphia. Journal of Mammalian Evolution 24: 373–414.

Beck, R.M.D., and M.L. Taglioretti. 2019. A nearly complete juvenile skull of the marsupial *Sparassocynus derivatus* from the Pliocene of Argentina, the affinities of "sparassocynids", and the diversification of opossums (Mammalia; Didelphimorphia; Didelphidae). Journal of Mammalian Evolution doi:10.1007/s10914-019-09471-y.

Beck, R.M.D., H. Godthelp, V. Weisbecker, M. Archer, and S.J. Hand. 2008. Australia's oldest marsupial fossils and their biogeographical implications. PLoS ONE 3(3): 1–8.

Beck-King, H., O. von Halversen, and R. Beck-King. 1999. Home range, population density, and food resources of *Agouti paca* (Rodentia: Agoutidae) in Costa Rica: a study using alternative methods. Biotropica 31: 675–685.

Belov, K., C.E. Sanderson, J.E. Deakin, E.S.W. Wong, D. Assange, K.A. McColl, A. Gout, et al. 2007. Characterization of the opossum immune genome provides insights into the evolution of the mammalian immune system. Genome Research 17: 982–991.

Berdoy, M., J.P. Webster, and D.W. Macdonald. 2000. Fatal attraction in rats infected with *Toxoplasma gondii*. Proceedings of the Royal Society of London B 267: 1591–1594.

Bergallo, H.G., and R. Cerqueira. 1994. Reproduction and growth of the opossum *Monodelphis domestica* (Mammalia: Didelphidae) in northeastern Brazil. Journal of Zoology London 232: 551–563.

Bhullar, B.-A.S., A.R. Manafzadeh, J.A. Miyamae, E.A. Hoffman, E.L. Brainerd, C. Musinsky, and A.W. Crompton. 2019. Rolling of the jaw is essential for mammalian chewing and tribosphenic molar function. Nature 566: 528–532.

Bi, S., X. Zheng, X. Wang, N.E. Cignetti, S. Yang, and J.R. Wible. 2018. An Early Cretaceous eutherian and the placental-marsupial dichotomy. Nature 558: 390–395.

Bianchi, R.C., and S.L. Mendes. 2007. Ocelot (*Felis pardalis*) predation on primates in Caratinga Biological Station, southeast Brazil. American Journal of Primatology 69: 1173–1178.

Bigio, N.C., and R.S. Secco. 2012. As espécies de *Pera* (Euphorbiaceae s.s.) na Amazônia brasileira. Rodriguésia 63: 163–207.

Birney, E.C., J.A. Montjeau, C.J. Phillips, R.S. Sikes, and I. Kim. 1996a. *Lestodelphys halli*: new information on a poorly known Argentine marsupial. Mastozoología Neotropical 3: 171–181.

Birney, E.C., R.S. Sikes, J.A. Monjeau, N. Guthmann, and C.J. Phillips. 1996b. Comments on Patagonian marsupials from Argentina. Pages 149–154 *in* H.H. Genoways and R.J. Baker, eds. Contributions in mammalogy: a memorial volume honoring Dr. J. Knox Jones, Jr. Lubbock, TX: Museum of Texas Tech University.

Bisbal, F.J. 1986. Food habits of some Neotropical carnivores in Venezuela (Mammalia, Carnivora). Mammalia 50: 329–339.

Black, K.H., M. Archer, S.J. Hand, and H. Godthelp. 2012. The rise of Australian marsupials: a synopsis of biostratigraphic, phylogenetic, palaeoecologic, and palaeobiogeographic understanding. Pages 983–1078 *in* J.A. Talent, ed. Earth and life: global biodiversity, extinction intervals, and biogeographic perturbations through time. Dordrecht: Springer Netherlands.

Bloch, J.I., E.D. Woodruff, A.R. Wood, A.F. Rincon, A.R. Harrington, G.S. Morgan, D.A. Foster, et al. 2016. First North American fossil monkey and early Miocene tropical biotic exchange. Nature 533: 243–246.

Bó, M.S., J.P. Isacch, A.I. Malizia, and M.M. Martínez. 2002. Lista comentada de los mamíferos de la Reserva de Biósfera Mar Chiquita, provincia de Buenos Aires, Argentina. Mastozoología Neotropical 9: 5–11.

Boas, J.E.V. 1918. Zur Kenntnis des Hinterfusses der Marsupialier. Biologiske Meddelelser Det Kongelige Danske Videnskabernes Selskab 1(8): 1–23 + 2 pls.

Bocchiglieri, A., A.F. Mendonça, and J.B. Campos. 2010. Diet composition of *Gracilinanus agilis* (Didelphimorphia, Didelphidae) in dry woodland areas of Cerrado in central Brazil. Mammalia 74: 225–227.

Bochkov, A.V. 2011. Evolution of parasitism in mammal-associated mites of the Psoroptidia group (Acari: Astigmata). Entomological Review 91: 1206–1215.

Bochkov, A.V., and B.M. OConnor. 2008. Morphology and systematics of the mite subfamily Dromiciocoptinae (Acari: Myocoptidae), with descriptions of two new species. Annals of the Entomological Society of America 101: 289–296.

Bochkov, A.V., and G. Wauthy. 2009. A review of the genus *Listropsoralges* (Acari: Psoroptidae), with the descriptions of two new species. Acta Parasitologica 54: 269–275.

Bochkov, A.V., B.M. OConnor, and P. Grootaert. 2013. Revision of the family Listropsoralgidae Fain, 1865 (Acariformes: Sarcoptoidea)—skin parasites of marsupials and rodents. Zootaxa 3611(1): 1–69.

Bodmer, R.E. 1989. Frugivory in Amazonian Artiodactyla: evidence for the evolution of the ruminant stomach. Journal of Zoology London 219: 457–467.

Boinski, S., and N.L. Fowler. 1989. Seasonal patterns in a tropical lowland forest. Biotropica 21: 223–233.

Bonaparte, J.F. 1990. New Late Cretaceous mammals from the Los Alamitos Formation, northern Patagonia. National Geographic Research 6: 63–93.

Bond, M., M.F. Tejedor, K.E. Campbell, Jr., L. Chornogubsky, N. Novo, and F. Goin. 2015. Eocene primates of South America and the African origin of New World monkeys. Nature 520: 538–542.

Bossi, D.E.P., and H. de G. Bergallo. 1992. Parasitism by cuterebrid botflies (*Metacuterebra apicalis*) in *Oryzomys nitidus* (Rodentia: Cricetidae) and *Metachirus nudicaudatus* (Marsupialia: Didelphidae) in a southeastern Brazilian rain forest. Journal of Parasitology 1992: 142–145.

Boyer, D.M. 2008. Relief index of second mandibular molars is a correlate of diet among prosimian primates and other euarchontan mammals. Journal of Human Evolution 55: 1118–1137.

Bozinovic, F., G. Ruiz, A. Cotés, and M. Rosenmann. 2005. Energetics, thermoregulation and torpor in the Chilean mouse-opossum *Thylamys elegans* (Didelphidae). Revista Chilena de Historia Natural 78: 199–206.

Braithwaite, R.W., and A.K. Lee. 1979. A mammalian example of semelparity. American Naturalist 113: 151–155.

Braun, J.K., N.L. Pratt, and M.A. Mares. 2010. *Thylamys pallidior* (Didelphimorphia: Didelphidae). Mammalian Species 42: 90–98.

Brecht, M., B. Preilowski, and M.M. Merzenich. 1997. Functional architecture of the mystacial vibrissae. Behavioural Brain Research 84: 81–97.

Breer, H., J. Fleischer, and J. Strotmann. 2006. The sense of smell: multiple olfactory subsystems. Cellular and Molecular Life Sciences 63: 1465–1475.

Brener, Z. 1973. Biology of *Trypanosoma cruzi*. Annual Review of Microbiology 27: 347–382.

Brennan, J.M., and J.T. Reed. 1975. A list of Venezuelan chiggers, particularly of small mammalian hosts (Acari: Trombiculidae). Brigham Young University Science Bulletin (Biol. Ser.) 20(1): 45–75.

Breviglieri, C.P.B., and V.V. Kuhnen. 2016. Resource-defense behavior: first report of an agonistic interaction between the opossum *Didelphis aurita* and the bat *Artibeus lituratus*. Folia Zoologica 65: 243–247.

Brightsmith, D.J. 2005. Competition, predation, and nest niche shifts among tropical cavity nesters: ecological evidence. Journal of Avian Biology 36: 74–83.

Brinkmann, A. 1911. Om Hudens Bygning paa Haand og Fod hos *Chironectes variegatus*. Videnskabelige Meddelelser fra den naturhistoriske Forening i Kjøbenhavn 1910: 1–17, 1 pl.

Brodie, E.D., III, and E.D. Brodie, Jr. 1999. Costs of exploiting poisonous prey: evolutionary trade-offs in a predator-prey arms race. Evolution 53: 626–631.

Bronstein, J.L. 2001. The exploitation of mutualisms. Ecology Letters 4: 277–287.

Bronstein, J.L., I. Izhaki, R. Nathan, J.J. Tewksbury, O. Spiegel, A. Lotan, and O. Alstein. 2007. Fleshy-fruited plants and frugivores in desert ecosystems. Pages 148–177 *in* A.J. Dennis, E.W. Schupp, R.J. Green, and D.A. Westcott, eds. Seed dispersal: theory and its application in a changing world. Wallingford, UK: CAB International.

Brown, J.C. 1971. The description of mammals. 1. The external characters of the head. Mammal Review 1: 151–168.

Brown, J.C., and D.W. Yalden. 1973. The description of mammals. 2. Limbs and locomotion of terrestrial mammals. Mammal Review 3: 107–134.

Bucher, J.E., and H.I. Fritz. 1977. Behavior and maintenance of the woolly opossum (*Caluromys*) in captivity. Laboratory Animal Science 27: 1007–1012.

Bucher, J.E., and R.S. Hoffmann. 1980. *Caluromys derbianus*. Mammalian Species 140: 1–4.

Burnham, R.J., and K.R. Johnson. 2004. South American palaeobotany and the origins of Neotropical rainforests. Philosophical Transactions of the Royal Society of London B 359: 1595–1610.

Busch, M., and F.O. Kravetz. 1991. Diet composition of *Monodelphis dimidiata* (Marsupialia, Didelphidae). Mammalia 55: 619–621.

Busse, S., D. Lutter, G. Heldmaier, M. Jastroch, and C.W. Meyer. 2014. Torpor at high ambient temperature in a Neotropical didelphid, the grey short-tailed opossum (*Monodelphis domestica*). Naturwissenschaften 101: 1003–1006.

Byles, B., F. Catzeflis, R.P. Scheibel, and F.A. Jiménez. 2013. Gastrointestinal helminths of two species of mouse opossums (*Marmosa demerarae* and *Marmosa murina*) from French Guiana. Comparative Parasitology 80: 210–216.

Cáceres, N.C. 2002. Food habits and seed dispersal by the white-eared opossum, *Didelphis albiventris*, in southern Brazil. Studies on Neotropical Fauna and Environment 37: 97–104.

Cáceres, N.C. 2004. Diet of three didelphid marsupials (Mammalia, Didelphimorphia) in southern Brazil. Mammalian Biology 69: 430–433.

Cáceres, N.C. 2006. O papel de marsupiais na dispersão de sementes. Pages 255–269 *in* N.C. Cáceres and E.L.A. Monteiro-Filho, eds. Os marsupiais do Brasil: biologia, ecologia e evolução. Campo Grande: Universidade Federal de Mato Grosso do Sul.

Cáceres, N.C., and A.F. Machado. 2013. Spatial, dietary, and temporal niche dimensions in ecological segregation of two sympatric, congeneric marsupial species. Open Ecology Journal 6: 10–23.

Cáceres, N.C., and E.L.A. Monteiro-Filho. 2000. The common opossum, *Didelphis aurita*, as a seed disperser of several plants in southern Brazil. Ciência e Cultura 52: 41–44.

Cáceres, N.C., and E.L.A. Monteiro-Filho. 2001. Food habits, home range, and activity of *Didelphis aurita* (Mammalia, Marsupialia) in a forest fragment of southern Brazil. Studies on Neotropical Fauna and Environment 36: 85–92.

Cáceres, N.C., and M. Pichorim. 2003. Use of an abandoned mottled piculet *Picumnus nebulosus* (Aves, Picidae) nest by the Brazilian gracile mouse opossum *Gracilinanus microtarsus* (Mammalia, Didelphidae). Biociências 11(1): 97-99.

Cáceres, N.C., V.A.O. Dittrich, and E.L.A. Monteiro-Filho. 1999. Fruit consumption, distance of seed dispersal and germination of solanaceous plants ingested by common opossum (*Didelphis aurita*) in southern Brazil. Revue d'Ecologie (La Terre et la Vie) 54: 225–234.

Cáceres, N.C., I.R. Ghizoni, Jr., and M.E. Graipel. 2002. Diet of two marsupials, *Lutreolina crassicaudata* and *Micoureus demerarae*, in a coastal Atlantic Forest island of Brazil. Mammalia 66: 331–340.

Cáceres, N.C., R.P. Napoli, W.H. Lopes, J. Casella, and G.S. Gazeta. 2007. Natural history of the marsupial *Thylamys macrurus* (Mammalia, Didelphidae) in fragments of savannah in southwestern Brazil. Journal of Natural History 41: 1979–1988.

Calvete, J.J. 2010. Snake venomics, antivenomics, and venom phenotyping: the ménage à trois of proteomic tools aimed at understanding the biodiversity of venoms. Pages 45–72 *in* R.M. Kini, K.J. Clemetson, F.S. Markland, M.A. McLane, and T. Morita, eds. Toxins and Hemostasis. Dordrecht: Springer.

Calzada, J., M. Delibes, C. Keller, F. Palomares, W. Magnusson. 2008. First record of the bushy-tailed opossum, *Glironia venusta* Thomas, 1912, (Didelphimorphia) from Manaus, Amazonas, Brazil. Acta Amazonica 38: 807–810.

Camargo, N.F. de, R.M.S. Cruz, J.F. Ribeiro, and E.M. Vieira. 2011. Frugivoria e potencial dispersão de sementes pelo marsupial *Gracilinanus agilis* (Didelphidae: Didelphimorphia) em áreas de Cerrado no Brasil central. Acta Botanica Brasilica 25: 646–656.

Camargo, N.F. de, J.F. Ribeiro, A.J.A. de Camargo, and E.M. Vieira. 2014. Diet of the gracile mouse opossum *Gracilinanus agilis* (Didelphimorphia: Didelphidae) in a Neotropical savanna: intraspecific variation and resource selection. Acta Theriologica 59: 183–191.

Camargo, N.F. de, N.Y. Sano, and E.M. Vieira. 2017. Predation upon small mammals by *Caluromys lanatus* (Didelphimorphia: Didelphidae) and *Callithrix penicillata* (Primates: Callitrichidae) in the Brazilian savanna. Mammalia 81: 207–210.

Campbell, K.E., Jr., C.D. Frailey, and L.R. Pittman. 2000. The late Miocene gomphothere *Amahuacatherium peruvium* (Proboscidea: Gomphotheriidae) from Amazonian Peru: implications for the great American faunal interchange. Instituto Geológico Minero y Metalúrgico Boletin (ser. D) 23: 1–152.

Cañeda-Guzmán, I.C., A. de Chambrier, and T. Scholz. 2001. *Thaumasioscolex didelphidis* n. gen., n. sp. (Eucestoda: Proteocephalidae) from the black-eared opossum *Didelphis marsupialis* from Mexico, the first proteocephalidean tapeworm from a mammal. Journal of Parasitology 87: 639–646.

Cantú-Salazar, L., M.G. Hidalgo-Mihart, C.A. López-González, and A. González-Romero. 2005. Diet and food resource use by the pygmy skunk (*Spilogale pygmaea*) in the tropical dry forest of Chamela, Mexico. Journal of Zoology 267: 283–289.

Carmignotto, A.P., and T. Monfort. 2006. Taxonomy and distribution of the Brazilian species of *Thylamys* (Didelphimorphia: Didelphidae). Mammalia 2006: 126–144.

Carmona, E.R., and M.M. Rivadeneira. 2006. Food habits of the barn owl *Tyto alba* in the National Reserve Pampa del Tamarugal, Atacama desert, north Chile. Journal of Natural History 40: 473–483.

Carneiro, L.M. 2018. A new species of *Varalphadon* (Mammalia, Metatheria, Sparassodonta) from the upper Cenomanian of southern Utah, North America: phylogenetic and biogeographic insights. Cretaceous Research 84: 88–96.

Carneiro, L.M., E.V. Oliveira, and F.J. Goin. 2018. *Austropediomys marshalli* gen et sp. nov., a new Pediomyoidea (Mammalia, Metatheria) from the Paleogene of Brazil: paleobiogeographic implications. Revista Brasileira de Paleontologia 21: 120–131.

Caro, T. 2005. Antipredator defenses in birds and mammals. Chicago: University of Chicago Press.

Caro, T. 2011. The functions of black-and-white coloration in mammals: review and synthesis. Pages 298–329 *in* M. Stevens and S. Merilaita, eds. Animal camouflage, mechanisms and function. Cambridge, UK: Cambridge University Press.

Carreira, J.C.A., A.M. Jansen, M.N. Meirelles, F.C. Silva, and H.L. Lenzi. 2001. *Trypanosoma cruzi* in the scent glands of *Didelphis marsupialis*: the kinetics of colonization. Experimental Parasitology 97: 129–140.

Carrillo, J.D., A. Forasiepi, C. Jaramillo, and M.R. Sánchez-Villagra. 2015. Neotropical mammal diversity and the Great American Biotic Interchange: spatial and temporal variation in South America's fossil record. Frontiers in Genetics 5 (451): 1–11.

Cartelle, C., and W.C. Hartwig. 1996. A new extinct primate among the Pleistocene megafauna of Bahia, Brazil. Proceedings of the National Academy of Sciences USA 93: 6405–6409.

Cartmill, M. 1974. Pads and claws in arboreal locomotion. Pages 45–83 *in* F.A. Jenkins, Jr., ed. Primate locomotion. New York: Academic Press.

Cartmill, M. 1985. Climbing. Pages 73–88 *in* M. Hildebrand, D.M. Bramble, K.F. Liem, and D.B. Wake, eds. Functional vertebrate morphology. Cambridge, MA: Belknap Press

Cartmill, M., P. Lemelin, and D. Schmitt. 2002. Support polygons and symmetrical gaits in mammals. Zoological Journal of the Linnean Society 136: 401–420.

Carvalho, F.M.V., F.A.S. Fernandez, and J.L. Nessimian. 2005. Food habits of sympatric opossums coexisting in small Atlantic Forest fragments in Brazil. Mammalian Biology 70: 366–375.

Carvalho, R.F. de, D.C. Passos, and L.G. Lessa. 2019. Diet variations in short-tailed opossum *Monodelphis domestica* (Didelphimorphia, Didelphidae) due to seasonal and intersexual factors. Mastozoología Neotropical 26: 340–348.

Case, J.A., F.J. Goin, and M.O. Woodburne. 2005. "South American" marsupials from the Late Cretaceous of North America and the origin of marsupial cohorts. Journal of Mammalian Evolution 12: 461–494.

Casella, J., and N.C. Cáceres. 2006. Diet of four small mammal species from Atlantic Forest patches in south Brazil. Neotropical Biology and Conservation 1: 5–11.

Castilheiro, W.F.F., and M.S. Filho. 2013. Diet of *Monodelphis glirina* (Mammalia: Didelphidae) in forest fragments in southern Amazon. Zoologia 30: 249–254.

Castro, J.M., and L.H. Emmons. 2012. Variation in diet and resources. Pages 37–55 *in* L.H. Emmons, ed. The maned wolves of Noel Kempff Mercado National Park (Smithsonian Contributions to Zoology 639). Washington, DC: Smithsonian Institution Scholarly Press.

Catanese, J.J., and L.F. Kress. 1993. Opossum $\alpha_1$-proteinase inhibitor: purification, linear sequence, and resistance to inactivation by rattlesnake venom metalloproteinases. Biochemistry 32: 509–515.

Catts, E.P. 1982. Biology of New World bot flies: Cuterebridae. Annual Review of Entomology 27: 313–338.

Catzeflis, F. 2018. *Hyladelphys kalinowskii* in French Guiana: new observations and first notes on its nesting biology. Mammalia 82: 431–437.

Catzeflis, F., C. Richard-Hansen, C. Fournier-Chambrillon, A. Lavergne, and J. Vié. 1997. Biométrie, reproduction et sympatrie chez *Didelphis marsupialis* et *D. albiventris* en Guyane française (Didelphidae: Marsupialia). Mammalia 61: 231–243.

Catzeflis, F.M., B.K. Lim, and C.R. da Silva. 2019. Litter size and seasonality in reproduction for Guianan rodents and marsupials. Studies on Neotropical Fauna and Environment 54: 31–39.

Ceballos, G. 1990. Comparative natural history of small mammals from tropical forests in western Mexico. Journal of Mammalogy 71: 263–266.

Ceballos, G. 1995. Vertebrate diversity, ecology, and conservation in Neotropical dry forests. Pages 195–220 *in* S.H. Bullock, H.A. Mooney, and E. Medina, eds. Seasonally dry tropical forests. Cambridge, UK: Cambridge University Press.

Ceotto, P., R. Finotti, R. Santori, and R. Cerqueira. 2009. Diet variation of the marsupials *Didelphis aurita* and *Philander frenatus* (Didelphimorphia, Didelphidae) in a rural area of Rio de Janeiro state, Brazil. Mastozoología Neotropical 16: 49–58.

Cerqueira, R. 1985. The distribution of *Didelphis* in South America. Journal of Biogeography 12: 135–145.

Charbonnel, N., J.G. de Bellocq, and S. Morand. 2006. Immunogenetics of micromammal-macroparasite interactions. Pages 401–442 *in* S. Morand, B.R. Krasnov, and R. Poulin, eds. Micromammals and macroparasites. Tokyo: Springer.

Charles-Dominique, P. 1983. Ecology and social adaptations in didelphid marsupials: comparisons with eutherians of similar ecology. Pages 395–422 *in* J.F. Eisenberg and D.G. Kleiman, eds. Advances in the study of mammalian behavior (American Society of Mammalogists Special Publication 7). Shippensburg, PA: American Society of Mammalogists.

Charles-Dominique, P. 1986. Inter-relations between frugivorous vertebrates and pioneer plants: *Cecropia*, birds and bats in French Guyana. Pages 119–135 *in* A. Estrada and T.H. Fleming, eds. Frugivores and seed dispersal. Dordrecht: Dr. W. Junk.

Charles-Dominique, P., M. Atramentowicz, M. Charles-Dominique, H. Gérard, A. Hladik, C.M. Hladik, and M.F. Prévost. 1981. Les mammifères frugivores arboricoles nocturnes d'une forêt guyanaise: inter-relations plantes-animaux. Revue d'Ecologie (La Terre et la Vie) 35: 341–435.

Chemisquy, M.A., and F.J. Prevosti. 2014. It takes more than large canines to be a sabretooth predator. Mastozoología Neotropical 21: 27–36.

Chemisquy, M.A., F.J. Prevosti, G. Martin, and D.A. Flores. 2015. Evolution of molar shape in didelphid marsupials (Marsupialia: Didelphidae): analysis of the influence of ecological factors and phylogenetic legacy. Zoological Journal of the Linnean Society 173: 217–235.

Chemisquy, M.A., S.D. Tarquini, C.O. Romano Muñoz, and F.J. Prevosti. 2020. Form, function and evolution of the skull of didelphid marsupials (Didelphimorphia: Didelphidae). Journal of Mammalian Evolution doi:10.1007/s10914-019-09495-4.

Chinchilla, F.A. 1997. La dieta del jaguar (*Panthera onca*), el puma (*Felis concolor*) y el manigordo (*Felis pardalis*) (Carnivora: Felidae) en el Parque Nacional Corcovado, Costa Rica. Revista de Biología Tropical 45: 1223–1229.

Christian, D.P. 1983. Water balance in *Monodelphis domestica* (Didelphidae) from the semiarid Caatinga. Comparative Biochemistry and Physiology 74A: 665–669.

Chubb, J.C., M.A. Ball, and G.A. Parker. 2009. Living in intermediate hosts: evolutionary adaptations in larval helminths. Trends in Parasitology 26: 93–102.

Cifelli, R.L., J.J. Eberle, D.L. Lofgren, J.A. Lillegraven, and W.A. Clemens. 2004. Mammalian biochronology of the latest Cretaceous. Pages 21–42 *in* M.O. Woodburne, ed. Late Cretaceous and Cenozoic mammals of North America. New York: Columbia University Press.

Cione, A.L., E.P. Tonni, and L. Soibelzon. 2009. Did humans cause the late Pleistocene–early Holocene mammalian extinctions in South America in a context of shrinking open areas? Pages 125–144 *in* G. Hayes, ed. American megafaunal extinctions at the end of the Pleistocene. Dordrecht: Springer.

Coates, A.G., and R.F. Stallard. 2013. How old is the Isthmus of Panama? Bulletin of Marine Science 89: 801–813.

Cockburn, A. 1997. Living slow and dying young: senescence in marsupials. Pages 163–174 *in* N. Saunders and L. Hinds, eds. Marsupial biology: recent research, new perspectives. Sydney: UNSW Press.

Cole, L.C. 1954. The population consequences of life history phenomena. Quarterly Review of Biology 29: 103–137.

Collins, L.R. 1973. Monotremes and marsupials: a reference for zoological institutions. Washington, DC: Smithsonian Institution Press.

Colwell, D.D., M.J.R. Hall, and P.J. Scholl. 2006. The oestrid flies: biology, host-parasite relationships, impact and management. Wallingford, UK: CABI Publishing.

Constantino, P.J., and B.W. Wright. 2009. The importance of fallback foods in primate ecology and evolution. American Journal of Physical Anthropology 140: 599–602.

Cooper, C.E., P.C. Withers, and A.P. Cruz-Neto. 2009. Metabolic, ventilatory, and hygric physiology of the gracile mouse opossum (*Gracilinanus agilis*). Physiological and Biochemical Zoology 82: 153–162.

Cooper, C.E., P.C. Withers, and A.P. Cruz-Neto. 2010. Metabolic, ventilatory, and hygric physiology of a South American marsupial, the long-furred woolly mouse opossum. Journal of Mammalogy 91: 1–10.

Corbet, G.B. 1988. The family Erinaceidae: a synthesis of its taxonomy, phylogeny, ecology, and zoogeography. Mammal Review 18: 117–172.

Cordero R., G.A., and R.A. Nicolas B. 1987. Feeding habits of the opossum (*Didelphis marsupialis*) in northern Venezuela. Fieldiana Zoology (new series) 39: 125–131.

Costa, L.P., Y.L.R. Leite, and J.L. Patton. 2003. Phylogeography and systematic notes on two species of gracile mouse opossums, genus *Gracilinanus* (Marsupialia: Didelphidae) from Brazil. Proceedings of the Biological Society of Washington 116: 275–292.

Cothran, E.G., M.J. Aivaliotis, and J.L. Vandeberg. 1985. The effects of diet on growth and reproduction in gray short-tailed opossums (*Monodelphis domestica*). Journal of Experimental Zoology 236: 103–114.

Coulson, G. 1996. Anti-predator behaviour in marsupials. Pages 158–186 *in* D.B. Croft and U. Ganslosser, eds. Comparison of marsupial and placental behaviour. Fürth, Germany: Filander Verlag.

Coura, J.R., and A.C.V. Junqeira. 2015. Ecological diversity of *Trypanosoma cruzi* transmission in the Amazon Basin: the main scenarios [*sic*] in the Brazilian Amazon. Acta Tropica 151: 51–57.

Cramer, J.D., T. Gaetano, J.P. Gray, P. Grobler, J.G. Lorenz, N.B. Freimer, C.A. Schmitt, and T.R. Turner. 2013. Variation in scrotal color among widely distributed vervet monkey populations (*Chlorocebus aethiops pygerythrus* and *Chlorocebus aethiops sabaeus*). American Journal of Primatology 75: 752–762.

Creighton, G.K., and A.L. Gardner. 2008 ("2007"). Genus *Gracilinanus* Gardner and Creighton, 1989. Pages 43–50 *in* A.L. Gardner, ed. Mammals of South America, vol. 1. Marsupials, xenarthrans, shrews, and bats. Chicago: University of Chicago Press.

Croft, D.A., R.K. Engelman, T. Dolgushina, and G. Wesley. 2018. Diversity and disparity of sparassodonts (Metatheria) reveal non-analogue nature of ancient South American carnivore guilds. Proceedings of the Royal Society B 285: 20172012.

Croft, D.B. 2003. Behaviour of carnivorous marsupials. Pages 332–346 *in* M. Jones, C. Dickman, and M. Archer, eds. Predators with pouches, the biology of carnivorous marsupials. Collingwood, Australia: CSIRO Publishing.

Crompton, A.W., and K. Hiiemae. 1969. Functional occlusion in tribosphenic molars. Nature 222: 678–679.

Crompton, A.W., and K. Hiiemae. 1970. Molar occlusion and mandibular movements during occlusion in the American opossum, *Didelphis marsupialis* L. Zoological Journal of the Linnean Society 49: 21–47.

Crooks, K.R., and M.E. Soulé. 1999. Mesopredator release and avifaunal extinctions in a fragmented system. Nature 400: 563–566.

Crumpton, N., and R.S. Thompson. 2013. The holes of moles: osteological correlates of the trigeminal nerve in Talpidae. Journal of Mammalian Evolution 20: 213–225.

Cruz, L.D., F.R. Fernandes, and A.X. Linhares. 2009. Prevalence of larvae of the botfly *Cuterebra simulans* (Diptera: Oestridae) on *Gracilinanus microtarsus* (Didelphimorphia: Didelphidae) in southeastern Cerrado from Brazil. Revista Brasileira de Entomologia 53: 314–317.

Cummings, A.R., and J.M. Read. 2016. Drawing on traditional knowledge to identify and describe ecosystem services associated with northern Amazon's multiple-use plants. International Journal of Biodiversity Science, Ecosystem Services & Management 12: 39–56.

Cundall, D. 2002. Envenomation strategies, head form, and feeding ecology in vipers. Pages 149–161 *in* G.W. Schuett, M. Höggren, M.E. Douglas, and H.W. Greene, eds. Biology of the vipers. Eagle Mountain, UT: Eagle Mountain Publishing.

Cunha, A.A., and M.V. Vieira. 2002. Support diameter, incline, and vertical movements of four didelphid marsupials in the Atlantic Forest of Brazil. Journal of Zoology London 258: 419–426.

Cunningham, S.A. 1995. Ecological constraints on fruit initiation by *Calyptrogyne ghiesbreghtiana* (Arecaceae): floral herbivory, pollen availability, and visitation by pollinating bats. American Journal of Botany 82: 1527–1536.

Curay, J., F. Sornoza-Molina, and J. Brito. 2019. Predación de la raposa de cuatro ojos *Philander opossum* (Mammalia, Didelphidae) por el gavilán campestre *Rupornis magnirostris* (Aves, Accipitridae). Avances en Ciéncias e Ingenierías B 11: 222–227.

Cusens, J., S.D. Wright, P.D. McBride, and L.N. Gillman. 2012. What is the form of the productivity-animal-species-richness relationship? A critical review and meta-analysis. Ecology 93: 2241–2252.

Czekanski-Moir, J.E., and R.J. Rundell. 2019. The ecology of nonadaptive speciation and nonadaptive radiations. Trends in Ecology & Evolution 34: 400–415.

Dalloz, M.F., D. Loretto, B. Papi, P. Cobra, and M.V. Vieira. 2012. Positional behaviour and tail use by the bare-tailed woolly opossum *Caluromys philander* (Didelphimorphia, Didelphidae). Mammalian Biology 77: 307–313.

da Silva, M.N.F., and A. Langguth. 1989. A new record of *Glironia venusta* from the lower Amazon, Brazil. Journal of Mammalogy 70: 873–875.

Dávalos, L.M., P.M. Velazco, O.M. Warsi, P.D. Smits, and N.B. Simmons. 2014. Integrating incomplete fossils by isolating conflicting signal in saturated and non-independent morphological characters. Systematic Biology 63: 582–600.

Davis, D.E. 1947. Notes on the life histories of some Brazilian mammals. Boletim do Museu Nacional (nova série) Zoologia 76: 1–8.

Davis, J.A., Jr. 1966. Maverick opossums. Animal Kingdom 69: 112–117.

Davis, M.A. 2003. Biotic globalization: does competition from introduced species threaten biodiversity? BioScience 53: 481–489.

Dawson, T.J., and A.J. Hulbert. 1970. Standard metabolism, body temperature, and surface areas of Australian marsupials. American Journal of Physiology 218: 1233–1238.

Dawson, T.J., and J.M. Olson. 1988. Thermogenic capabilities of the opossum *Monodelphis domestica* when warm and cold acclimated: similarities between American and Australian marsupials. Comparative Biochemistry and Physiology 89A: 85–91.

Deane, M.P., H.L. Lenzi, and A. Jansen. 1984. *Trypanosoma cruzi*: vertebrate and invertebrate cycles in the same mammal host, the opossum *Didelphis marsupialis*. Memórias do Instituto Oswaldo Cruz 79: 513–515.

Degen, A.A. 1997. Ecophysiology of small desert mammals. Berlin: Springer.

Degrange, F.J., C.P. Tambussi, K. Moreno, L.M. Witmer, and S. Wroe. 2010. Mechanical analysis of feeding behavior in the extinct "terror bird" *Andalgalornis steulleti* (Gruiformes: Phorusrhacidae). PLoS ONE 5(8): e11856.

Delciellos, A.C., and M.V. Vieira. 2006. Arboreal walking performance in seven didelphid marsupials as an aspect of their fundamental niche. Austral Ecology 31: 449–457.

Delciellos, A.C., and M.V. Vieira. 2009a. Allometric, phylogenetic, and adaptive components of climbing performance in seven species of didelphid marsupials. Journal of Mammalogy 90: 104–113.

Delciellos, A.C., and M.V. Vieira. 2009b. Jumping ability in the arboreal locomotion of didelphid marsupials. Mastozoología Neotropical 16: 299–307.

Delciellos, A.C., D. Loretto, and M.V. Vieira. 2006. Novos métodos no estudo da estratificação vertical de marsupiais neotropicais. Oecologia Brasiliensis 10: 135–153.

Delgado-V., C.A., A. Arias-Alzate, S. Aristizábal-Arango, and J.D. Sánchez-Londoño. 2014. Uso de la cola y el marsupio en *Didelphis marsupialis* y *Metachirus nudicaudatus* (Didelphimorphia: Didelphidae) para transportar material de anidación. Mastozoología Neotropical 21: 129–134.

D'Elía, G., and J.A. Martínez. 2006. Registros uruguayos de *Gracilinanus* Gardner y Creighton, 1989 y *Cryptonanus* Voss, Lunde y Jansa, 2005 (Didelphimorphia, Didelphidae). Mastozoología Neotropical 13: 245–249.

Delupi, L.H., M.H. Carrera, and J.J. Bianchini. 1997 ("1995"). Morfología comparada de la musculatura craneal en *Lutreolina crassicaudata* (Desmarest, 1804) y *Didelphis albiventris* Lund, 1840 (Marsupialia: Didelphidae). Physis (C) 53: 19–28.

Dennis, A.J. 2002. The diet of the musky rat-kangaroo, *Hypsiprymnodon moschatus*, a rainforest specialist. Wildlife Research 29: 209–219.

Derryberry, E.P., S. Claramunt, G. Derryberry, R.T. Chesser, J. Cracraft, A. Aleixo, J. Pérez-Emán, J.V. Remsen, Jr., and R.T. Brumfield. 2011. Lineage diversification and morphological evolution in a large-scale continental radiation: the Neotropical ovenbirds and woodcreepers. Evolution 65: 2973–2986.

Díaz, G.B., R.A. Ojeda, and M. Dacar. 2001. Water conservation in the South American desert mouse opossum, *Thylamys pusillus* (Didelphimorphia, Didelphidae). Comparative Biochemistry and Physiology 130A: 323–330.

Díaz, M.M., and M.R. Willig. 2004. Nuevos registros de *Glironia venusta* y *Didelphis albiventris* (Didelphimorphia) para Perú. Mastozoología Neotropical 11: 185–192.

Díaz, M.M., S. Nava, and A.A. Guglielmone. 2009. The parasitism of *Ixodes luciae* (Acari: Ixodidae) on marsupials and rodents in Peruvian Amazonia. Acta Amazonica 39: 997–1002.

Díaz-Nieto, J.F., and R.S. Voss. 2016. A revision of the didelphid marsupial genus *Marmosops*, part 1. Species of the subgenus *Sciophanes*. Bulletin of the American Museum of Natural History 402: 1–70.

Dice, L.R. 1947. Effectiveness of selection by owls of deer mice (*Peromyscus maniculatus*). Contributions from the Laboratory of Vertebrate Zoology University of Michigan 34: 1–20.

Dickman, C.R., and C. Huang. 1988. The reliability of fecal analysis as a method for determining the diet of insectivorous mammals. Journal of Mammalogy 69: 108–113.

Donadio, E., S. di Martino, M. Aubone, and A.J. Novaro. 2004. Feeding ecology of the Andean hog-nosed skunk (*Conepatus chinga*) in areas under different land use in north-western Patagonia. Journal of Arid environments 56: 709–718.

Donoghue, M.J., and M.J. Sanderson. 2015. Confluence, synnovation, and depauperons in plant diversification. New Phytologist 207: 260–274.

Dooley, J.C., H.M. Nguyen, A.M.H. Seelke, and L. Krubitzer. 2012. Visual acuity in the short-tailed opossum (*Monodelphis domestica*). Neuroscience 223: 124–130.

Drabeck, D.H., A.M. Dean, and S.A. Jansa. 2015. Why the honey badger don't care: convergent evolution of venom-targeted nicotinic acetylcholine receptors in mammals that survive venomous snake bites. Toxicon 99: 68–72.

Dubost, G. 1988. Ecology and social life of the red acouchy, *Myoprocta exilis*; comparison with the orange-rumped agouti, *Dasyprocta leporina*. Journal of Zoology London 214: 107–123.

Dubost, G., and O. Henry. 2006. Comparison of diets of the acouchy, agouti, and paca, the three largest terrestrial rodents of French Guiana. Journal of Tropical Ecology 22: 641–651.

Duchêne, D.A., J.G. Bragg, S. Duchêne, L.E. Neaves, S. Potter, C. Moritz, R.N. Johnson, S.Y.W. Ho, and M.D.B. Eldridge. 2018. Analysis of phylogenomic tree space resolves relationships among marsupial families. Systematic Biology 67: 400–412.

Duda, R., and L.P. Costa. 2015. Morphological, morphometric, and genetic variation among cryptic and sympatric species of southeastern South American three-striped opossums (*Monodelphis*: Mammalia: Didelphidae). Zootaxa 3936: 485–506.

Duellman, W.E. 1990. Herpetofaunas in Neotropical rainforests: comparative composition, history, and resource use. Pages 455–505 *in* A.H. Gentry, ed. Four Neotropical rainforests. New Haven, CT: Yale University Press.

Dulmen, A. van. 2001. Pollination and phenology of flowers in the canopy of two contrasting rain forest types in Amazonia, Colombia. Plant Ecology 153: 73–85.

Dumont, E.R., K. Samadevam, I. Grosse, O.M. Warsi, B. **Baird, and** L.M. Dávalos. 2014. Selection for mechanical advantage underlies multiple **cranial optima** in New World leaf-nosed bats. Evolution 68: 1436–1449.

Duncan, R.B., Jr., C.R. Reinemeyer, and R.S. Funk. 1989. Fatal lungworm infection in an opossum. Journal of Wildlife Diseases 25: 266–269.

Durant, P. 2002. Notes on white-eared opossum *Didelphis albiventris* Lund, 1840 from Mérida Andes, Venezuela. Revista de Ecología Lantinoamericana 9: 1–7.

Durden, L.A. 2006. Taxonomy, host associations, life cycles, and vectorial importance of ticks parasitizing small mammals. Pages 91–102 *in* S. Morand, B.R. Krasnov, and R. Poulin, eds. Micromammals and macroparasites. Tokyo: Springer.

Durden, L.A., and N. Wilson. 1990. Ectoparasitic and phoretic arthropods of Virginia opossums (*Didelphis virginiana*) in central Tennessee. Journal of Parasitology 76: 581–583.

Eden, R. 1555. The Decades of the newe worlde or west India, conteynyng the navigations and conquests of the Spanyardes, with the particular description of the moste ryche and large landes and Ilandes lately founde in the west Ocean perteynyng to the inheritaunce of the kinges of Spayne [etc.]. Londini: In aedibus Guilhelmi Powell. [Reprinted and edited by Edward Arber (1885) in "The first three English books on America" (Edinburgh: Turnbull & Spears).]

Eisenberg, J.F., and D.W. Wilson. 1981. Relative brain size and demographic strategies in didelphid marsupials. American Naturalist 118: 1–15.

Eiten, G. 1972. The Cerrado vegetation of Brazil. Botanical Review 38: 201–341.

Eldridge, M.D.B., R.M.D. Beck, D.A. Croft, K.J. Travouillon, and B.J. Fox. 2019. An emerging consensus in the evolution, phylogeny, and systematics of marsupials and their fossil relatives (Metatheria). Journal of Mammalogy 100: 802–837.

Ellis, R.D., O.J. Pung, and D.J. Richardson. 1999. Site selection by intestinal helminths of the Virginia opossum (*Didelphis virginiana*). Journal of Parasitology 85: 1–5.

Elton, C.S. 1973. The structure of invertebrate populations inside Neotropical rain forest. Journal of Animal Ecology 42: 55–104.

Emerson, K.C., and R.D. Price. 1975. Mallophaga of Venezuelan mammals. Brigham Young University Science Bulletin (Biol. Ser.) 20(3): 1–77.

Emmons, L.H. 1984. Geographic variation in densities and diversities of non-flying mammals in Amazonia. Biotropica 16: 210–222.

Emmons, L.H. 1987. Comparative feeding ecology of felids in a Neotropical rainforest. Behavioral Ecology and Sociobiology 20: 271–283.

Emmons, L.H. 1997. Neotropical rainforest mammals (2nd ed.). Chicago: University of Chicago Press.

Emmons, L.H. 2000. Tupai: a field study of Bornean treeshrews. Berkeley: University of California Press.

Emmons, L.H. 2008 ("2007"). Genus *Caluromysiops* Sanborn, 1951. Pages 11–12 *in* A.L. Gardner, ed. Mammals of South America, volume 1. Marsupials, xenarthrans, shrews, and bats. Chicago: University of Chicago Press.

Emmons, L.H., V. Chávez, N. Rocha, B. Phillips, I. Phillips, L.F. del Aguila, and M.J. Swarner. 2006. The non-flying mammals of Noel Kempff Mercado National Park (Bolivia). Revista Boliviana de Ecología y Conservación Ambiental 19: 23–46.

Enders, R.K. 1935. Mammalian life histories from Barro Colorado Island, Panama. Bulletin of the Museum of Comparative Zoology 78: 385–502, pls. 1–5.

Enders, R.K. 1937. Panniculus carnosus and the formation of the pouch in didelphids. Journal of Morphology 61: 1–26.

Endler, J.A. 1993. The color of light in forests and its implications. Ecological Monographs 63: 1–27.

Engelman, R.K., and D.A. Croft. 2014. A new species of small-bodied sparassodont (Mammalia: Metatheria) from the middle Miocene locality of Quebrada Honda, Bolivia. Journal of Vertebrate Paleontology 34: 672–688.

Estrada, A., R. Coates-Estrada, and C. Vazquez-Yanes. 1984. Observations on fruiting and dispersers of *Cecropia obtusifolia* at Los Tuxtlas, Mexico. Biotropica 16: 315–318.

Ewer, R.F. 1968. Ethology of mammals. New York: Plenum Press.

Facure, K.G., and V.N. Ramos. 2011. Food habits of the thick-tailed opossum *Lutreolina crassicaudata* (Didelphimorphia, Didelphidae) in two urban areas of southeastern Brazil. Mammalian Biology 76: 234–236.

Fadem, B.H. 1987. Activation of estrus by pheromones in a marsupial: stimulus control and endocrine factors. Biology of Reproduction 36: 328–332.

Fadem, B.H., D.B. Kraus, and R.H. Sheffet. 1986. Nest-building in gray short-tailed opossums: temperature effects and sex differences. Physiology and Behavior 36: 667–670.

Fain, A. 1994. Adaptation, specificity, and host-parasite coevolution in mites (Acari). International Journal for Parasitology 24: 1273–1283.

Fain, A., A.W.A.M. de Cock, and F.S. Lukoschus. 1972. Parasitic mites of Surinam XVII. Description and life cycle of *Marsupialichus marsupialis* sp. n. from *Didelphis marsupialis* (Glycyphagidae: Sarcoptiformes). Acarologia 14: 81–93.

Faurby, S., and J.-C. Svenning. 2016. The asymmetry in the Great American Biotic Interchange in mammals is consistent with differential susceptibility to mammalian predation. Global Ecology and Biogeography 25: 1443–1453.

Faurby, S., W.L. Eiserhardt, and J.-C. Svenning. 2016. Strong effects of variation in taxonomic opinion on diversification analyses. Methods in Ecology and Evolution 7: 4–13.

Fedriani, J.M., M. Zywiec, and M. Delibes. 2012. Thieves or mutualists? Pulp feeders enhance endozoochore local recruitment. Ecology 93: 575–587.

Fenster, C.B., W.S. Armbruster, P. Wilson, M.R. Dudash, and J.D. Thomson. 2004. Pollination syndromes and floral specialization. Annual Review of Ecology, Evolution, and Systematics 35: 375–403.

Fernandes, F.R., L.D. Cruz, E.G. Martins, and S.F. dos Reis. 2010. Growth and home range size of the gracile mouse opossum *Gracilinanus microtarsus* (Marsupialia: Didelphidae) in Brazilian Cerrado. Journal of Tropical Ecology 26: 185–192.

Ferreira, G.A., E. Nakano-Oliveira, G. Genaro, and A.K. Lacerda-Chaves. 2013. Diet of the coati *Nasua nasua* (Carnivora: Procyonidae) in an area of woodland inserted in an urban environment in Brazil. Revista Chilena de Historia Natural 86: 95–102.

Ferreira, M.S., M. Kajin, M.V. Vieira, P.L. Zangrandi, R. Cerqueira, and R. Gentile. 2013. Life history of a Neotropical marsupial: evaluating potential contributions of survival and reproduction to population growth rate. Mammalian Biology 78: 406–411.

Ferreira, M.S., M.V. Vieira, R. Cerqueira, and C.R. Dickman. 2016. Seasonal dynamics with compensatory effects regulate populations of tropical forest marsupials: a 16-year study. Oecologia 182: 1095–1106.

Ficken, R.W., P.E. Matthiae, and R. Horwich. 1971. Eye marks in vertebrates: aids to vision. Science 173: 936–939.

Finotti, R., R. Cerqueira, and M.V. Vieira. 2018. Frugivory vs. insectivory in marsupials of the Atlantic Forest: trade-offs in the use of vertical strata. Oecologia Australis 22: 191–200.

Firestein, S. 2001. How the olfactory system makes sense of scents. Nature 413: 211–218.

Fish, F.E. 1993. Comparison of swimming kinematics between terrestrial and semiaquatic opossums. Journal of Mammalogy 74: 275–284.

Fisher, D.O., I.P.F. Owens, and C.N. Johnson. 2001. The ecological basis of life history variation in marsupials. Ecology 82: 3531–3540.

Fisher, D.O., M.C. Double, S.P. Blomburg, M.D. Jennions, and A. Cockburn. 2006. Post-mating sexual selection increases lifetime fitness of polyandrous females in the wild. Nature 444: 89–92.

Fisher, D.O., C.R. Dickman, M.E. Jones, and S.P. Blomberg. 2013. Sperm competition drives the evolution of suicidal reproduction in mammals. Proceedings of the National Academy of Sciences 110: 17910–17914.

Fitch, H.S., and H.W. Shirer. 1970. A radiotelemetric study of spatial relationships in the opossum. American Midland Naturalist 84: 170–186.

Fleck, D.W., and J.D. Harder. 1995. Ecology of marsupials in two Amazonian rain forests in northeastern Peru. Journal of Mammalogy 76: 809–818.

Fleming, T.H. 1972. Aspects of the population dynamics of three species of opossums in the Panama Canal Zone. Journal of Mammalogy 53: 619–623.

Fleming, T.H. 1973. The reproductive cycles of three species of opossums and other mammals in the Panama Canal Zone. Journal of Mammalogy 54: 439–455.

Fleming, T.H. 1981. Fecundity, fruiting pattern, and seed dispersal in *Piper amalago* (Piperaceae), a bat-dispersed tropical shrub. Oecologia 51: 42–46.

Fleming, T.H. 1986. Opportunism versus specialization: the evolution of feeding strategies in frugivorous bats. Pages 105–118 *in* A. Estrada and T.H. Fleming, eds. Frugivores and seed dispersal. Dordrecht: Dr. W. Junk.

Fleming, T.H., R. Breitwisch, and G.H. Whitesides. 1987. Patterns of tropical vertebrate frugivore diversity. Annual Review of Ecology and Systematics 18: 91–109.

Fleming, T.H., C. Geiselman, and W.J. Kress. 2009. The evolution of bat pollination: a phylogenetic perspective. Annals of Botany 104: 1017–1043.

Flores, D.A. 2009. Phylogenetic analyses of postcranial skeletal morphology in didelphid marsupials. Bulletin of the American Museum of Natural History 320: 1–81.

Flores, D.A., N. Giannini, and F. Abdala. 2018. Evolution of post-weaning skull ontogeny in New World opossums (Didelphidae). Organisms, Diversity & Evolution 18: 367–382.

Flynn, J.J., and C.C. Swisher, III. 1995. Cenozoic South American land mammal ages: correlations to global geochronologies. Society for Sedimentary Geology Special Publication 54: 317–333.

Flynn, J.J., R. Charrier, D.A. Croft, and A.R. Wyss. 2012. Cenozoic Andean faunas[:] shedding new light on South American mammal evolution, biogeography, environments, and tectonics. Pages 51–75 *in* B.D. Patterson and L.P. Costa, eds. Bones, clones, and biomes[:] the history and geography of Recent Neotropical mammals. Chicago: University of Chicago Press.

Fogel, R., and J.M. Trappe. 1978. Fungus consumption (mycophagy) by small animals. Northwest Science 52: 1–31.

Fonseca, G.A.B., and M.C.M. Kierulff. 1989. Biology and natural history of Brazilian Atlantic Forest small mammals. Bulletin of the Florida State Museum (Biological Sciences) 34: 99–152.

Fontúrbel, F.E., M. Franco, M.A. Rodríguez-Cabal, M. Daniela-Rivarola, and G.C. Amico. 2012. Ecological consistency across space: a synthesis of the ecological aspects of *Dromiciops gliroides* in Argentina and Chile. Naturwissenschaften 99: 873–881.

Forasiepi, A.M. 2009. Osteology of *Arctodictis sinclairi* (Mammalia, Metatheria, Sparassodonta) and phylogeny of Cenozoic metatherian carnivores from South America. Monografías del Museo Argentino de Ciencias Naturales (n.s.) 6: 1–174.

Forasiepi, A.M., F. Goin, and A.G. Martinelli. 2009. Contribution to the knowledge of the Sparassocynidae (Mammalia, Metatheria, Didelphoidea), with comments on the age of the Aisol Formation (Neogene), Mendoza Province, Argentina. Journal of Vertebrate Paleontology 29: 1252–1263.

Forasiepi, A.M., M.J. Babot, and N. Zimicz. 2015. *Australohyaena antiqua* (Mammalia: Metatheria, Sparassodonta), a large predator from the Late Oligocene of Patagonia. Journal of Systematic Palaeontology 13: 503–525.

Formoso, A.E., D.E. Udrizar Sauthier, P. Teta, and U.F.J. Pardiñas. 2011. Dense sampling reveals a complex distributional pattern between the southernmost marsupials *Lestodelphys* and *Thylamys* in Patagonia, Argentina. Mammalia 75: 371–379.

Formoso, A.E., G.M. Martin, P. Teta, A.E. Carbajo, D.E. Udrizar Sauthier, and U.F.J. Pardiñas. 2015. Regional extinctions and Quaternary shifts in the geographic range of *Lestodelphys halli*, the southernmost living marsupial: clues for its conservation. PLoS ONE doi:10.1371/journal. pone.0132130.

Foster, A.O. 1939. Some helminths of the woolly opossum in Panama. Transactions of the American Microscopical Society 58: 185–198.

Fox, J.W., and S.M.T. Serrano. 2008. Exploring snake venom proteomes: multifaceted analyses for complex toxin mixtures. Proteomics 8: 909–920.

Fragoso, C., and P. Lavelle. 1992. Earthworm communities of tropical rain forests. Soil Biology and Biochemistry 24: 1397–1408.

Frailey, C.D., and K.E. Campbell, Jr. 2012. Two new genera of peccaries (Mammalia, Artiodactyla, Tayassuidae) from upper Miocene deposits of the Amazon Basin. Journal of Paleontology 86: 852–877.

Franq, E.N. 1969. Behavioral aspects of feigned death in the opossum *Didelphis marsupialis*. American Midland Naturalist 81: 556–568.

Friedman, H. 1967. Colour vision in the Virginia opossum. Nature 213: 835–836.

Frost, S.B., and R.B. Masterton. 1994. Hearing in primitive mammals: *Monodelphis domestica* and *Marmosa elegans*. Hearing Research 76: 67–72.

Furman, D.P. 1972. Laelapid mites (Laelapidae: Laelapinae) of Venezuela. Brigham Young University Science Bulletin (Biol. Ser.) 17(3): 1–58.

Gabbert, S.L. 1998. Basicranial anatomy of *Herpetotherium* (Marsupialia: Didelphimorphia) from the Eocene of Wyoming. American Museum Novitates 3235: 1–13.

Gabrielsen, G.W., and E.N. Smith. 1985. Physiological responses associated with feigned death in the American opossum. Acta Physiologica Scandinavica 123: 393–398.

Galliez, M., M.S. Leite, T.L. Queiroz, and F.A.S. Fernandez. 2009. Ecology of the water opossum *Chironectes minimus* in Atlantic Forest streams of southeastern Brazil. Journal of Mammalogy 90: 93–103.

Gallus, S., A. Janke, V. Kumar, and M.A. Nilsson. 2015. Disentangling the relationship of the Australian marsupial orders using retrotransposon and evolutionary network analyses. Genome Biology and Evolution 7(4): 985–992.

Garber, P.A., and L.M. Porter. 2010. The ecology of exudate production and exudate feeding in *Saguinus* and *Callimico*. Pages 89–108 in A.M. Burrows and L.T. Nash, eds. The evolution of exudativory in Primates. New York: Springer.

García-Márquez, L.J., J.L. Vázquez-García, D. Osorio-Sarabia, V. León-Régagnon, L. García-Prieto, R. Lamothe-Argumedo, and F. Constantino-Casas. 2010. Lung lesions in opossums (*Didelphis virginiana*) naturally infected by *Paragonimus mexicanus* in Colima, Mexico. Veterinaria México 41: 65–70.

García-Prieto, L., J. Falcón-Ordaz, and C. Guzmán-Cornejo. 2012. Helminth parasites of wild Mexican mammals: list of species, hosts, and geographical distribution. Zootaxa 3290: 1–92.

Gardner, A.L. 1973. The systematics of the genus *Didelphis* (Marsupialia: Didelphidae) in North and Middle America. Special Publications of the Museum of Texas Tech University 4: 1–81.

Gardner, A.L. 1982. Virginia opossum. Pages 3–36 *in* J.A. Chapman and G.A. Feldhamer, eds. Wild mammals of North America. Baltimore, MD: Johns Hopkins University Press.

Gardner, A.L. 2008 ("2007"). Mammals of South America, volume 1. Marsupials, xenarthrans, shrews, and bats. Chicago: University of Chicago Press.

Garland, T., Jr. 1983. The relation between maximal running speed and body mass in terrestrial mammals. Journal of Zoology London 199: 157–170.

Garrett, C.M., and D.M. Boyer. 1993. *Bufo marinus* (cane toad). Predation. Herpetological Review 24: 148.

Gasch, T., M. Schott, C. Wehrenfennig, R.-A. Düring, and A. Vilcinskas. 2013. Multifunctional weaponry: the chemical defenses of earwigs. Journal of Insect Physiology 59: 1186–1193.

Gatti, A., R. Bianchi, C.R.X. Rosa, and S.L. Mendes. 2006. Diet of two sympatric carnivores, *Cerdocyon thous* and *Procyon cancrivorus*, in a restinga area of Espirito Santo state, Brazil. Journal of Tropical Ecology 22: 227–230.

Gaunt, M., and M. Miles. 2000. The ecotypes and evolution of triatomine bugs (Triatominae) and their associated trypanosomes. Memórias do Instituto Oswaldo Cruz 95: 557–565.

Gayot, M., O. Henry, G. Dubost, and D. Sabatier. 2004. Comparative diet of the two forest cervids of the genus *Mazama* in French Guiana. Journal of Tropical Ecology 20: 31–43.

Geise, L., and D. Astúa. 2009. Distribution extension and sympatric occurrence of *Gracilinanus agilis* and *G. microtarsus* (Didelphimorphia, Didelphidae), with cytogenetic notes. Biota Neotropica 9: 269–276.

Geiser, F., and R.M. Brigham. 2012. The other functions of torpor. Pages 109–121 *in* T. Ruf, C. Bieber, W. Arnold, and E. Milesi, eds. Living in a seasonal world. Berlin: Springer.

Geiser, F., and G.M. Martin. 2013. Torpor in the Patagonian opossum (*Lestodelphys halli*): implications for the evolution of daily torpor. Naturwissenschaften 100: 975–981.

Gelfo, J.N., F.J. Goin, M.O. Woodburne, and C. de Muizon. 2009. Biochronological relationships of the earliest South American Paleogene mammalian faunas. Palaeontology 52: 251–269.

Gentile, R., P.S. D'Andrea, R. Cerqueira, and L.S. Maroja. 2000. Population dynamics and reproduction of marsupials and rodents in a Brazilian rural area: a five-year study. Studies of Neotropical Fauna and Environment 35: 1–9.

Gentles, A.J., M.J. Wakefield, O. Kahany, W. Gu, M.A. Batzer, D.D. Pollock, and J. Jurka. 2007. Evolutionary dynamics of transposable elements in the short-tailed opossum *Monodelphis domestica*. Genome Research 17: 992–1004.

Gentry, A.H. 1995. Diversity and floristic composition of Neotropical dry forests. Pages 146–194 *in* S.H. Bullock, H.A. Mooney, and E. Medina, eds. Seasonally dry tropical forests. Cambridge, UK: Cambridge University Press.

Gettinger, D., F. Martins-Hitano, M. Lareschi, and J. Malcolm. 2005. Laelapine mites (Acari: Laelapidae) associated with small mammals from Amazonas, Brazil, including a new species from marsupials. Journal of Parasitology 91: 45–48.

Giannini, N.P., and E.K.V. Kalko. 2004. Trophic structure in a large assemblage of phyllostomid bats in Panama. Oikos 105: 209–220.

Giannini, N.P., P. Gaudioso, D.A. Flores, and T.J. Gaudin. 2011. A possible function for an enigmatic synapomorphy of *Didelphis*. Mammalian Biology 76: 512–514.

Giarla, T.C., R.S. Voss, and S.A. Jansa. 2010. Species limits and phylogenetic relationships in the didelphid marsupial genus *Thylamys* based on mitochondrial DNA sequences and morphology. Bulletin of the American Museum of Natural History 346: 1–67.

Gill, G.G., Jr., R.T. Fowler, and S.A. Mori. 1998. Pollination biology of *Symphonia globulifera* (Clusiaceae) in central French Guiana. Biotropica 30: 139–144.

Gillette, L.N. 1980. Movement patterns of radio-tagged opossums in Wisconsin. American Midland Naturalist 104: 1–12.

Gittenberger, E. 1991. What about nonadaptive radiations? Biological Journal of the Linnean Society 43: 263–272.

Givnish, T.J. 2015. Adaptive radiation versus "radiation" and "explosive diversification": why conceptual distinctions are fundamental to understanding evolution. New Phytologist 207: 297–303.

Glor, R.E. 2010. Phylogenetic insights on adaptive radiations. Annual Review of Ecology and Systematics 41: 251–270.

Godfrey, G.K. 1975. A study of oestrus and fecundity in a laboratory colony of mouse opossums (*Marmosa robinsoni*). Journal of Zoology London 175: 541–555.

Goeldi, E.A. 1894. Critical gleanings on the Didelphyidae of the Serra dos Orgãos, Brazil. Proceedings of the Zoological Society of London. 1894: 457–467.

Goin, F.J. 1997. New clues for understanding Neogene marsupial radiations. Pages 187–206 *in* R.F. Kay, R.H. Madden, R.L. Cifelli, and J.J. Flynn, eds. Vertebrate paleontology in the Neotropics: the Miocene fauna of La Venta, Colombia. Washington, DC: Smithsonian Institution Press.

Goin, F.J. 2003. Early marsupial radiations in South America. Pages 30–42 *in* M. Jones, C. Dickman, and M. Archer, eds. Predators with pouches: the biology of carnivorous marsupials. Collingwood, Australia: CSIRO Publishing.

Goin, F.J. 2006. A review of the Caroloameghiniidae, Paleogene South American "primate-like" marsupials (?Didelphimorphia, Peradectoidea). Paleontographica Abteilung A 278: 57–67, pl. 1.

Goin, F.J., and U.F.J. Pardiñas. 1996. Revisión de las especies del género *Hyperdidelphys* Ameghino, 1904 (Mammalia, Marsupialia, Didelphidae). Su significación filogenética, estratigráfica, y adaptativa en el Neogeno del Cono Sur suadmericano. Estudios Geológicos 52: 327–359.

Goin, F.J., A.M. Candela, M.A. Abello, and E.V. Oliveira. 2009a. Earliest South America paucituberculatans and their significance in the understanding of "pseudodiprotodont" marsupial radiations. Zoological Journal of the Linnean Society 155: 867–884.

Goin, F.J., N. Zimicz, M. de los Reyes, and L. Soibelzon. 2009b. A new large didelphid of the genus *Thylophorops* (Mammalia: Didelphimorphia: Didelphidae) from the late Tertiary of the Pampean Region, Argentina. Zootaxa 2005: 35–46.

Goin, F.J., M.A. Abello, and L. Chornogubsky. 2010. Middle Tertiary marsupials from central Patagonia (early Oligocene of Gran Barranca): understanding South America's *Grande Coupure*. Pages 69–105 *in* R.H. Madden, A.A. Carlini, M.G. Vucetich, and R.F. Kay, eds. The paleontology of Gran Barranca. Cambridge, UK: Cambridge University Press.

Goin, F.J., J.N. Gelfo, L. Chornogubsky, M.O. Woodburne, and T. Martin. 2012. Origins, radiations, and distribution of South American mammals. Pages 20–50 *in* B.D. Patterson and L.P. Costa, eds. Bones, clones, and biomes: the history and geography of Recent Neotropical mammals. Chicago: University of Chicago Press.

Goin, F.J., M.O. Woodburne, A.N. Zimicz, G.M. Martin, and L. Chornogubsky. 2016. A brief history of South American metatherians: evolutionary contexts and intercontinental dispersals. Dordrecht: Springer.

Gomes, A.C., G. Lessa, C. Cartelle, and L. Kerber. 2019. New fossil remains of Quaternary capybaras (Rodentia: Caviomorpha: Caviidae) from the intertropical region of Brazil: morphology and taxonomy. Journal of South American Earth Sciences 91: 36–46.

Gomes, D.C., R.P. da Cruz, J.J. Vicente, and R.M. Pinto. 2003. Nematode parasites of marsupials and small rodents from the Brazilian Atlantic Forest in the state of Rio de Janeiro, Brazil. Revista Brasileira de Zoologia 20: 699–707.

Gómez de Silva, H., M.P. Pérez-Villafaña, and J.A. Santos-Moreno. 1997. Diet of the spectacled owl (*Pulsatrix perspicillata*) during the rainy season in northern Oaxaca, Mexico. Journal of Raptor Research 31: 385–387.

Gómez-Martínez, M.J., A. Gutierrez, and F. DeClerck. 2008. Four-eyed opossum (*Philander opossum*) predation on a coral snake. Mammalia 72: 350–351.

González, E.M., and S. Claramunt. 2000. Behaviors of captive short-tailed opossums, *Monodelphis dimidiata* (Wagner, 1847) (Didelphimorphia, Didelphidae). Mammalia 64: 271–285.

Goodstadt, L., A. Heger, C. Webber, and C.P. Ponting. 2007. An analysis of the gene complement of a marsupial, *Monodelphis domestica*: evolution of lineage-specific genes and giant chromosomes. Genome Research 17: 969–981.

Goris, R.C. 2011. Infrared organs of snakes: an integral part of vision. Journal of Herpetology 45: 2–14.

Gould, S.J. 1980. Sticking up for marsupials. Pages 289–295 *in* S.J. Gould, ed. The panda's thumb, more reflections in natural history. New York: W.W. Norton.

Granatosky, M.C., P. Lemelin, S.G.B. Chester, J.D. Pampush, and D. Schmitt. 2014. Functional and evolutionary aspects of axial stability in euarchontans and other mammals. Journal of Morphology 275: 313–327.

Grand, T.I. 1983. Body weight: its relationship to tissue composition, segmental distribution of mass, and motor function III. The Didelphidae of French Guiana. Australian Journal of Zoology 31: 299–312.

Grant, R.A., S. Haidarliu, N.J. Kennerley, and T.J. Prescott. 2013. The evolution of active vibrissal sensing in mammals: evidence from vibrissal musculature and function in the marsupial opossum *Monodelphis domestica*. Journal of Experimental Biology 216: 3483–3494.

Granzinolli, M.A.M., and J.C. Motta, Jr. 2006. Small mammal selection by the white-tailed hawk in southeastern Brazil. Wilson Journal of Ornithology 118: 91–98.

Gray, J.B., and R.C. Anderson. 1982. Observations on *Turgida turgida* (Rudolphi, 1819) (Nematoda: Physalopteroidea) in the American opossum (*Didelphis virginiana*). Journal of Wildlife Diseases 18: 279–285.

Greene, H.W. 1988. Species richness in tropical predators. Pages 259–280 *in* F. Almeda and C.M. Pringle, eds. Diversity and conservation of tropical rainforests. San Francisco: California Academy of Science.

Greene, H.W., and F.M. Jaksic. 1983. Food-niche relationships among sympatric predators: effects of level of prey identification. Oikos 40: 151–154.

Gregory, T., F.C. Rueda, J. Deichmann, J. Kolowski, and A. Alonso. 2014. Arboreal camera trapping: taking a proven method to new heights. Methods in Ecology and Evolution 5: 443–451.

Gribel, R. 1988. Visits of *Caluromys lanatus* (Didelphidae) to flowers of *Pseudobombax tomentosum* (Bombacaceae): a probable case of pollination by marsupials in central Brazil. Biotropica 20: 344–347.

Gribel, R., P.E. Gibbs, and A.L. Queiróz. 1999. Flowering phenology and pollination biology of *Ceiba pentandra* (Bombacaceae) in central Amazonia. Journal of Tropical Ecology 15: 247–263.

Grubb, P.J. 1977. Control of forest growth and distribution on wet tropical mountains: with special reference to mineral nutrition. Annual Review of Ecology and Systematics 8: 83–107.

Grubb, P.J., J.R. Lloyd, T.D. Pennington, and T.C. Whitmore. 1963. A comparison of montane and lowland rain forest in Ecuador. I. The forest structure, physiognomy, and floristics. Journal of Ecology 51: 567–601.

Guariguata, M.R., and R. Ostertag. 2001. Neotropical secondary forest succession: changes in structural and functional characteristics. Forest Ecology and Management 148: 185–206.

Guglielmone, A.A., S. Nava, and M.M. Díaz. 2011. Relationships of South American marsupials (Didelphimorphia, Microbiotheria and Paucituberculata) and hard ticks (Acari: Ixodidae) with distribution of four species of *Ixodes*. Zootaxa 3086: 1–30.

Guillemin, M.-L., M. Atramentowicz, and D. Julien-Laferrière. 2001. The marsupial community. Pages 121–128 *in* F. Bongers, P. Charles-Dominique, P.-M. Forget, and M. Théry, eds. Nouragues: dynamics and plant-animal interactions in a Neotropical rainforest. Dordrecht: Kluwer.

Guillotin, M. 1982. Rythmes d'activité et régimes alimentaires de *Proechimys cuvieri* et d'*Oryzomys capito velutinus* (Rodentia) en forêt Guyanaise. Rev. Ecol. (Terre Vie) 36: 337–371.

Guillotin, M., G. Dubost, and D. Sabatier. 1994. Food choice and food competition among the three major primate species of French Guiana. Journal of Zoology London 233: 551–579.

Gutierrez, E.A., B.M. Pegoraro, B. Magalhães-Castro, and V.F. Pessoa. 2011. Behavioral evidence of dichromacy in a species of South American marsupial. Animal Behaviour 81: 1049–1054.

Gutiérrez, E.E., S.A. Jansa, and R.S. Voss. 2010. Molecular systematics of mouse opossums (Didelphidae: *Marmosa*): assessing species limits using mitochondrial DNA sequences, with comments on phylogenetic relationships and biogeography. American Museum Novitates 3692: 1–22.

Halenar, L.B., and A.L. Rosenberger. 2013. A closer look at the "*Protopithecus*" fossil assemblages: new genus and species from Bahia, Brazil. Journal of Human Evolution 65: 374–390.

Hall, E.R., and W.W. Dalquest. 1963. The mammals of Veracruz. University of Kansas Publications Museum of Natural History 14: 165–362.

Hallé, F., R.A.A. Oldeman, and P.B. Tomlinson. 1978. Tropical trees and forests: an architectural analysis. Berlin: Springer-Verlag.

Halliday, T.J.D., P. Upchurch, and A. Goswami. 2017. Resolving the relationships of Paleocene placental mammals. Biological Reviews 92: 521–550.

Hamilton, W.J. 1936. Seasonal food of skunks in New York. Journal of Mammalogy 17: 240–246.

Hamilton, W.J., Jr. 1958. Life history and economic relations of the opossum (*Didelphis marsupialis virginiana*) in New York State. Cornell University Agricultural Experimental Station Memoir 354: 1–48.

Hamrick, M.W. 2001. Morphological diversity in digital skin microstructure of didelphid marsupials. Journal of Anatomy 198: 683–688.

Handley, C.O., Jr. 1976. Mammals of the Smithsonian Venezuelan Project. Brigham Young University Science Bulletin (biol. ser.) 20(5): 1–89.

Hanson, A.M., M.B. Hall, L.M. Porter, and B. Lintzenich. 2006. Composition and nutritional characteristics of fungi consumed by *Callimico goeldii* in Pando, Bolivia. International Journal of Primatology 27: 323–346.

Harder, J.D., and D.W. Fleck. 1997. Reproductive ecology of New World marsupials. Pages 175–203 *in* N. Saunders and L. Hinds, eds. Marsupial biology: recent research, new perspectives. Sydney: UNSW Press.

Harder, J.D., and L.M. Jackson. 1999. Opossums. Encyclopedia of Reproduction 3: 515–523.

Harder, J.D., and L.M. Jackson. 2010. Chemical communication and reproduction in the gray short-tailed opossum (*Monodelphis domestica*). Vitamins and Hormones 83: 373–399.

Hart, B.L. 1992. Behavioral adaptations to parasites: an ethological approach. Journal of Parasitology 78: 256–265.

Hart, S.L., M.M. Spicer, T. Wrynn, T.L. Chapman, and K.L. Spivey. 2019. Palatability and predator avoidance behavior of salamanders in response to the Virginia opossum (*Didelphis virginiana*). American Midland Naturalist 181: 245–258.

Hartman, C.G. 1952. Possums. Austin, TX: University of Texas Press.

Hay, W.W., R.M. Deconto, C.N. Wold, K.M. Wilson, S. Voigt, M. Schulz, A.R. Wold, et al. 1999. Alternative global Cretaceous paleogeography. Geological Society of America Special Paper 332: 1–47.

Hayden, S., M. Bekaert, T.A. Crider, S. Mariani, W.J. Murphy, and E.C. Teeling. 2010. Ecological adaptation determines functional mammalian olfactory subgenomes. Genome Research 20: 1–9.

Hayes, R.A., M.R. Crossland, M. Hagman, R.J. Capon, and R. Shine. 2009. Ontogenetic variation in the chemical defenses of cane toads (*Bufo marinus*): toxin profiles and effects on predators. Journal of Chemical Ecology 35: 391–399.

Hayssen, V., and R.C. Lacey. 1985. Basal metabolic rates in mammals: taxonomic differences in the allometry of BMR and body mass. Comparative Biochemistry and Physiology 81A: 741–754.

Hayward, J.S., and P.A. Lisson. 1992. Evolution of brown fat: its absence in marsupials and monotremes. Canadian Journal of Zoology 70: 171–179.

Heath, T.A., D.J. Zwickl, J. Kim, and D.M. Hillis. 2008. Taxon sampling affects inferences of macroevolutionary processes from phylogenetic trees. Systematic Biology 57: 160–166.

Heffner, R.S. 1997. Comparative study of sound localization and its anatomical correlates in mammals. Acta Otolaryngologica (suppl.) 532: 46–53.

Heffner, R.S., and H.E. Heffner. 1992. Evolution of sound localization in mammals. Pages 691–715 *in* D.B. Webster, R.R. Fay, and A.N. Popper, eds. The evolutionary biology of hearing. New York: Springer-Verlag.

Heithaus, E.R., and T.H. Fleming. 1978. Foraging movements of a frugivorous bat, *Carollia perspicillata* (Phyllostomatidae). Ecological Monographs 48: 127–143.

Henderson, R.W., and M.J. Pauers. 2012. On the diets of Neotropical treeboas (Squamata: Boidae: *Corallus*). South American Journal of Herpetology 7: 172–180.

Henry, O., F. Feer, and D. Sabatier. 2000. Diet of the lowland tapir (*Tapirus terrestris* L.) in French Guiana. Biotropica 32: 364–368.

Henry, V.G. 1969. Predation on dummy nests of ground-nesting birds in the southern Appalachians. Journal of Wildlife Management 33: 169–172.

Hershkovitz, P. 1992. The South American gracile mouse opossums, genus *Gracilinanus* Gardner and Creighton, 1989 (Marmosidae, Marsupialia): a taxonomic review with notes on general morphology and relationships. Fieldiana Zoology (new series) 39: i–vi, 1–56.

Hershkovitz, P. 1999. *Dromiciops gliroides* Thomas, 1894, last of the Microbiotheria (Marsupialia), with a review of the family Microbiotheriidae. Fieldiana Zoology (New Series) 93: 1–60.

Hibert, F., D. Sabatier, J. Andrivot, C. Scotti-Saintagne, S. Gonzalez, M.-F. Prévost, P. Grenand, J. Chave, H. Caron, and C. Richard-Hansen. 2011. Botany, genetics, and ethnobotany: a crossed investigation on the elusive tapir's diet in French Guiana. PLoS ONE 6 (10): e25850.

Hice, C.L., and P.M. Velazco. 2012. The nonvolant mammals of the Reserva Nacional Allpahuayo-Mishana, Loreto, Peru. Special Publications of the Museum of Texas Tech University 60: [i–ii], 1–135.

Hickman, G.C. 1979. The mammalian tail: a review of functions. Mammal Review 9: 143–157.

Hiiemae, K., and A.W. Crompton. 1971. A cinefluorographic study of feeding in the American opossum, *Didelphis marsupialis*. Pages 299–334 *in* A.A. Dahlberg, ed. Dental morphology and evolution. Chicago: University of Chicago Press.

Hilário, R.R., and S.F. Ferrari. 2010. Feeding ecology of a group of buffy-headed marmosets (*Callithrix flaviceps*): fungi as a preferred food resource. American Journal of Primatology 72: 515–521.

Hildebrand, M. 1961. Body proportions of didelphid (and some other) marsupials, with emphasis on variability. American Journal of Anatomy 109: 239–249.

Hildebrand, M. 1980. The adaptive significance of tetrapod gait selection. American Zoologist 20: 255–267.

Hingst, E., P.S. D'Andrea, R. Santori, and R. Cerqueira. 1998. Breeding of *Philander frenata* (Didelphimorphia, Didelphidae) in captivity. Laboratory Animals 32: 434–438.

Hirsch, B.T. 2009. Seasonal variation in the diet of ring-tailed coatis (*Nasua nasua*) in Iguazu, Argentina. Journal of Mammalogy 90: 136–143.

Hoogstraal, H., and K.C. Kim. 1985. Tick and mammal coevolution, with emphasis on *Haemaphysalis*. Pages 505–568 *in* K.C. Kim, ed. Coevolution of parasitic arthropods and mammals. New York: John Wiley & Sons.

Hopkins, H.C. 1984. Floral biology and pollination ecology of the Neotropical species of *Parkia*. Journal of Ecology 72: 1–23.

Horovitz, I., S. Ladevèze, C. Argot, T.E. Macrini, T. Martin, J.J. Hooker, C. Kurz, C. de Muizon, and M. Sánchez-Villagra. 2008. The anatomy of *Herpetotherium* cf. *fugax* Cope, 1873, a metatherian from the Oligocene of North America. Palaeontographica (Abteilung A) 284: 109–141.

Horovitz, I., T. Martin, J. Bloch, S. Ladevèze, C. Kurz, and M.R. Sánchez-Villagra. 2009. Cranial anatomy of the earliest marsupials and the origin of opossums. PLoS ONE 4(12): 1–9.

Howe, H.F., and J. Smallwood. 1982. Ecology of seed dispersal. Annual Review of Ecology and Systematics 13: 201–228.

Hsu, M.J., D.W. Garton, and J.D. Harder. 1999. Energetics of offspring production: a comparison of a marsupial (*Monodelphis domestica*) and a eutherian (*Mesocricetus auratus*). Journal of Comparative Physiology B 169: 67–76.

Huck, M., C.P. Juarez, M.A. Rotundo, V.M. Dávalos, and E. Fernandez-Duque. 2017a. Mammals and their activity patterns in a forest area in the Humid Chaco, northern Argentina. Check List 13: 363–378.

Huck, M., C.P. Juárez, and E. Fernández-Duque. 2017b. Relationship between moonlight and nightly activity patterns of the ocelot (*Leopardus pardalis*) and some of its prey species in Formosa, northern Argentina. Mammalian Biology 82: 57–64.

Hudson, W.H. 1892. The naturalist in La Plata. London: Chapman and Hall.

Hume, I.D. 1999. Marsupial nutrition. Cambridge, UK: Cambridge University Press.

Humphreys, R.K., and G.D. Ruxton. 2018. A review of thanatosis (death feigning) as an anti-predator behaviour. Behavioral Ecology and Sociobiology 72(22): 1–16.

Hunsaker, D., III. 1977. The biology of marsupials. New York: Academic Press.

Hunt, D.M., J. Chan, L.S. Carvalho, J.N. Hokoc, M.C. Ferguson, C.A. Arrese, and L.D. Beazley. 2009. Cone visual pigments in two species of South American marsupials. Gene 433: 50–55.

Hunt, G., and G. Slater. 2016. Integrating paleontological and phylogenetic approaches to macroevolution. Annual Review of Ecology and Systematics 47: 189–213.

Ibarra-Cerdeña, C., V. Sánchez-Cordero, P. Ibarra-López, and L.I. Iñiguez-Dávalos. 2007. Noteworthy record of *Musonycteris harrisoni* and *Tlacuatzin canescens* pollinating a columnar cactus in west-central Mexico. International Journal of Zoological Research 3: 223–226.

Izor, R.J. 1985. Sloths and other mammalian prey of the harpy eagle. Pages 343–346 *in* G.G. Montgomery, ed. The evolution and ecology of armadillos, sloths, and vermilinguas. Washington, DC: Smithsonian Institution Press.

Izor, R.J., and R.H. Pine. 1987. Notes on the black-shouldered opossum, *Caluromysiops irrupta*. Fieldiana Zoology (New Series) 39: 117–124.

Jackson, J.B.C., and A. O'Dea. 2013. Timing of the oceanographic and biological isolation of the Caribbean Sea from the tropical eastern Pacific Ocean. Bulletin of Marine Science 89: 779–800.

Jacobs, G.H. 2009. Evolution of colour vision in mammals. Philosophical Transactions of the Royal Society B 364: 2957–2967.

Jacobs, G.H. 2010. Recent progress in understanding mammalian color vision. Ophthalmic and Physiological Optics 30: 422–434.

Jacobs, G.H., and M.P. Rowe. 2004. Evolution of vertebrate colour vision. Clinical and Experimental Optometry 87 (4–5): 206–216.

Jacobs, G.H., and G.A. Williams. 2010. Cone pigments in a North American marsupial, the opossum (*Didelphis virginiana*). Journal of Comparative Physiology A 196: 379–384.

Jaksic, F.M., H.W. Greene, and J.L. Yanez. 1981. The guild structure of a community of predatory vertebrates in central Chile. Oecologia 49: 21–28.

Jaksic, F.M., P.L. Meserve, J.R. Gutierrez, and E.L. Tabilo. 1993. The components of predation on small mammals in semiarid Chile: preliminary results. Revista Chilena de Historia Natural 66: 305–321.

Janos, D.P., C.T. Sahley, and L.H. Emmons. 1995. Rodent dispersal of vesicular-arbuscular mycorrhizal fungi in Amazonian Peru. Ecology 76: 1852–1858.

Jansa, S.A., and R.S. Voss. 2005. Phylogenetic relationships of the marsupial genus *Hyladelphys* based on nuclear gene sequences and morphology. Journal of Mammalogy 86: 853–865.

Jansa, S.A., and R.S. Voss. 2011. Adaptive evolution of the venom-targeted vWF protein in opossums that eat pitvipers. PLoS ONE 6(6): e20997.

Jansa, S.A., F.K. Barker, and R.S. Voss. 2014. The early diversification history of didelphid marsupials: a window into South America's "splendid isolation." Evolution 68: 684–695.

Jansen, A.M., S.C.C. Xavier, and A.L.R. Roque. 2015. The multiple and complex and changeable scenarios of the *Trypanosoma cruzi* transmission cycle in the sylvatic environment. Acta Tropica 151: 1–15.

Janson, C.H., and L.H. Emmons. 1990. Ecological structure of the nonflying mammal community at Cocha Cashu Biological Station, Manu National Park, Peru. Pages 314–338 *in* A.H. Gentry, ed. Four Neotropical rainforests. New Haven, CT: Yale University Press.

Janson, C.H., J. Terborgh, and L.H. Emmons. 1981. Non-flying mammals as pollinating agents in the Amazonian forest. Biotropica 13(2, suppl.): 1–6.

Janzen, D.H. 1973a. Sweep samples of tropical foliage insects: effects of seasons, vegetation types, elevation, time of day, and insularity. Ecology 54: 687–708.

Janzen, D.H. 1973b. Rate of regeneration after a tropical high elevation fire. Biotropica 5: 117–122.

Janzen, D.H., M. Ataroff, M. Fariñas, S. Reyes, N. Rincon, A. Soler, P. Soriano, and M. Vera. 1976. Changes in the arthropod community along an elevational transect in the Venezuelan Andes. Biotropica 8: 193–203.

Jenkins, F.A., Jr. 1971. Limb posture and locomotion in the Virginia opossum (*Didelphis marsupialis*) and in other non-cursorial mammals. Journal of Zoology London 165: 303–315.

Jenkins, F.A., Jr., and D. McClearn. 1984. Mechanisms of hind foot reversal in climbing mammals. Journal of Morphology 182: 197–219.

Jerison, H.J. 1973. Evolution of the brain and intelligence. New York: Academic Press.

Jerison, H.J. 1985. Animal intelligence as encephalization. Philosophical Transactions of the Royal Society of London B 308: 21–35.

Jiménez, F.A., J.K. Braun, M.L. Campbell, and S.L. Gardner. 2008. Endoparasites of fat-tailed mouse opossums (*Thylamys*: Didelphidae) from northwestern Argentina and southern Bolivia, with the description of a new species of tapeworm. Journal of Parasitology 94: 1098–1102.

Jiménez, F.A., F. Catzeflis, and S.L. Gardner. 2011. Structure of parasite component communities of didelphid marsupials: insights from a comparative study. Journal of Parasitology 97: 779–787.

Johansen, K. 1962. Buoyancy and insulation in the muskrat. Journal of Mammalogy 43: 64–68.

Johnson, J.I., Jr. 1977. Central nervous system of marsupials. Pages 157–278 *in* D. Hunsaker, III, ed. The biology of marsupials. New York: Academic Press.

Johnson, M.A., P.M. Saraiva, and D. Coelho. 1999. The role of gallery forests in the distribution of Cerrado mammals. Revista Brasileira de Biologia 59: 421–427.

Jones, E.K., and C.M. Clifford. 1972. The systematics of the subfamily Ornithodorinae (Acarina: Argasidae). V. A revised key to larval Argasidae of the western hemisphere and description of seven new species of *Ornithodoros*. Annals of the Entomological Society of America 65: 730–740.

Jones, E.K., C.M. Clifford, J.E. Keirans, and G.M. Kohls. 1972. The ticks of Venezuela (Acarina: Ixodoidea) with a key to the species of *Amblyomma* in the western hemisphere. Brigham Young University Science Bulletin (Biol. Ser.) 17(4): 1–40.

Jordano, P. 2000. Fruits and frugivory. Pages 125–164 *in* M. Fenner, ed. Seeds: the ecology of regeneration in plant communities (2nd ed.). Wallingford, UK: CABI Publishing.

Julien-Laferrière, D. 1991. Organisation du peuplement de marsupiaux en Guyane française. Revue d'Ecologie (La Terre et la Vie) 46: 125–144.

Julien-Laferrière, D. 1995. Use of space by the woolly opossum *Caluromys philander* (Marsupialia, Didelphidae) in French Guiana. Canadian Journal of Zoology 73: 1280–1289.

Julien-Laferrière, D. 1997. The influence of moonlight on activity of woolly opossums (*Caluromys philander*). Journal of Mammalogy 78: 251–255.

Julien-Laferrière, D., and M. Atramentowicz. 1990. Feeding and reproduction of three didelphid marsupials in two Neotropical forests (French Guiana). Biotropica 22: 404–415.

Julien-Laferrière, J. 1999. Foraging strategies and food partitioning in the Neotropical frugivorous mammals *Caluromys philander* and *Potos flavus*. Journal of Zoology London 247: 71–80.

Julien-Laferrière, J. 2001. Frugivory and seed dispersal by kinkajous. Pages 217–225 *in* F. Bongers, P. Charles-Dominique, P.-M. Forget, and M. Théry, eds. Nouragues, dynamics and plant-animal interactions in a Neotropical rainforest. Dordrecht: Kluwer Academic.

Jurgelski, W., Jr., and M.E. Porter. 1974. The opossum (*Didelphis virginiana* Kerr) as a biomedical model. III. Breeding the opossum in captivity: methods. Laboratory Animal Science 24: 412–425.

Jurgilas, P.B., A.G.C. Neves-Ferreira, G.B. Domont, and J. Perales. 2003. PO41, a snake venom metalloproteinase inhibitor isolated from *Philander opossum* serum. Toxicon 42: 621–628.

Kajin, M., R. Cerqueira, M.V. Vieira, and R. Gentile. 2008. Nine-year demography of the black-eared opossum *Didelphis aurita* (Didelphimorphia: Didelphidae) using life tables. Revista Brasileira de Zoologia 25: 206–213.

Kalka, M., and E.K.V. Kalko. 2006. Gleaning bats as underestimated predators of herbivorous insects: diet of *Micronycteris microtis* (Phyllostomidae) in Panama. Journal of Tropical Ecology 22: 1–10.

Kamler, J.E., and P.S. Gipson. 2004. Survival and cause-specific mortality among furbearers in a protected area. American Midland Naturalist 151: 27–34.

Kanda, L.L., T.K. Fuller, P.R. Sievert, and R.L. Kellogg. 2009. Seasonal source-sink dynamics at the edge of a species' range. Ecology 90: 1574–1585.

Kasparian, M.A., E.C. Hellgren, and S.M. Ginger. 2002. Food habits of the Virginia opossum during raccoon removal in the Cross Timbers Ecoregion, Oklahoma. Proceedings of the Oklahoma Academy of Sciences 82: 73–78.

Kasparian, M.A., E.C. Hellgren, S.M. Ginger, L.P. Levesque, J.E. Clark, and D.L. Winkelman. 2004. Population characteristics of Virginia opossum in the Cross Timbers during raccoon reduction. American Midland Naturalist 151: 154–163.

Kaufman, D.W. 1974. Adaptive coloration in *Peromyscus polionotus*: experimental selection by owls. Journal of Mammalogy 55: 271–283.

Kaufmann, J.H. 1962. Ecology and social behavior of the coati, *Nasua narica*, on Barro Colorado Island, Panama. University of California Publications in Zoology 60: 95–222.

Kavaliers, M., and D.D. Colwell. 1995. Discrimination by female mice between the odours of parasitized and non-parasitized males. Proceedings of the Royal Society of London B 261: 31–35.

Kay, R.F., and E.C. Kirk. 2000. Osteological evidence for the evolution of activity pattern and visual acuity in primates. American Journal of Physical Anthropology 113: 235–262.

Kay, R.F., and R.H. Madden. 1997. Paleogeography and paleoecology. Pages 520–550 *in* R.F. Kay, R.H. Madden, R.L. Cifelli, and J.J. Flynn, eds. Vertebrate paleontology in the Neotropics: the Miocene fauna of La Venta, Colombia. Washington, DC: Smithsonian Institution Press.

Kay, R.F., R.H. Madden, C. van Shaik, and D. Higdon. 1997a. Primate species richness is determined by plant productivity: implications for conservation. Proceedings of the National Academy of Sciences USA 94: 13023–13027.

Kay, R.F., R.H. Madden, R.L. Cifelli, and J.J. Flynn (eds.). 1997b. Vertebrate paleontology in the Neotropics: the Miocene fauna of La Venta, Colombia. Washington, DC: Smithsonian Institution Press.

Kay, R.F., S.F. Vizcaíno, and M.S. Bargo. 2012. A review of the paleoenvironment and paleoecology of the Miocene Santa Cruz Formation. Pages 331–365 *in* S.F. Vizcaíno, R.F. Kay, and M.S. Bargo, eds. Early Miocene paleobiology in Patagonia: high-latitude paleocommunities of the Santa Cruz Formation. Cambridge, UK: Cambridge University Press.

Kays, R.W. 1999. Food preferences of kinkajous (*Potos flavus*): a frugivorous carnivore. Journal of Mammalogy 80: 589–599.

Kays, R.W. 2000. The behavior and ecology of olingos (*Bassaricyon gabbii*) and their competition with kinkajous (*Potos flavus*). Mammalia 64: 1–10.

Kays, R.W., M.E. Rodríguez, L.M. Valencia, R. Horan, A.R. Smith, and C. Ziegler. 2012. Animal visitation and pollination of flowering balsa trees (*Ochroma pyramidale*) in Panama. Mesoamericana 16: 56–70.

Kearn, G.C. 1998. Parasitism and the platyhelminths. London: Chapman & Hall.

Keesing, F., J. Brunner, S. Duerr, M. Killilea, K. LoGiudice, K. Schmidt, H. Vuong, and R.S. Ostfeld. 2009. Hosts as ecological traps for the vector of Lyme disease. Proceedings of the Royal Society B 276: 3911–3919.

Kennedy, M.L., G.D. Schnell, M.L. Romero-Almaraz, B.S. Malakouti, C. Sánchez-Hernández, T.L. Best, and M.C. Wooten. 2013. Demographic features, distribution, and habitat selection of the gray mouse opossum (*Tlacuatzin canescens*) in Colima, Mexico. Acta Theriologica 58: 285–298.

Kerber, L., E.L. Mayer, A.M. Ribeiro, and M.G. Vucetich. 2016. Late Quaternary caviomorph rodents (Rodentia: Hystricognathi) from the Serra da Capivara, northeastern Brazil, with descriptions of a new taxon. Historical Biology 28: 439–458.

Kielan-Jaworowska, Z., R.L. Cifelli, and Z.-X. Luo. 2004. Mammals from the age of dinosaurs. New York: Columbia Univ. Press.

Killeen, T.J., A. Jardim, F. Mamani, and N. Rojas. 1998. Diversity, composition, and structure of a tropical semideciduous forest in the Chiquitanía region of Santa Cruz, Bolivia. Journal of Tropical Ecology 14: 803–827.

Kiltie, R.A. 2000. Scaling of visual acuity with body size in mammals and birds. Functional Ecology 14: 226–234.

Kimble, D.P. 1997. Didelphid behavior. Neuroscience and Biobehavioral Reviews 21: 361–369.

Kimmel, T.M., L.M. do Nascimento, D. Piechowski, E.V.S.B. Sampaio, M.J.N. Rodal, and G. Gottsberger. 2010. Pollination and seed dispersal modes of woody species of 12-year-old secondary forest in the Atlantic Forest region of Pernambuco, NE Brazil. Flora 205: 540–547.

Kirkendall, L.R., and N.C. Stenseth. 1985. On defining "breeding once." American Naturalist 125: 189–204.

Kissell, R.E., and M.L. Kennedy. 1992. Ecologic relationships of co-occurring populations of opossums (*Didelphis virginiana*) and raccoons (*Procyon lotor*) in Tennessee. Journal of Mammalogy 73: 808–813.

Kitchings, J.T., and B.T. Walton. 1991. Fauna of the North American temperate deciduous forest. Pages 345–375 in E. Röhrig and B. Ulrich, eds. Temperate deciduous forests (Ecosystems of the world, vol. 7). Amsterdam: Elsevier.

Klein, B.C., L.H. Harper, R.O. Bierregaard, and G.V.N. Powell. 1988. The nesting and feeding behavior of the ornate hawk-eagle near Manaus, Brazil. Condor 90: 239–241.

Kluge, A.G. 1981. The life history, social organization, and parental behavior of *Hyla rosenbergi* Boulenger, a nest-building gladiator frog. Miscellaneous Publications Museum of Zoology University of Michigan 100: 1–170.

Kolb, H., and H.H. Wang. 1985. The distribution of photoreceptors, dopaminergic amacrine cells, and ganglion cells in the retina of the North American opossum (*Didelphis virginiana*). Vision Research 25: 1207–1209.

Konecny, M.J. 1989. Movement patterns and food habits of four sympatric carnivore species in Belize, Central America. Pages 243–264 in K.H. Redford and J.F. Eisenberg, eds. Advances in Neotropical mammalogy. Gainesville, FL: Sandhill Crane Press.

Kraaijeveld, K., F.J.L. Kraaijeveld-Smit, and G.J. Adcock. 2003. Does female mortality drive male semelparity in dasyurid marsupials? Proceedings of the Royal Society of London B (suppl.) 270: S251–S253.

Kress, W.J., and D.E. Stone. 1993. Morphology and floral biology of *Phenakospermum* (Strelitziaceae), an arborescent herb of the Neotropics. Biotropica 25: 290–300.

Kurz, C. 2007. The opossum-like marsupials (Didelphimorphia and Peradectia, Marsupialia, Mammalia) from the Eocene of Messel and Geiseltal—ecomorphology, diversity, and palaeogeography. Kaupia 15: 3–65.

Labruna, M.B., R.R. Cabrera, and A. Pinter. 2009. Life cycle of *Ixodes luciae* (Acari: Ixodidae) in the laboratory. Parasitology Research 105: 1749–1753.

Lainson, R. 2010. The Neotropical *Leishmania* species: a brief historical review of their discovery, ecology and taxonomy. Revista Pan-Amazônica de Saúde 1(2): 13–32.

Lambert, T.D., J.R. Malcolm, and B.L. Zimmerman. 2005. Variation in small mammal species richness by trap height and trap type in southeastern Amazonia. Journal of Mammalogy 86: 982–990.

Lammers, A.R., and A.R. Biknevicius. 2004. The biodynamics of arboreal locomotion: the effects of substrate diameter on locomotor kinetics in the gray short-tailed opossum (*Monodelphis domestica*). Journal of Experimental Biology 207: 4325–4336.

Lareschi, M., J.P. Sanchez, M. C. Ezquiaga, A.G. Autino, M.M. Díaz, and R.M. Barquez. 2010. Fleas associated with mammals from northwestern Argentina, with new distributional reports. Comparative Parasitology 77: 207–213.

Laurance, S.G., and W.E. Laurance. 2007. *Chaunus marinus* (cane toad). Predation. Herpetological Review 38: 320–321.

Laver, P.N., and M.J. Kelly. 2008. A critical review of home range studies. Journal of Wildlife Management 72: 290–298.

Lawson Handley, L.J., and N. Perrin. 2007. Advances in our understanding of mammalian sex-biased dispersal. Molecular Ecology 16: 1559–1578.

Lee, C.H., D.K. Jones, C. Ahern, M.F. Sarhan, and P.C. Ruben. 2011. Biophysical costs associated with tetrodotoxin resistance in the sodium channel pore of the garter snake, *Thamnophis sirtalis*. Journal of Comparative Physiology A 197: 33–54.

Leigh, E.G., A. O'Dea, and G.J. Vermeij. 2014. Historical biogeography of the Isthmus of Panama. Biological Reviews 89: 148–172.

Leiner, N.O., and W.R. Silva. 2007a. Effects of resource availability on the use of space by the mouse opossum *Marmosops paulensis* (Didelphidae) in a montane Atlantic Forest area in southeastern Brazil. Acta Theriologica 52: 197–204.

Leiner, N.O., and W.R. Silva. 2007b. Seasonal variation in the diet of the Brazilian slender opossum (*Marmosops paulensis*) in a montane Atlantic Forest area, southeastern Brazil. Journal of Mammalogy 88: 158–164.

Leiner, N.O., and W.R. Silva. 2009. Territoriality in females of the slender opossum (*Marmosops paulensis*) in the Atlantic Forest of Brazil. Journal of Tropical Ecology 25: 671–675.

Leiner, N.O., E.Z.F. Setz, and W.R. Silva. 2008. Semelparity and factors affecting the reproductive activity of the Brazilian slender opossum (*Marmosops paulensis*) in southeastern Brazil. Journal of Mammalogy 89: 153–158.

Leiner, N.O., C.R. Dickman, and W.R. Silva. 2010. Multiscale habitat selection by slender opossums (*Marmosops* spp.) in the Atlantic Forest of Brazil. Journal of Mammalogy 91: 561–565.

Leite, M.S., T.L. Queiroz, M. Galliez, P.P. de Mendonça, and F.A.S. Fernandez. 2013. Activity patterns of the water opossum *Chironectes minimus* in Atlantic Forest rivers of south-eastern Brazil. Journal of Tropical Ecology 29: 261–264.

Leite, M.S., M. Galliez, T.L. Queiroz, and F.A.S. Fernandez. 2016. Spatial ecology of the water opossum *Chironectes minimus* in Atlantic Forest streams. Mammalian Biology 81: 480–487.

Leite, Y.L.R., L.P. Costa, and J.R. Stallings. 1996. Diet and vertical space use of three sympatric opossums in a Brazilian Atlantic Forest reserve. Journal of Tropical Ecology 12: 435–440.

Lemelin, P. 1999. Morphological correlates of substrate use in didelphid marsupials: implications for primate origins. Journal of Zoology London 247: 165–175.

Lemelin, P., and D. Schmitt. 2007. Origins of grasping and locomotor adaptations in primates: comparative and experimental approaches using an opossum model. Pages 329–380 *in* M.J. Ravosa and M. Dagosto, eds. Primate origins: adaptations and evolution. New York: Springer.

Lemelin, P., D. Schmitt, and M. Cartmill. 2003. Footfall patterns and interlimb co-ordination in opossums (family Didelphidae): evidence for the evolution of diagonal-sequence walking gates in primates. Journal of Zoology London 260: 423–429.

Lenzi, H.L., A.M. Jansen, and M.P. Deane. 1984. The recent discovery of what might be a primordial escape mechanism for *Trypanosoma cruzi*. Memórias do Instituto Oswaldo Cruz (supl.) 79: 13–18.

Leone, M.F., A.C. Loss, R.G. Rocha, R.D. Paes, and L.P. Costa. 2019. To stripe or not to stripe? Natural selection and disruptive coloration in two sympatric species of Neotropical marsupials from the genus *Monodelphis* (Mammalia, Didelphidae). Boletim da Sociedade Brasileira de Mastozoologia 85: 86–94.

Lessa, E.P., and R.A. Fariña. 1996. Reassessment of extinction patterns among the late Pleistocene mammals of South America. Palaeontology 39: 651–662.

Lessa, E.P., B. van Valkenburgh, and R.A. Fariña. 1997. Testing hypotheses of differential mammalian extinctions subsequent to the Great American Biotic Interchange. Palaeogeography, Palaeoclimatology, Palaeoecology 135: 157–162.

Lessa, L.G., and F.N. da Costa. 2010. Diet and seed dispersal by five marsupials (Didelphimorphia: Didelphidae) in a Brazilian cerrado reserve. Mammalian Biology 75: 10–16.

Lessa, L.G., and L. Geise. 2014. Food habits and carnivory by a small size opossum, *Gracilinanus agilis* (Didelphimorphia: Didelphidae). Mastozoología Neotropical 21: 139–143.

Lessa, L.G., L. Geise, and F.N. da Costa. 2013. Effects of gut passage on the germination of seeds ingested by didelphid marsupials in a Neotropical savanna. Acta Botanica Brasilica 27: 519–525.

Levey, D.J. 1990. Habitat-dependent fruiting behaviour of an understorey tree, *Miconia centrodesma*, and tropical treefall gaps as keystone habitats for frugivores in Costa Rica. Journal of Tropical Ecology 6: 409–420.

Levey, D.J., T.C. Moermond, and J.S. Denslow. 1994. Frugivory: an overview. Pages 282–294 *in* L.A. McDade, K.S. Bawa, H.A. Hespenheide, and G.S. Hartshorn, eds. La Selva, ecology and natural history of a Neotropical rainforest. Chicago: Chicago University Press.

Levings, S.C., and D.M. Windsor. 1996. Seasonal and annual variation in litter arthropod populations. Pages 355–387 *in* E.G. Leigh, Jr., A.S. Rand, and D.M. Windsor, eds. The ecology of a tropical forest (2nd ed.). Washington, DC: Smithsonian Institution.

Lieberman, D., M. Lieberman, R. Peralta, and G.S. Hartshorn. 1996. Tropical forest structure and composition on a large-scale altitudinal gradient in Costa Rica. Journal of Ecology 84: 137–152.

Liem, K.F. 1980. Adaptive significance of intra- and interspecific differences in the feeding repertoires of cichlid fishes. American Zoologist 20: 295–314.

Lima, B.S., F. Dantas-Torres, M.R. de Carvalho, J.F. Marinho-Junior, E.L. de Almeida, M.E.F. Brito, F. Gomes, and S.P. Brandão-Filho. 2013. Small mammals as hosts of *Leishmania* spp. in a highly endemic area for zoonotic leishmaniasis in north-eastern Brazil. Transactions of the Royal Society of Tropical Medicine and Hygiene 107: 592–597.

Lima, M., N.C. Stenseth, N.G. Yoccoz, and F.M. Jaksic. 2001. Demography and population dynamics of the mouse opossum (*Thylamys elegans*) in semi-arid Chile: seasonality, feedback structure, and climate. Proceedings of the Royal Society B 268: 2053–2064.

Lima, R.B.S., P.A. Oliveira, and A.G. Chiarello. 2010. Diet of the thin-spined porcupine (*Chaetomys subspinosus*), an Atlantic Forest endemic threatened with extinction in southeastern Brazil. Mammalian Biology 75: 538–546.

Linardi, P.M. 2006. Os ectoparasitos de marsupiais brasileiros. Pages 35–52 *in* N.C. Cáceres and E.L.A. Monteiro-Filho, eds. Os marsupiais do Brasil. Campo Grande: Editora UFMS.

Lindenfors, P., J.L. Gittleman, and K.E. Jones. 2007. Sexual size dimorphism in mammals. Pages 16–26 *in* D.J. Fairbairn, W.U. Blanckenhorn, and T. Székely, eds. Sex, size, and gender roles: evolutionary studies of sexual size dimorphism. New York: Oxford University Press.

Lira, P.K., and F.A.S. Fernandez. 2009. A comparison of trapping- and telemetry-based estimates of home range of the Neotropical opossum *Philander frenatus*. Mammalian Biology 74: 1–8.

Lira, P.K., A.S. Pires, H.S.A. Carlos, P.L. Curzio, and F.A.S. Fernandez. 2018. Resting sites of opossums (Didelphimorphia, Didelphidae) in Atlantic Forest fragments. Mammalia 82: 62–64.

Lobova, T.A., C.K. Geiselman, and S.A. Mori. 2009. Seed dispersal by bats in the Neotropics. Memoirs of the New York Botanical Garden 101: i–xii, 1–471.

Loo, S.-K., and Z. Halata. 1991. Innervation of hairs in the facial skin of marsupial mammals. Journal of Anatomy 174: 207–219.

Lopes, G.P., and N.O. Leiner. 2015. Semelparity in a population of *Gracilinanus agilis* (Didelphimorphia: Didelphidae) inhabiting the Brazilian Cerrado. Mammalian Biology 80: 1–6.

Lopes-Torres, E.J., A. Maldonado, Jr., and R. Marisa-Lanfredi. 2009. Spirurids from *Gracilinanus agilis* (Marsupialia: Didelphidae) in Brazilian Pantanal wetlands, with a new species of *Physaloptera* (Nematoda: Spirurida). Veterinary Parasitology 163: 87–92.

López-Aguirre, C.L., S.J. Hand, S.W. Laffan, and M. Archer. 2018. Phylogenetic diversity, types of endemism, and the evolutionary history of New World bats. Ecography 41: 1955–1966.

López-Arévalo, H., O. Montenegro-Díaz, and A. Cadena. 1993. Ecología de los pequeños mamíferos de la Reserva Biológica Carpanta, en la Cordillera Oriental colombiana. Studies on Neotropical Fauna and Environment 28: 193–210.

Loretto, D., and M.V. Vieira. 2008. Use of space by the marsupial *Marmosops incanus* (Didelphimorphia, Didelphidae) in the Atlantic Forest, Brazil. Mammalian Biology 73: 255–261.

Loretto, D., and M.V. Vieira. 2011. Artificial nests as an alternative to studies of arboreal small mammal populations: a five-year study in the Atlantic Forest, Brazil. Zoologia 28: 388–394.

Loretto, D., E. Ramalho, and M.V. Vieira. 2005. Defense behavior and nest architecture of *Metachirus nudicaudatus* Desmarest, 1817 (Marsupialia, Didelphidae). Mammalia 69: 417–419.

Lorini, M.L., J.A. de Oliveira, and V.G. Persson. 1994. Annual age structure and reproductive patterns in *Marmosa incana* (Lund, 1841) (Didelphidae, Marsupialia). Zeitshrift für Säugetierkunde 59: 65–73.

Losos, J.B., R.E. Glor, J.J. Kolbe, and K. Nicholson. 2006. Adaptation, speciation, and convergence: a hierarchical analysis of adaptive radiation in Caribbean *Anolis* lizards. Annals of the Missouri Botanical Gardens 93: 24–33.

Loughry, W.J., and C.M. McDonough. 2013. The nine-banded armadillo: a natural history. Norman, OK: University of Oklahoma Press.

Lukeš, J., T. Skalický, J. Týč, J. Votýpka, and V. Yurchenko. 2014. Evolution of parasitism in kinetoplastid flagellates. Molecular & Biochemical Parasitology 195: 115–122.

Lunaschi, L.I., and F.B. Drago. 2007. Checklist of digenean parasites of wild mammals from Argentina. Zootaxa 1580: 35–50.

Lunde, D.P., and W.A. Schutt. 1999. The peculiar carpal tubercles of male *Marmosops parvidens* and *Marmosa robinsoni* (Didelphidae: Didelphinae). Mammalia 63: 495–504.

Luo, Z.-X., Q. Ji, J.R. Wible, and C.-X. Yuan. 2003. An Early Cretaceous tribosphenic mammal and metatherian evolution. Science 302: 1934–1940.

Lutz, N.D., E. Lemes, L. Krubitzer, S.P. Collin, S. Haverkamp, and L. Peichl. 2018. The rod signaling pathway in marsupial retinae. PLoS ONE 13(8): e0202089.

Lydekker, R. 1894. A hand-book to the Marsupialia and Monotremata. London: W.H. Allen & Co.

Lyne, A.G. 1959. The systematic and adaptive significance of the vibrissae in the Marsupialia. Proceedings of the Zoological Society of London 133: 79–132.

MacArthur, R.H. 1972. Geographical ecology: patterns in the distribution of species. New York: Harper and Row.

Macedo, L., F.A.S. Fernandez, J.L. Nessimian. 2010. Feeding ecology of the marsupial *Philander frenatus* in a fragmented landscape in southeastern Brazil. Mammalian Biology 75: 363–369.

MacFadden, B.J. 2006. Extinct mammalian biodiversity of the ancient New World. Trends in Ecology & Evolution 21: 157–165.

Mackessy, S.P. 2010. The field of reptile toxinology: snakes, lizards, and their venoms. Pages 3–23 *in* S.P. Mackessy, ed. Handbook of venoms and toxins of reptiles. Boca Raton: CRC Press.

MacMillen, R.E., and J.E. Nelson. 1969. Bioenergetics and body size in dasyurid marsupials. American Journal of Physiology 217: 1246–1251.

Macrini, T.E. 2012. Comparative morphology of the internal nasal skeleton of adult marsupials based on X-ray computed tomography. Bulletin of the American Museum of Natural History 365: 1–91.

Maga, A.M., and R.M.D. Beck. 2017. Skeleton of an unusual, cat-sized marsupial relative (Metatheria: Marsupialiformes) from the middle Eocene (Lutetian: 44–43 million years ago) of Turkey. PLoS ONE 12(8): e0181712.

Magnusson, W.E., E.V. da Silva, and A.P. Lima. 1987. Diets of Amazonian crocodilians. Journal of Herpetology 21: 85–95.

Malcolm, J.R. 1990. Estimation of mammalian densities in continuous forest north of Manaus. Pages 339–357 *in* A.H. Gentry, ed. Four Neotropical rainforests. New Haven, CT: Yale University Press.

Malcolm, J.R. 1991. Comparative abundances of Neotropical small mammals by trap height. Journal of Mammalogy 72: 188–192.

Mangan, S.A., and G.H. Adler. 2000. Consumption of arbuscular mycorrhizal fungi by terrestrial and arboreal small mammals in a Panamanian cloud forest. Journal of Mammalogy 81: 563–570.

Mangan, S.A., and G.H. Adler. 2002. Seasonal dispersal of arbuscular mycorrhizal fungi by spiny rats in a Neotropical forest. Oecologia 131: 587–597.

Mani, M.S. 1968. Ecology and biogeography of high altitude insects (Series Entomologica 4). The Hague: Dr. W. Junk.

Mann, G. 1978. Los pequeños mamíferos de Chile. Gayana (Zoología) 40: 1–342.

Mares, M.A., J.K. Braun, and D. Gettinger. 1989. Observations on the distribution and ecology of the mammals of the Cerrado grasslands of central Brazil. Annals of Carnegie Museum 58: 1–60.

Marshall, A.J., and R.W. Wrangham. 2007. Evolutionary consequences of fallback foods. International Journal of Primatology 28: 1219–1235.

Marshall, L.G. 1978a. *Lutreolina crassicaudata*. Mammalian Species 91: 1–4.

Marshall, L.G. 1978b. *Chironectes minimus*. Mammalian Species 109: 1–6.

Marshall, L.G., and C. de Muizon. 1988. The dawn of the age of mammals in South America. National Geographic Research 4: 23–55.

Marshall, L.G., S.D. Webb, J.J. Seposki, Jr., and D.M. Raup. 1982. Mammalian evolution and the Great American Interchange. Science 215: 1351–1357.

Marshall, L.G., T. Sempere, and R.F. Butler. 1997. Chronostratigraphy of the mammal-bearing Paleocene of South America. Journal of South American Earth Sciences 10: 49–70.

Martin, G.M., and D.E. Udrizar Sauthier. 2011. Observations on the captive behavior of the rare Patagonian opossum *Lestodelphys halli* (Thomas, 1921) (Marsupialia, Didelphimorphia, Didelphidae). Mammalia 75: 281–286.

Martínez, J.I.Z., A. Travaini, S.Z.D. Procopio, and M.A. Santillán. 2012. The ecological role of native and introduced species in the diet of the puma *Puma concolor* in southern Patagonia. Oryx 46: 106–111.

Martínez-Lanfranco, J.A., D. Flores, J.P. Jayat, and G. D'Elía. 2014. A new species of lutrine opossum, genus *Lutreolina* Thomas (Didelphidae), from the South American Yungas. Journal of Mammalogy 95: 225–240.

Martins, E.G., and V. Bonato. 2004. On the diet of *Gracilinanus microtarsus* (Marsupialia, Didelphidae) in an Atlantic rainforest fragment in southeastern Brazil. Mammalian Biology 69: 58–60.

Martins, E.G., V. Bonato, H.P. Pinheiro, and S.F. dos Reis. 2006a. Diet of the gracile mouse opossum (*Gracilinanus microtarsus*) (Didelphimorphia: Didelphidae) in a Brazilian cerrado: patterns of food consumption and intrapopulation variation. Journal of Zoology 269: 21–28.

Martins, E.G., V. Bonato, C.Q. da Silva, and S.F. dos Reis. 2006b. Seasonality in reproduction, age structure, and density of the gracile mouse opossum *Gracilinanus microtarsus* (Marsupialia: Didelphidae) in a Brazilian Cerrado. Journal of Tropical Ecology 22: 461–468.

Martins, E.G., V. Bonato, C.Q. da Silva, and S.F. dos Reis. 2006c. Partial semelparity in the Neotropical didelphid marsupial *Gracilinanus microtarsus*. Journal of Mammalogy 87: 915–920.

Martins, R.L., and R. Gribel. 2007. Polinização de *Caryocar villosum* (Aubl.) Pers. (Caryocaraceae) uma árvore emergente da Amazônia central. Revista Brasiliera de Botânica 30: 37–45.

Martins-Hatano, F., D. Gettinger, and H.G. Bergallo. 2002. Ecology and host specificity of laelapine mites (Acari: Laelapidae) of small mammals in an Atlantic Forest area of Brazil. Journal of Parasitology 88: 36–40.

Maslin, M., Y. Malhi, O. Phillips, and S. Cowling. 2005. New views on an old forest: assessing the longevity, resilience, and future of the Amazon rainforest. Transactions of the Institute of British Geographers 30: 477–499.

Massoia, E., and A. Fornes. 1972. Presencia y rasgos ecológicos de *Marmosa agilis chacoensis* Tate en las provincias de Buenos Aires, Entre Ríos y Misiones (Mammalia-Marsupialia-Didelphidae). Revista de Investigaciones Agropecuarias (ser. 1) 9(2): 71–82.

Mathews, A.G.A. 1977. Studies on termites from the Mato Grosso state, Brazil. Rio de Janeiro: Academia Brasileira de Ciências.

May, M.R., S. Höhne, and B.R. Moore. 2016. A Bayesian approach for detecting the impact of mass-extinction events on molecular phylogenies when rates of lineage diversification may vary. Methods in Ecology and Evolution 7: 947–959.

Mayer, E.L., A. Hubbe, L. Kerber, P.M. Haddad-Martim, and W. Neves. 2016. Taxonomic, biogeographic, and taphonomic reassessment of a large extinct species of paca from the Quaternary of Brazil. Acta Palaeontologica Polonica 61: 743–758.

McKenna, M.C. 1980. Early history and biogeography of South America's extinct land mammals. Pages 43–77 in R.L. Ciochon and A.B. Chiarelli, eds. Evolutionary biology of the New World monkeys and continental drift. New York: Plenum Press.

McKenna, M.C., and S.K. Bell. 1997. Classification of mammals above the species level. New York: Colombia University Press.

McManus, J.J. 1970. Behavior of captive opossums, *Didelphis marsupialis virginiana*. American Midland Naturalist 84: 144–169.

McManus, J.J. 1971. Activity of captive *Didelphis marsupialis*. Journal of Mammalogy 52: 846–848.

McNab, B.K. 1978. The comparative energetics of Neotropical marsupials. Journal of Comparative Physiology 125: 115–128.

McNab, B.K. 1986. Food habits, energetics, and the reproduction of marsupials. Journal of Zoology 2008: 595–614.

Mebs, D., O. Arakawa, and M. Yotsu-Yamashita. 2010. Tissue distribution of tetrodotoxin in the red-spotted newt *Notophthalmus viridescens*. Toxicon 55: 1353–1357.

Medellín, R.A. 1994. Seed dispersal of *Cecropia obtusifolia* by two species of opossums in the Selva Lacandona, Chiapas, Mexico. Biotropica 26: 400–407.

Menezes, J.C.T., and M.A. Marini. 2017. Predators of bird nests in the Neotropics: a review. Journal of Field Ornithology 88: 99–114.

Meredith, R.W., M. Westerman, J.A. Case, and M.S. Springer. 2008. A phylogeny and timescale for marsupial evolution based on sequences for five nuclear genes. Journal of Mammalian Evolution 15: 1–36.

Meredith, R.W., C. Crajewski, M. Westerman, and M.S. Springer. 2009. Relationships and divergence times among the orders and families of Marsupialia. Museum of Northern Arizona Bulletin 65: 383–406.

Meredith, R.W., J.E. Janecka, J. Gatesy, O.A. Ryder, C.A. Fisher, E.C. Teeling, A. Goodbla, et al. 2011. Impacts of the Cretaceous terrestrial revolution and KPg extinction on mammalian diversification. Science 334: 521–524.

Merilaita, S., and M. Stevens. 2011. Crypsis through background-matching. Pages 17–33 *in* M. Stevens and S. Merilaita, eds. Animal camouflage, mechanisms and function. Cambridge, UK: Cambridge University Press.

Meserve, P.L. 1981. Trophic relationships among small mammals in a Chilean semiarid thorn scrub community. Journal of Mammalogy 62: 304–314.

Mesquita, P.C.M.D. de, D.M. Borges-Nojosa, D.C. Passos, and C.H. Bezerra. 2011. Ecology of *Philodryas nattereri* in the Brazilian semi-arid region. Herpetological Journal 21: 193–198.

Meyer-Lucht, Y., C. Otten, T. Püttker, R. Pardini, J.P. Metzger, and S. Sommer. 2010. Variety matters: adaptive genetic diversity and parasite load in two mouse opossums from the Brazilian Atlantic Forest. Conservation Genetics 11: 2001–2013.

Meza, A.V., E.M. Meyer, and C.A. L. González. 2002. Ocelot (*Leopardus pardalis*) food habits in a tropical deciduous forest of Jalisco, Mexico. American Midland Naturalist 148: 146–154.

Mikkelson, T.S., M.J. Wakefield, B. Aken, C.T. Amemiya, J.L. Chang, S. Duke, M. Garber, et al. 2007. Genome of the marsupial *Monodelphis domestica* reveals innovation in non-coding sequences. Nature 447: 167–177.

Milano, M.Z., and E.L.A. Monteiro-Filho. 2009. Predation on small mammals by capuchin monkeys, *Cebus cay.* Neotropical Primates 16: 78–80.

Miles, M.A., A.A. de Souza, and M.M. Póvoa. 1981. Mammal tracking and nest location in Brazilian forest with an improved spool-and-line device. Journal of Zoology London 195: 331–347.

Mills, J.R.E. 1967. A comparison of lateral jaw movements in some mammals from wear facets on the teeth. Archives of Oral Biology 12: 645–661.

Mitchell, K.J., R.C. Pratt, L.N. Watson, G.C. Gibb, B. Llamas, M. Kasper, J. Edson, et al. 2014. Molecular phylogeny, biogeography, and habitat preference evolution of marsupials. Molecular Biology and Evolution 31: 2322–2330.

Mitchinson, B., R.A. Grant, K. Arkley, V. Rankov, I. Perkon, and T.J. Prescott. 2011. Active vibrissal sensing in rodents and marsupials. Philosophical Transactions of the Royal Society B 366: 3037–3048.

Mittermeier, R.A., and M.G.M. van Roosmalen. 1981. Preliminary observations on habitat utilization and diet in eight Surinam monkeys. Folia Primatologica 36: 1–39.

Mondolfi, E., and G.M. Padilla. 1957. Contribución al conocimiento del "perrito de agua." Memoria de la Sociedad Ciencias Naturales La Salle 17: 141–155.

Monet-Mendoza, A., D. Osorio-Sarabia, and L. García-Prieto. 2005. Helminths of the Virginia opossum *Didelphis virginiana* (Mammalia: Didelphidae) in Mexico. Journal of Parasitology 91: 213–219.

Monteiro-Filho, E.L.A., and V.S. Dias. 1990. Observações sobre a biologia de *Lutreolina crassicaudata* (Mammalia: Marsupialia). Revista Brasileira de Biologia 50: 393–399.

Monticelli-Almada, P.F., and A. Gasco. 2018. Nesting behavior of *Didelphis aurita*: twenty days of continuous recording of a female in a coati nest. Biota Neotropica 18(3): e20180550.

Mooney, H.A., and E.E. Cleland. 2001. The evolutionary impact of invasive species. Proceedings of the National Academy of Sciences 98: 5446–5451.

Moore, S.J., and G.D. Sanson. 1995. A comparison of the molar efficiency of two insect-eating mammals. Journal of Zoology 235: 175–192.

Moraes, E.A., Jr., and A.G. Chiarello. 2005a. A radio tracking study of home range and movements of the marsupial *Micoureus demerarae* (Thomas) (Mammalia, Didelphidae) in the Atlantic Forest of south-eastern Brazil. Revista Brasileira de Zoologia 22: 85–91.

Moraes, E.A., Jr., and A.G. Chiarello. 2005b. Sleeping sites of woolly mouse opossum *Micoureus demerarae* (Thomas) (Didelphimorphia, Didelphidae) in the Atlantic Forest of southeastern Brazil. Revista Brasileira de Zoologia 22: 839–843.

Morales, V.R., and R.W. McDiarmid. 1996. Annotated checklist of the amphibians and reptiles of Pakitza, Manu National Park Reserved Zone, with comments on the herpetofauna of Madre de Dios, Peru. Pages 503–522 *in* D.E. Wilson and A. Sandoval, eds. Manu, the biodiversity of southern Peru. Washington, DC: Smithsonian Institution.

Morand, S., B.R. Krasnov, and R. Poulin (eds.). 2006. Micromammals and macroparasites: from evolutionary ecology to management. Tokyo: Springer.

Mori, S.A., G. Cremers, C.A. Gracie, J.-J. de Granville, M. Hoff, and J.D. Mitchell. 1997. Guide to the vascular plants of central French Guiana. Part 1. Pteridophytes, gymnosperms, and monocotyledons. Memoirs of the New York Botanical Garden 76(1): i–viii + 1–422.

Mori, S.A., G. Cremers, C.A. Gracie, J.-J. de Granville, S.V. Heald, M. Hoff, and J.D. Mitchell. 2002. Guide to the vascular plants of central French Guiana. Part 2. Dicotyledons. Memoirs of the New York Botanical Garden 76(2): 1–776.

Morton, S.R. 1980. Ecological correlates of caudal fat storage in small mammals. Australian Mammalogy 3: 81–86.

Mothé, D., and L. Avilla. 2015. Mythbusting evolutionary issues on South American Gomphotheriidae (Mammalia: Proboscidea). Quaternary Science Reviews 110: 23–35.

Motta, J.C., Jr. 2006. Relações tróficas entre cinco Strigiformes simpátricas na região central do Estado de São Paulo, Brasil. Revista Brasileira de Ornitologia 14: 359–377.

Motta, J.C., Jr., S.A. Talamoni, J.A. Lombardi, and K. Simokomaki. 1996. Diet of the maned wolf, *Chrysocyon brachyurus*, in central Brazil. Journal of Zoology, London 240: 277–284.

Mourelle, C., and E. Ezcurra. 1996. Species richness of Argentine cacti: a test of biogeographic hypotheses. Journal of Vegetation Science 7: 667–680.

Muizon, C. de. 1994. A new carnivorous marsupial from the Palaeocene of Bolivia and the problem of marsupial monophyly. Nature 370: 208–211.

Muizon, C. de. 1998. *Mayulestes ferox*, a borhyaenoid (Metatheria, Mammalia) from the early Paleocene of Bolivia. Phylogenetic and paleobiologic implications. Geodiversitas 20: 19–142.

Muizon, C. de., and R.L. Cifelli. 2001. A new basal "didelphoid" (Marsupialia, Mammalia) from the early Paleocene of Tiupampa, Bolivia. Journal of Vertebrate Paleontology 21: 87–97.

Muizon, C. de, G. Billet, C. Argot, S. Ladevèze, and S. Goussard. 2015. *Alcidedorbignya inopinata*, a basal pantodont (Eutheria, Mammalia) from the early Paleocene of Bolivia: anatomy, phylogeny, and paleobiology. Geodiversitas 37: 397–634.

Muizon, C. de, S. Ladevèze, C. Selva, R. Vignaud, and S. Goussard. 2018. *Allqokirus australis* (Sparassodonta, Metatheria) from the early Paleocene of Tiupampa (Bolivia) and the rise of the metatherian carnivorous radiation in South America. Geodiversitas 40: 363–459.

Murphy, J.C., and R.W. Henderson. 1997. Tales of giant snakes: a natural history of anacondas and pythons. Malabar, FL: Krieger.

Murphy, P.G., and A.E. Lugo. 1986. Ecology of tropical dry forest. Annual Review of Ecology and Systematics 17: 67–88.

Murray, K.G., S. Kinsman, and J.L. Bronstein. 2000. Plant-animal interactions. Pages 245–302 *in* N.M. Nadkarni and N.T. Wheelwright, eds. Monteverde, ecology and conservation of a tropical cloud forest. New York: Oxford University Press.

Muschetto, E., G.R. Cueto, and O.V. Suárez. 2011. New data on the natural history and morphometrics of *Lutreolina crassicaudata* (Didelphimorphia) from central-eastern Argentina. Mastozoología Neotropical 18: 73–79.

Mustrangi, M.A., and J.L. Patton. 1997. Phylogeography and systematics of the slender mouse opossum *Marmosops* (Marsupialia, Didelphidae). University of California Publications (Zoology) 130: i–x, 1–86.

Nadkarni, N.M., and N.T. Wheelwright. 2000. Monteverde: ecology and conservation of a tropical cloud forest. New York: Oxford University Press.

Naranjo, M.E., C. Rengifo, and P.J. Soriano. 2003. Effect of ingestion by bats and birds on seed germination of *Stenocereus griseus* and *Subpilocereus repandus* (Cactaceae). Journal of Tropical Ecology 19: 19–25.

Nee, S., R.M. May, and P.H. Harvey. 1994. The reconstructed evolutionary process. Philosophical Transactions of the Royal Society B 344: 305–311.

Nelson, B.W., V. Kapos, J.B. Adams, W.J. Oliveira, O.P.G. Braun, and I.L. do Amaral. 1994. Forest disturbance by large blowdowns in the Brazilian Amazon. Ecology 75: 853–858.

Nettles, V.F., A.K. Prestwood, and W.R. Davidson. 1975. Severe parasitism in an opossum. Journal of Wildlife Diseases 11: 419–420.

Neves-Ferreira, A.G.C., R.H. Valente, J. Perales, and G.B. Domont. 2010. Natural inhibitors: innate immunity to snake venoms. Pages 259–284 *in* S.P. Mackessy, ed. Handbook of venoms and toxins of reptiles. Boca Raton, FL: CRC Press.

Newman, C., C.D. Buesching, and J.O. Wolff. 2005. The function of facial masks in "midguild" carnivores. Oikos 108: 623–633.

Nickle, D.A., and J.L. Castner. 1995. Strategies utilized by katydids (Orthoptera: Tettigoniidae) against diurnal predators in rainforests of northeastern Peru. Journal of Orthoptera Research 4: 75–88.

Nickle, D.A., and E.W. Heymann. 1996. Predation on Orthoptera and other orders of insects by tamarin monkeys, *Saguinus mystax mystax* and *Saguinus fuscicollis nigrifrons* (Primates: Callitrichidae), in north-eastern Peru. Journal of Zoology London 239: 799–819.

Nickol, B.B. 1985. Epizootiology. Pages 307–346 *in* D.W.T. Crompton and B.B. Nickol, eds. Biology of the Acanthocephala. Cambridge, UK: Cambridge University Press.

Nicolson, S.W., and R.W. Thornburg. 2007. Nectar chemistry. Pages 215–264 *in* S.W. Nicholson, M. Nepi, and E. Pacini, eds. Nectaries and nectar. Dordrecht: Springer.

Nilsson, M.A., G. Churakov, M. Sommer, N.V. Tran, A. Zemann, J. Brosius, and J. Schmitz. 2010. Tracking marsupial evolution using archaic genomic retrotransposon insertions. PLoS Biology 8(7): e1000436.

Nitikman, L.Z., and M.A. Mares. 1987. Ecology of small mammals in a gallery forest of central Brazil. Annals of Carnegie Museum 56: 75–95.

Noireau, F., P. Diosque, and A.M. Jansen. 2009. *Trypanosoma cruzi*: adaptation to its vectors and its hosts. Veterinary Research 40: 26.

Norris, D., F. Michalski, and C.A. Peres. 2010. Habitat patch size modulates terrestrial mammal activity patterns in Amazonian forest fragments. Journal of Mammalogy 91: 551–560.

Norton, A.C., A.V. Beran, and G.A. Misrahy. 1964. Electroencephalograph during "feigned" sleep in the opossum. Nature 204: 162–163.

Nowack, J., S. Giroud, W. Arnold, and T. Ruf. 2017. Muscle non-shivering thermogenesis and its role in the evolution of endothermy. Frontiers in Physiology 8(889): 1–13.

Nutting, W.B. 1985. Prostigmata-Mammalia, validation of coevolutionary phylogenies. Pages 569–640 *in* K.C. Kim, ed. Coevolution of parasitic arthropods and mammals. New York: John Wiley & Sons.

Nutting, W.B., F.S. Lukoschus, and L.E. Desch. 1980. Parasitic mites of Surinam, XXXVII. *Demodex marsupiali* sp. nov. from *Didelphis marsupialis*: adaptation to glandular habitat. Zoologische Mededelingen 56: 83–90.

O'Connell, M.A. 1979. Ecology of didelphid marsupials from northern Venezuela. Pages 73–87 *in* J.F. Eisenberg, ed. Vertebrate ecology in the northern Neotropics. Washington, DC: Smithsonian Institution Press.

O'Connell, M.A. 1989. Population dynamics of Neotropical small mammals in seasonal habitats. Journal of Mammalogy 70: 532–548.

OConnor, B.M. 1982. Evolutionary ecology of astigmatid mites. Annual Review of Entomology 27: 385–409.

OConnor, B.M., F.S. Lukoschus, and K.T.M. Giesen. 1982. Two new species of *Marmosopus* (Acari: Astigmata) from rodents of the genus *Scotinomys* (Cricetidae) in Central America. Occasional Papers of the Museum of Zoology University of Michigan 703: 1–22.

O'Dea, A., H.A. Lessios, A.G. Coates, R.I. Eytan, S.A. Restrepo-Moreno, A.L. Cione, L.S. Collins, et al. 2016. Formation of the Isthmus of Panama. Science Advances 2(8): e1600883.

Oelkrug, R., E.T. Polymeropoulos, and M. Jastroch. 2015. Brown adipose tissue: physiological function and evolutionary significance. Journal of Comparative Physiology B 185: 587–606.

Ojala-Barbour, R., C.M. Pinto, J. Brito M., L. Albuja V., T.E. Lee, Jr., and B.D. Patterson. 2013. A new species of shrew-opossum (Paucituberculata: Caenolestidae) with a phylogeny of extant caenolestids. Journal of Mammalogy 94: 967–982.

Ojeda, R.A. [no date]. Lista de los mamíferos de Ñacuñán (34° 02' S, 67° 58' W), Mendoza, Argentina. Mendoza: IADIZA/CONICET.

Ojeda, R.A., and S. Tabeni. 2009. The mammals of the Monte Desert revisited. Journal of Arid Environments 73: 173–181.

Oliveira, E.V., and F.J. Goin. 2015. A new species of *Gaylordia* Paula Couto (Mammalia, Metatheria) from Itaboraí, Brazil. Revista Brasileira de Paleontologia 18: 97–108.

Oliveira, E.V., P.V. Nova, F.J. Goin, and L.D.S. Avilla. 2011. A new hyladelphine marsupial (Didelphimorphia, Didelphidae) from cave deposits of northern Brazil. Zootaxa 3041: 51–62.

Oliveira, M.E., and R.T. Santori. 1999. Predatory behavior of the opossum *Didelphis albiventris* on the pitviper *Bothrops jararaca*. Studies of Neotropical Fauna and Environment. 34: 72–75.

Oliveira-Filho, A.T., and J.A. Ratter. 2002. Vegetation physiognomies and woody flora of the Cerrado biome. Pages 91–120 *in* P.S. Oliveira and R.J. Marquis, eds. The cerrados of Brazil, ecology and natural history of a Neotropical savanna. New York: Columbia University Press.

Oliveira-Santos, L.G.R., M.A. Tortato, and M.E. Graipel. 2008. Activity patterns of Atlantic Forest small arboreal mammals as revealed by camera traps. Journal of Tropical Ecology 24: 563–567.

Oliver, W.L.R. 1976. The management of yapoks (*Chironectes minimus*) at Jersey Zoo, with observations on their behavior. Jersey Wildlife Preservation Trust 13: 32–36.

Olrog, C.C. 1979. Los mamíferos de la selva húmeda, Cerro Calilegua, Jujuy. Acta Zoológica Lilloana 33: 9–14.

Olson, D.M. 1994. The distribution of leaf litter invertebrates along a Neotropical altitudinal gradient. Journal of Tropical Ecology 10: 129–150.

O'Meara, R.N., and R.S. Thompson. 2014. Were there Miocene meridiolestidans? Assessing the phylogenetic placement of *Necrolestes patagonensis* and the presence of a 40 million year meridiolestidan ghost lineage. Journal of Mammalian Evolution 21: 271–284.

Ortolani, A. 1999. Spots, stripes, tail tips and dark eyes: Predicting the function of carnivore colour patterns using the comparative method. Biological Journal of the Linnean Society 67: 433–476.

Ostfeld, R.S. 1990. The ecology of territoriality in small mammals. Trends in Ecology and Evolution 5: 411–415.

Oswaldo-Cruz, E., J.N. Hokoc, and A.P.B. Sousa. 1979. A schematic eye for the opossum. Vision Research 19: 263–276.

Owen, R.D., H. Sánchez, L. Rodríguez, and C.B. Jonsson. 2018. Composition and characteristics of a diverse didelphid community (Mammalia: Didelphimorphia) in sub-tropical South America. Occasional Papers Museum of Texas Tech University 358: 1–18.

Palacios, A.G., F. Bozinovic, A. Vielma, C.A. Arrese, D.M. Hunt, and L. Peichl. 2010. Retinal photoreceptor arrangement, SWS1 and LWS opsin sequence, and electroretinography in the South American marsupial *Thylamys elegans* (Waterhouse, 1839). Journal of Comparative Neurology 518: 1589–1602.

Palmeirim, A.F., M.S. Leite, M. Santos-Reis, and F.A.S. Fernandez. 2014. Habitat selection for resting sites by the water opossum (*Chironectes minimus*) in the Brazilian Atlantic Forest. Studies on Neotropical Fauna and Environment 49: 231–238.

Pardini, R. 1998. Feeding ecology of the Neotropical river otter *Lontra longicaudis* in an Atlantic Forest stream, south-eastern Brazil. Journal of Zoology London 245: 385–391.

Park, O., A. Barden, and E. Williams. 1940. Studies in nocturnal ecology, IX. Further analysis of activity of Panama rain forest animals. Ecology 21: 122–134.

Parker, G.A. 1974. Assessment strategy and the evolution of fighting behaviour. Journal of Theoretical Biology 47: 223–243.

Parrish, S.S. 1997. The female opossum and the nature of the New World. William and Mary Quarterly 54: 475–514.

Paruelo, J.M., E.G. Jobbágy, M. Oesterheld, R.A. Golluscio, and M.R. Aguiar. 2007. The grasslands and steppes of Patagonia and the Río de la Plata plains. Pages 232–248 *in* T.T. Veblen, K.R. Young, and A.R. Orme, eds. The physical geography of South America. Oxford: Oxford University Press.

Passamani, M., and A.B. Rylands. 2000. Feeding behavior of Geoffroy's marmoset (*Callithrix geoffroyi*) in an Atlantic Forest fragment of south-eastern Brazil. Primates 41: 27–38.

Patel, B.A., and S.A. Maiolino. 2016. Morphological diversity in the digital rays of primate hands. Pages 55–100 *in* T.L. Kivell, P. Lemelin, B.G. Richmond, and D. Schmitt, eds. The evolution of the primate hand. New York: Springer.

Patrizi, G., and B.L. Munger. 1966. The ultrastructure and innervation of rat vibrissae. Journal of Comparative Neurology 126: 423–435.

Patrón, J.C., U.H. Salinas, A.R. Bautista, S. Lozano-Trejo, and F. Marini-Zúñiga. 2011. *Masticophis* (= *Coluber*) *mentovarius* (Neotropical whipsnake). Diet. Herpetological Review 42: 293.

Patterson, B., and R. Pascual. 1972. The fossil mammal fauna of South America. Pages 247–309 *in* A. Keast, F.C. Erk, and B. Glass, eds. Evolution, mammals, and southern continents. Albany, NY: State University of New York Press.

Patterson, B.D. 2008 ("2007"). Order Paucituberculata Ameghino, 1894. Pages 119–127 *in* A.L. Gardner, ed. Mammals of South America, volume 1. Marsupials, xenarthrans, shrews, and bats. Chicago: University of Chicago Press.

Patterson, B.D, and M.A. Rogers. 2008 ("2007"). Order Microbiotheria Ameghino, 1889. Pages 117–119 *in* A.L. Gardner, ed. Mammals of South America, volume 1. Marsupials, xenarthrans, shrews, and bats. Chicago: University of Chicago Press.

Patton, J.L., M.N.F. da Silva, and J.R. Malcolm. 2000. Mammals of the Rio Juruá and the evolutionary and ecological diversification of Amazonia. Bulletin of the American Museum of Natural History 244: 1–306.

Pavan, S.E., and R.S. Voss. 2016. A revised subgeneric classification of short-tailed opossums (Didelphidae: *Monodelphis*). American Museum Novitates 3868: 1–44.

Pavan, S.E., S.A. Jansa, and R.S. Voss. 2014. Molecular phylogeny of short-tailed opossums (Didelphidae: *Monodelphis*): taxonomic implications and tests of evolutionary hypotheses. Molecular Phylogenetics and Evolution 79: 199–214.

Pearson, O.P. 2008 ("2007"). Genus *Lestodelphys*. Pages 50–51 *in* A.L. Gardner, ed. Mammals of South America, volume 1. Marsupials, xenarthrans, shrews, and bats. Chicago: University of Chicago Press.

Pedersen, A.B., and T.J. Greives. 2008. The interaction of parasites and resources cause crashes in a wild mouse population. Journal of Animal Ecology 77: 370–377.

Peichl, L. 2005. Diversity of mammalian photoreceptor properties: adaptations to habitat and lifestyle? Anatomical Record 287A: 1001–1012.

Penny, N.D., and J.R. Arias. 1982. Insects of an Amazon forest. New York: Columbia University Press.

Perales, J., R. Muñoz, and H. Moussatché. 1986. Isolation and partial characterization of a protein fraction from the opossum (*Didelphis marsupialis*) serum, with protecting properties against the *Bothrops jararaca* snake venom. Anais da Academia Brasileira de Ciências 58: 155–162.

Perales, J., H. Moussatché, S. Marangoni, B. Oliveira, and G.B. Domont. 1994. Isolation and partial characterization of an anti-bothropic complex from the serum of South American Didelphidae. Toxicon 32: 1237–1249.

Peres, C.A. 1993. Diet and feeding ecology of saddle-back (*Saguinus fuscicollis*) and moustached (*S. mystax*) tamarins in an Amazonian forest. Journal of Zoology 230: 567–592.

Pernetta, J.C. 1976. Diets of the shrews *Sorex araneus* L. and *Sorex minutus* L. in Wytham grassland. Journal of Animal Ecology 45: 899–912.

Perret, M., and S.B. M'Barek. 1991. Male influence on oestrous cycles in female woolly opossum (*Caluromys philander*). Journal of Reproduction and Fertility 91: 557–566.

Peters, R.H., and J.V. Raelson. 1984. Relations between individual size and mammalian population density. American Naturalist 124: 498–517.

Pheasey, H., P. Smith, and K. Atkinson. 2018. Observations on the diet and behavior of captive *Monodelphis* (*Mygalodelphys*) *kunsi* Pine, 1975 (Didelphimorphia: Didelphidae) from Paraguay. Notas Mastozoológicas 4(2): 38–45.

Phillimore, A.B., and T.D. Price. 2008. Density-dependent cladogenesis in birds. PLoS Biology 6(3): e71.

Piana, R.P. 2007. Anidamiento y dieta de *Harpia harpyja* Linnaeus en la comunidad nativa de Infierno, Madre de Dios, Perú. Revista Peruana de Biología 14: 135–138.

Pianka, E.R. 1974. Niche overlap and diffuse competition. Proceedings of the National Academy of Sciences USA 71: 2141–2145.

Pigot, A.L., A.B. Phillimore, I.P.F. Owens, and C.D.L. Orme. 2010. The shape and temporal dynamics of phylogenetic trees arising from geographic speciation. Systematic Biology 59: 660–673.

Pilatti, P., and D. Astúa. 2016. Orbit orientation in didelphid marsupials (Didelphimorphia: Didelphidae). Current Zoology 63: 403–415.

Pine, R.H, P.L. Dalby, and J.O. Matson. 1985. Ecology, postnatal development, morphometrics, and taxonomic status of the short-tailed opossum, *Monodelphis dimidiata*, an apparently semelparous annual marsupial. Annals of Carnegie Museum 54: 195–231.

Pineda-Munoz, S., and J. Alroy. 2014. Dietary characterization of terrestrial mammals. Proceedings of the Royal Society B 281: 20141173.

Pinheiro, P.S., F.M.V. Carvalho, F.A.S. Fernandez, and J.L. Nessimian. 2002. Diet of the marsupial *Micoureus demerarae* in small fragments of Atlantic Forest in southeastern Brazil. Studies on Neotropical Fauna and Environment 37: 213–218.

Pinotti, B.T., L. Naxara, and R. Pardini. 2011. Diet and food selection by small mammals in an old-growth Atlantic Forest of south-eastern Brazil. Studies on Neotropical Fauna and Environment 46: 1–9.

Pinto, C.D., and T. de Lema. 2002. Comportamento alimentar e dieta de serpentes, gêneros *Boiruna* e *Clelia* (Serpentes, Colubridae). Iheringia (ser. zool.) 92: 9–19.

Pinto, I. del S., J.R. Botelho, L.P. Costa, Y.L.R. Leite, and P.M. Linardi. 2009. Siphonaptera associated with wild mammals from the central Atlantic Forest biodiversity corridor in southeastern Brazil. Journal of Medical Entomology 46: 1146–1151.

Pinto, R.M., M. Knoff, D.C. Gomes, and D. Noronha. 2011. Nematodes from mammals in Brazil: an updating. Neotropical Helminthology 5: 139–183.

Pires, A.S., and F.A.S. Fernandez. 1999. Use of space by the marsupial *Micoureus demerarae* in small Atlantic Forest fragments in south-eastern Brazil. Journal of Tropical Ecology 15: 279–290.

Pires, J.M., and G.T. Prance. 1985. The vegetation types of the Brazilian Amazon. Pages 109–145 *in* G.T. Prance and T.E. Lovejoy, eds. Amazonia (Key Environments series). Oxford: Pergamon Press.

Pires, M.M., E.G. Martins, L.D. Cruz, F.R. Fernandes, R.B.G. Clemente-Carvalho, M.N.F. da Silva, and S.F. dos Reis. 2010a. Young didelphid consumption by *Micoureus paraguayanus* (Didelphimorphia: Didelphidae) in southeastern Brazil. Mastozoología Neotropical 17: 183–187.

Pires, M.M., E.G. Martins, M.N.F. da Silva, and S.F. dos Reis. 2010b. *Gracilinanus microtarsus* (Didelphimorphia: Didelphidae). Mammalian Species 42 (851): 33–40.

Pires, M.M., E.G. Martins, M.S. Araújo, and S.F. dos Reis. 2012. Between-individual variation drives the seasonal dynamics in the trophic niche of a Neotropical marsupial. Austral Ecology 38: 664–671.

Pizzatto, L., O.A.V. Marques, and K. Facure. 2009. Food habits of Brazilian boid snakes: overview and new data, with special reference to *Corallus hortulanus*. Amphibia-Reptilia 30: 533–544.

Plavcan, J.M., and C.P. van Schaik. 1992. Intrasexual competition and canine dimorphism in anthropoid primates. American Journal of Physical Anthropology 87: 461–477.

Pol, R., and J.L. de Casenave. 2004. Activity patterns of harvester ants *Pogonomyrmex pronotalis* and *Pogonomyrmex rastratus* in the central Monte Desert, Argentina. Journal of Insect Behavior 17: 647–661.

Pollock, K.H., J.D. Nichols, C. Brownie, and J.E. Hines. 1990. Statistical inference for capture-recapture experiments. Wildlife Monographs 107: 3–97.

Potkay, S. 1970. Diseases of the opossum (*Didelphis marsupialis*): a review. Laboratory Animal Care 20: 502–511.

Poulin, R. 2007. Evolutionary ecology of parasites (2nd ed.). Princeton, NJ: Princeton University Press.

Power, M.L. 2010. Nutritional and digestive challenges to being a gum-feeding primate. Pages 25–44 *in* A.M. Burrows and L.T. Nash, eds. The evolution of exudativory in primates. New York: Springer.

Prance, G.T. 1979. Notes on the vegetation of Amazonia III. The terminology of Amazonian forest types subject to inundation. Brittonia 31: 26–38.

Pratas-Santiago, L.P., A.L.S. Gonçalves, A.M.V.M. Soares, and W.R. Spironello. 2016. The moon cycle effect on the activity patterns of ocelots and their prey. Journal of Zoology 299: 275–283.

Pratas-Santiago, L.P., A.L.S. Gonçalves, A.J.A. Nogueira, and W.R. Spironello. 2017. Dodging the moon: the moon effect on activity allocation of prey in the presence of predators. Ethology 123: 467–474.

Prevedello, J.A., R.G. Rodrigues, and E.L.A. Monteiro-Filho. 2009. Vertical use of space by the marsupial *Micoureus paraguayanus* (Didelphimorphia, Didelphidae) in the Atlantic Forest of Brazil. Acta Theriologica 54: 259–266.

Prévost, M.-F. 1983. Les fruits et les graines des espèces végétales pionnières de Guyane Française. Revue d'Ecologie (La Terre et la Vie) 38: 121–145.

Prevosti, F.J., E.P. Tonni, and J.C. Bidegain. 2009. Stratigraphic range of the large canids (Carnivora, Canidae) in South America, and its relevance to Quaternary biostratigraphy. Quaternary International 2010: 76–81.

Prevosti, F.J., A.M. Forasiepi, M.D. Ercoli, and G.F. Turazzini. 2012. Paleoecology of the mammalian carnivores (Metatheria, Sparassodonta) of the Santa Cruz Formation (late Early Miocene). Pages 173–193 *in* S.F. Vizcaíno, R.F. Kay, and M.S. Bargo, eds. Early Miocene paleobiology in Patagonia: high-latitude paleocommunities of the Santa Cruz Formation. Cambridge, UK: Cambridge University Press.

Price, R.D., and K.C. Emerson. 1986. New species of *Cummingsia* Ferris (Mallophaga: Trimenoponidae) from Peru and Venezuela. Proceedings of the Biological Society of Washington 99: 748–752.

Pridmore, P.A. 1992. Trunk movements during locomotion in the marsupial *Monodelphis domestica*. Journal of Morphology 211: 137–146.

Prugh, L.R., and C.D. Golden. 2014. Does moonlight increase predation risk? Meta-analysis reveals divergent responses of nocturnal mammals to lunar cycles. Journal of Animal Ecology 83: 504–514.

Prum, R.O., and R.H. Torres. 2004. Structural coloration of mammalian skin: convergent evolution of coherently scattering dermal collagen arrays. Journal of Experimental Biology 207: 2157–2172.

Puhakka, M., and R. Kalliola. 1995. Floodplain vegetation mosaics in western Amazonia. Biogeographica 71: 1–14.

Pung, O.J., L.A. Durden, C.W. Banks, and D.N. Jones. 1994. Ectoparasites of opossums and raccoons in southeastern Georgia. Journal of Medical Entomology 31: 915–919.

Quadros, J., and E.L.A. Monteiro-Filho. 2001. Diet of the Neotropical otter, *Lontra longicaudis*, in an Atlantic Forest area, Santa Catarina State, southern Brazil. Studies on Neotropical Fauna and Environment 36: 15–21.

Quaresma, P.F., F.D. Rêgo, H.A. Botelho, S.R. da Silva, A.J. Moura, Jr., R.G.T. Neto, et al. 2011. Wild, synanthropic and domestic hosts of *Leishmania* in an endemic area of cutaneous leishmaniasis in Minas Gerais state, Brazil. Transactions of the Royal Society of Tropical Medicine and Hygiene 105: 579–585.

Quintela, F.M., M.B. Santos, A. Gava, and A.U. Christoff. 2011. Notas sobre morfologia, distribucão geográfica, história natural y citogenética de *Cryptonanus guahybae* (Didelphimorphia: Didelphidae). Mastozoología Neotropical 18: 247–257.

Quintela, F.M., G. Iob, and L.G.S. Artioli. 2014. Diet of *Procyon cancrivorus* (Carnivora, Procyonidae) in restinga and estuarine environments of southern Brazil. Iheringia (ser. Zoologia) 104: 143–149.

Rabosky, D.L. 2010. Extinction rates should not be estimated from molecular phylogenies. Evolution 64: 1816–1824.

Rademaker, V., and R. Cerqueira. 2006. Variation in the latitudinal reproductive patterns of the genus *Didelphis* (Didelphimorphia: Didelphidae). Austral Ecology 31: 337–342.

Radovsky, F.J. 1985. Evolution of mammalian mesostigmate mites. Pages 441–504 *in* K.C. Kim, ed. Coevolution of parasitic arthropods and mammals. New York: John Wiley & Sons.

Radovsky, F.J. 1994. The evolution of parasitism and the distribution of some dermanyssid mites (Mesostigmata) on vertebrate hosts. Pages 186–217 *in* M.A. Houck, ed. Mites, ecological and evolutionary analyses of life-history patterns. New York: Chapman & Hall.

Ramamurthy, D.L., and L.A. Krubitzer. 2016. The evolution of whisker-mediated somatosensation in mammals: sensory processing in barrelless S1 cortex of a marsupial, *Monodelphis domestica*. Journal of Comparative Neurology 524: 3587–3612.

Ramírez-Cañas, S.A., M. George-Nascimento, L. García-Prieto, and R. Mata-López. 2019. Helminth community structure of the gray four-eyed opossum *Philander opossum* (Mammalia: Didelphidae) in the Neotropical portion of Mexico. Journal of Parasitology 105: 624–629.

Rasmussen, D.T. 1990. Primate origins: lessons from a Neotropical marsupial. American Journal of Primatology 22: 263–277.

Ravizza, R.J., H.E. Hefner, and B. Masterton. 1969. Hearing in primitive mammals, 1: Opossum (*Didelphis virginianus* [sic]). Journal of Auditory Research 9: 1–7.

Redford, K.H., and J.G. Dorea. 1984. The nutritional value of invertebrates with emphasis on ants and termites as food for mammals. Journal of Zoology London 203: 385–395.

Redford, K.H., and G.A.B. da Fonseca. 1986. The role of gallery forests in the zoogeography of the Cerrado's non-volant mammalian fauna. Biotropica 18: 126–135.

Redwood, S.D. 2019. The geology of the Panama-Chocó Arc. Pages 901–932 *in* F. Cediel and R.P. Shaw, eds. Geology and tectonics of northwestern South America. Cham, Switzerland: Springer International.

Reed, R.R. 2004. After the holy grail: establishing a molecular basis for mammalian olfaction. Cell 116: 329–336.

Regidor, H.A., and M. Gorostiague. 1996. Reproduction in the white eared opossum (*Didelphis albiventris*) under temperate conditions in Argentina. Studies of Neotropical Fauna & Environment 31: 133–136.

Reguero, M.A., and A.M. Candela. 2011. Late Cenozoic mammals from the northwest of Argentina. Pages 411–426 *in* J.A. Salfity and R.A. Marquillas, eds. Cenozoic geology of the central Andes of Argentina. Salta: SCS Publisher.

Reguero, M.A., J.N. Gelfo, G.M. López, M. Bond, A. Abello, S.N. Santillana, and S.A. Marenssi. 2014. Final Gondwana breakup: the Paleogene South American native ungulates and the demise of the South American-Antarctica land connection. Global and Planetary Change 123 (part B): 400–413.

Reig, O.A., J.A.W. Kirsch, and L.G. Marshall. 1987. Systematic relationships of the living and Neocenozoic American "opossum-like" marsupials (suborder Didelphimorphia), with comments on the classification of these and of Cretaceous and Paleogene New World and European metatherians. Pages 1–89 *in* M. Archer, ed. Possums and opossums: studies in evolution, vol. 1. Sydney: Surrey Beatty.

Resende, B.D., V.L.G. Greco, E.B. Ottoni, and P. Izar. 2003. Some observations on the predation of small mammals by tufted capuchin monkeys (*Cebus apella*). Neotropical Primates 11: 103–104.

Reynolds, H.C. 1945. Some aspects of the life history and ecology of the opossum in central Missouri. Journal of Mammalogy 26: 361–379.

Reynolds, H.C. 1952. Studies on reproduction in the opossum (*Didelphis virginiana virginiana*). University of California Publications in Zoology 52: [i–iv], 223–283.

Ribeiro, M.C.P., and J.E.P.W. Bicudo. 2007. Oxygen consumption and thermoregulatory responses in three species of South American marsupials. Comparative Biochemistry and Physiology A 147: 658–664.

Richards, P.W. 1952. The tropical rain forest, an ecological study. Cambridge, UK: Cambridge University Press.

Richardson, D.J., W.B. Owen, and D.E. Snyder. 1992. Helminth parasites of the raccoon (*Procyon lotor*) from north-central Arkansas. Journal of Parasitology 78: 163–166.

Rico-Guevara, A., and K.J. Hurme. 2019. Intrasexually selected weapons. Biological Reviews 94: 60–101.

Riek, A., and F. Geiser. 2014. Heterothermy in pouched mammals—a review. Journal of Mammalogy 292: 74–85.

Riley, J. 1986. The biology of pentastomids. Advances in Parasitology 25: 45–128.

Robinson, B.W., and D.S. Wilson. 1998. Optimal foraging, specialization, and a solution to Liem's paradox. American Naturalist 151: 223–235.

Robinson, J.G., and K.H. Redford. 1986. Body size, diet, and population density of Neotropical forest mammals. American Naturalist 128: 665–680.

Robinson, S.K. 1994. Habitat selection and foraging ecology of raptors in Amazonian Peru. Biotropica 26: 443–458.

Rocha-Mendes, F., S.B. Mikich, J. Quadros, and W.A. Pedro. 2010. Feeding ecology of carnivores (Mammalia: Carnivora) in Atlantic Forest remnants, southern Brazil. Biota Neotropica 10: 21–30.

Rodrigues, F.H.G. 2005. *Bothrops neuwiedi pauloensis* (Jararaca Rabo-de-osso). Predation. Herpetological Review 36: 67–68.

Rodrigues, F.H.G., L. Silveira, A.T.A. Jácomo, A.P. Carmignotto, A.M.R. Bezerra, D.C. Coelho, H. Garbogini, J. Pagnozzi, and A. Hass. 2002. Composição e caracterização da fauna de mamíferos do Parque Nacional das Emas, Goiás, Brasil. Revista Brasileira de Zoologia 19: 589–600.

Rodriguez, L.B., and J.E. Cadle. 1990. A preliminary overview of the herpetofauna of Cocha Cashu, Manu National Park, Peru. Pages 410–425 in A.H. Gentry, ed. Four Neotropical rainforests. New Haven, CT: Yale University Press.

Rogers, S.M., and S.J. Simpson. 2014. Thanatosis. Current Biology 24: R1031–R1033.

Rojas, D., O.M. Warsi, and L.M. Dávalos. 2016. Bats (Chiroptera: Noctilionoidea) challenge a recent origin of extant Neotropical diversity. Systematic Biology 65: 432–448.

Roosmalen, M.G.M. van. 1985a. Fruits of the Guianan flora. Utrecht: Institute of Systematic Botany.

Roosmalen, M.G.M. van. 1985b. Habitat preferences, diet, feeding strategy, and social organization of the black spider monkey (*Ateles paniscus* Linnaeus, 1758) in Surinam. Acta Amazonica 15 (3/4, suppl.): 1–238.

Rose, K.D. 2006. The beginning of the Age of Mammals. Baltimore, MD: Johns Hopkins University Press.

Rosenberger, A.L. 1992. Evolution of feeding niches in New World monkeys. American Journal of Physical Anthropology 88: 525–562.

Rosenthal, M.A. 1975. Observations on the water opossum or yapok in captivity. International Zoo Yearbook 15: 4–6.

Rosenzweig, M.L., and R.D. McCord. 1991. Incumbent replacement: evidence for long-term evolutionary progress. Paleobiology 17: 202–213.

Rossi, R.V. 2005. Revisão taxonômica de *Marmosa* Gray, 1821 (Didelphimorphia, Didelphidae). PhD dissertation, Universidade de São Paulo.

Rossi, R.V., R.S. Voss, and D.P. Lunde. 2010. A revision of the didelphid marsupial genus *Marmosa*, part 1. The species in Tate's "Mexicana" and "Mitis" sections and other closely related forms. Bulletin of the American Museum of Natural History 334: 1–83.

Rotureau, B. 2006. Ecology of the *Leishmania* species in the Guianan ecoregion complex. American Journal of Tropical Medicine and Hygiene 74: 81–96.

Rougier, G.W., J.R. Wible, and M.J. Novacek. 1998. Implications of *Deltatheridium* specimens for early marsupial history. Nature 396: 458–463.

Rougier, G.W., J.R. Wible, R.M.D. Beck, and S. Apesteguía. 2012. The Miocene mammal *Necrolestes* demonstrates the survival of a Mesozoic nontherian lineage into the late Cenozoic of South America. Proceedings of the National Academy of Sciences USA 109: 20053–20058.

Rowe, T.B., T.P. Eiting, T.E. Macrini, and R.A. Ketcham. 2005. Organization of the olfactory and respiratory skeleton in the nose of the gray short-tailed opossum *Monodelphis domestica*. Journal of Mammalian Evolution 12: 303–336.

Rundell, R.J., and T.D. Price. 2009. Adaptive radiation, nonadaptive radiation, ecological speciation, and nonecological speciation. Trends in Ecology and Evolution 24: 394–399.

Russell, E.M. 1982. Patterns of parental care and parental investment in marsupials. Biological Reviews 57: 423–486.

Russell, E.M. 1984. Social behaviour and social organization of marsupials. Mammal Review 14: 101–154.

Ruxton, G.D., T.N. Sheratt, and M.P. Speed. 2004. Avoiding predation: the evolutionary ecology of crypsis, warning signals, and mimicry. Oxford: Oxford University Press.

Ryan, M.J., M.D. Tuttle, and L.K. Taft. 1981. The costs and benefits of frog chorusing behavior. Behavioral Ecology and Sociobiology 8: 273–278.

Rylands, A.B. 1993. Marmosets and tamarins: systematics, behaviour, and ecology. Oxford: Oxford University Press.

Ryser, J. 1990. The mating system, ecology, and biology of the Virginia opossum, *Didelphis virginiana*, in north-central Florida. PhD dissertation, University of Berne.

Ryser, J. 1992. The mating system and male mating success of the Virginia opossum (*Didelphis virginiana*) in Florida. Journal of Zoology London 228: 127–139.

Ryser, J. 1995. Activity, movement, and home range of Virginia opossum (*Didelphis virginiana*) in Florida. Bulletin of the Florida Museum of Natural History 38: 177–194.

Sabat, P., F. Bozinovic, and F. Zambrano. 1995. Role of dietary substrates on intestinal disaccharides, digestibility, and energetics in the insectivorous mouse-opossum (*Thylamys elegans*). Journal of Mammalogy 76: 603–611.

Salo, P., E. Korpimäki, P.B. Banks, M. Nordström, and C.R. Dickman. 2007. Alien predators are more dangerous than native predators to prey populations. Proceedings of the Royal Society B 274: 1237–1243.

Saltré, F., M. Rodríguez-Rey, B.W. Brook, C.N. Johnson, C.S.M. Turney, J. Alroy, et al. 2016. Climate change not to blame for late Quaternary megafauna extinctions in Australia. Nature Communications 7:10511.

Samollow, P.B. 2008. The opossum genome: insights and opportunities from an alternative mammal. Genome Research 18: 1199–1215.

Sampaio, E.M., E.K.V. Kalko, E. Bernard, B. Rodríguez-Herrera, and C.O. Handley, Jr. 2003. A biodiversity assessment of bats (Chiroptera) in a tropical lowland rainforest of central Amazonia, including methodological and conservation considerations. Studies on Neotropical Fauna and Environment 38: 17–31.

Sánchez-Villagra, M.R. 2001. The phylogenetic relationships of argyrolagid marsupials. Zoological Journal of the Linnean Society 131: 481–496.

Sánchez-Villagra, M.R., and R.J. Asher. 2002. Cranio-sensory adaptations in small faunivorous semiaquatic mammals, with special reference to olfaction and the trigeminal system. Mammalia 66: 93–109.

Sánchez-Villagra, M.R., and R.F. Kay. 1997. A skull of *Proargyrolagus*, the oldest argyrolagid (late Oligocene Salla beds, Bolivia), with brief comments concerning its paleobiology. Journal of Vertebrate Paleontology 17: 717–724.

Sandom, C., S. Faurby, B. Sandel, and J.-C. Svenning. 2014. Global late Quaternary megafauna extinctions linked to humans, not climate change. Proceedings of the Royal Society B 281: 20133254.

Santana, S.E., S. Strait, and E.R. Dumont. 2011. The better to eat you with: functional correlates of tooth structure in bats. Functional Ecology 25: 839–847.

Santori, R.T., D. Astúa de Moraes, and R. Cerqueira. 1995. Diet composition of *Metachirus nudicaudatus* and *Didelphis aurita* (Marsupialia, Didelphoidea) in southeastern Brazil. Mammalia 59: 511–516.

Santori, R.T., D. Astúa de Moraes, C.E.V. Grelle, and R. Cerqueira. 1997. Natural diet at a restinga forest and laboratory food preferences of the opossum *Philander frenata* in Brazil. Studies of Neotropical Fauna and Environment 32: 12–16.

Santori, R.T., D. Astúa de Moraes, and R. Cerqueira. 2004. Comparative gross morphology of the digestive tract in ten Didelphidae marsupial species. Mammalia 68: 27–36

Santori, R.T., O. Rocha-Barbosa, M.V. Vieira, J.A. Magnan-Neto, and M.F.C. Loguercio. 2005. Locomotion in aquatic, terrestrial, and arboreal habitats of thick-tailed opossum, *Lutreolina crassicaudata* (Desmarest, 1804). Journal of Mammalogy 86: 902–908.

Santori, R.T., D. Astúa, and M. Martins. 2016. An additional record for the rare black-shouldered opossum *Caluromysiops irrupta* Sanborn, 1951 (Didelphimorphia: Didelphidae) in northwestern Brazil. Check List 12(3): 1–3.

Sarmiento, G. 1976. Evolution of arid vegetation in tropical America. Pages 65–99 *in* D.W. Goodall, ed. Evolution of desert biota. Austin, TX: University of Texas Press.

Sarmiento, L. 1954. *Gigantorhynchus ortizi* n. sp., an acanthocephalan from *Metachirus nudicaudatus*. Journal of Parasitology 40: 448–452.

Saunders, R.C. 1975. Venezuelan Macronyssidae (Acarina: Mesostigmata). Brigham Young University Science Bulletin (Biol. Ser.) 20(2): 75–90.

Sawaya, R.J., O.A.V. Marques, and M. Martins. 2008. Composition and natural history of a cerrado snake assemblage at Itirapina, São Paulo state, southeastern Brazil. Biota Neotropica 8(2): 127–149.

Sazima, I. 1992. Natural history of the jararaca pitviper, *Bothrops jararaca*, in southeastern Brazil. Pages 199–216 *in* J.A. Campbell and E.D. Brodie, Jr., eds. Biology of the pitvipers. Tyler, TX: Selva.

Scheibler, D.R. 2007. Food partitioning between breeding white-tailed kites (*Elanus leucurus*; Aves; Accipitridae) and barn owls (*Tyto alba*; Aves; Tytonidae) in southern Brazil. Brazilian Journal of Biology 67: 65–71.

Scher, H.D., and E.E. Martin. 2006. Timing and climatic consequences of the opening of Drake Passage. Science 312: 428–430.

Schluter, D. 2000. Ecology of adaptive radiation. Oxford: Oxford University Press.

Schmitt, D., and P. Lemelin. 2002. Origins of primate locomotion: gait mechanics of the woolly opossum. American Journal of Physical Anthropology 118: 231–238.

Schupp, E.W. 1993. Quantity, quality, and the effectiveness of seed dispersal by animals. Vegetatio 107/108: 15–29.

Schweizer, M., S.T. Hertwig, and O. Seehausen. 2014. Diversity versus disparity and the role of ecological opportunity in a continental bird radiation. Journal of Biogeography 41: 1301–1312.

Seelke, A.H.M., J.C. Dooley, and L.A. Krubitzer. 2014. Photic preference of the short-tailed opossum (*Monodelphis domestica*). Neuroscience 269: 273–280.

Selig, K.R., E.J. Sargis, and M.T. Silcox. 2019. The frugivorous insectivores? Functional morphological analysis of molar topography for inferring diet in extant treeshrews (Scandentia). Journal of Mammalogy 100: 1901–1917.

Semedo, T.B.F., M.V. Brandão, A.P. Carmignotto, M.S. Nunes, I.P. Farias, M.N.F. da Silva, and R.V. Rossi. 2015. Taxonomic status and phylogenetic relationships of *Marmosa agilis peruana* Tate, 1931 (Didelphimorphia: Didelphidae), with comments on the morphological variation of *Gracilinanus* from central-western Brazil. Zoological Journal of the Linnean Society 173: 190–216.

Seton, M., R.D. Müller, S. Zahirovic, C. Gaina, T. Torsvik, G. Shephard, A. Talsma, et al. 2012. Global continental and ocean basin reconstructions since 200 Ma. Earth-Science Reviews 113: 212–270.

Shattuck, M.R., and S.A. Williams. 2010. Arboreality has allowed for the evolution of increased longevity in mammals. Proceedings of the National Academy of Sciences 107: 4635–4639.

Shibuya, P.S., G.L. Melo, and N.C. Cáceres. 2018. Determinants of home range size and spatial overlap of *Gracilinanus agilis* (Mammalia: Didelphidae) in central-western Brazil. Mammalia 82: 328–337.

Shine, R. 1989. Ecological causes for the evolution of sexual dimorphism: a review of the evidence. Quarterly Review of Biology 64: 419–461.

Shuster, S.M., W.R. Briggs, and P.A. Dennis. 2013. How multiple matings by females affects sexual selection. Philosophical Transactions of the Royal Society B 368: 20120046.

Sih, A., G. Eglund, and D. Wooster. 1998. Emergent impacts of multiple predators on prey. TREE 13: 350–355.

Sih, A., D.I. Bolnick, B. Luttberg, J.L. Orrock, S.D. Peacor, L.M. Pintor, E. Preisser, J.S. Rehage, and J.R. Vonesh. 2010. Predator-prey naïveté, antipredator behavior, and the ecology of predator invasions. Oikos 119: 610–621.

Silva, M.G.Q., and H.M.A. Costa. 1999. Helminths of the white-bellied opossum from Brazil. Journal of Wildlife Diseases 35: 371–374.

Silveira, T.B. da, F.R de Melo, and J.E.P. Lima. 2014. New field data on reproduction, diet, and activity of *Glironia venusta* Thomas, 1912 (Didelphimorphia, Didelphidae) in northern Brazil. Mammalia 78: 217–222.

Silvestro, D., M.F. Tejedor, M.L. Serrano-Serrano, O. Loiseau, V. Rossier, J. Rolland, A. Zizka, S. Höhna, A. Antonelli, and N. Salamin. 2019. Early arrival and climatically linked geographic expansion of New World monkeys from tiny African ancestors. Systematic Biology 68: 78–92.

Simmen, B., and D. Sabatier. 1996. Diets of some French Guianan primates: food composition and food choices. International Journal of Primatology 17: 661–693.

Simmons, N.B., and R.S. Voss. 1998. The mammals of Paracou, French Guiana: a Neotropical lowland rainforest fauna. Part 1. Bats. Bulletin of the American Museum of Natural History 237: 1–219.

Simonetti, J.A., A. Poiani, and K.J. Raedeke. 1984. Food habits of *Dusicyon griseus* in northern Chile. Journal of Mammalogy 65: 515–517.

Simpson, G.G. 1945. The principles of classification and a classification of mammals. Bulletin of the American Museum of Natural History 85: i–xvi, 1–350.

Simpson, G.G. 1970. The Argyrolagidae, extinct South American marsupials. Bulletin of the Museum of Comparative Zoology 139: 1–86.

Simpson, G.G. 1971. The evolution of marsupials in South America. Anais da Academia Brasileira de Ciências 43 (suppl.) 103–118.

Simpson, G.G. 1980. Splendid isolation, the curious history of South American mammals. New Haven, CT: Yale University Press.

Smith, P., and R.D. Owen. 2015. The subgenus *Micoureus* (Didelphidae: *Marmosa*) in Paraguay: morphometrics, distributions, and habitat associations. Mammalia 79: 463–471.

Smith, P., H. Pheasey, K. Atkinson, J. Ramakers, and J. Sarvary. 2012. The Didelphimorphia (Didelphidae) of Reserva Natural Laguna Blanca, Departamento de San Pedro, Paraguay. Acta Zoológica Lilloana 56: 141–153.

Smythe, N. 1978. The natural history of the Central American agouti (*Dasyprocta punctata*). Smithsonian Contributions to Zoology 257: i–iv, 1–52.

Smythe, N. 1986. Competition and resource partitioning in the guild of Neotropical terrestrial frugivorous mammals. Annual Review of Ecology and Systematics 17: 169–188.

Smythe, N. 1996. The seasonal abundance of night-flying insects in a Neotropical forest. Pages 309–318 *in* E.G. Leigh, Jr., A.S. Rand, and D.M. Windsor, eds. The ecology of a tropical forest (2nd ed.). Washington, DC: Smithsonian Institution.

Snyder, D.E., A.N. Hamir, C.A. Hanlon, and C.E. Rupprecht. 1991. Lung lesions in an opossum (*Didelphis virginiana*) associated with *Capillaria didelphis*. Journal of Wildlife Diseases 27: 175–177.

Sofroniew, N.J., and K. Svoboda. 2015. Whisking. Current Biology 25: R137–R140.

Soibelzon, L.H. 2011. First description of milk teeth of fossil South American procyonid from the lower Chapadmalalan (late Miocene–early Pliocene) of "Farola Monte Hermoso," Argentina: paleoecological considerations. Paläontologische Zeitschrift 85: 83–89.

Solari, S., V. Pacheco, L. Luna, P.M. Velazco, and B.D. Patterson. 2006. Mammals of the Manu Biosphere Reserve. Fieldiana Zoology 110: 13–22.

Solari, S., V. Pacheco, E. Vivar, and L.H. Emmons. 2012. A new species of *Monodelphis* (Mammalia: Didelphimorphia: Didelphidae) from the montane forests of central Peru. Proceedings of the Biological Society of Washington 125: 295–307.

Soong, T.W., and B. Venkatesh. 2006. Adaptive evolution of tetrodotoxin resistance in animals. Trends in Genetics 22: 621–626.

Sperr, E.B., E.A. Fronhofer, and M. Tschapka. 2008. The Mexican mouse opossum (*Marmosa mexicana*) as a flower visitor at a Neotropical palm. Mammalian Biology 74: 76–80.

Steiner, K.E. 1981. Nectarivory and potential pollination by a Neotropical marsupial. Annals of the Missouri Botanical Garden. 68: 505–513.

Steiner, K.E. 1983. Pollination of *Mabea occidentalis* (Euphorbiaceae) in Panama. Systematic Botany 8: 105–117.

Stonerook, M.J., and J.D. Harder. 1992. Sexual maturation in female gray short-tailed opossums, *Monodelphis domestica*, is dependent upon male stimuli. Biology of Reproduction 46: 290–294.

Streilein, K.E. 1982a. Ecology of small mammals in the semiarid Brazilian caatinga. I. Climate and faunal composition. Annals of Carnegie Museum 51: 79–107.

Streilein, K.E. 1982b. Ecology of small mammals in the semiarid Brazilian caatinga. III. Reproductive biology and population ecology. Annals of Carnegie Museum 51: 251–269.

Streilein, K.E. 1982c. The ecology of small mammals in the semiarid Brazilian caatinga. V. Agonistic behavior and overview. Annals of Carnegie Museum 51: 345–369.

Striedter, G.F. 2005. Principles of brain evolution. Sunderland, MA: Sinauer Associates.

Stroud, J.T., and J.B. Losos. 2016. Ecological opportunity and adaptive radiation. Annual Review of Ecology, Evolution, and Systematics 47: 507–532.

Suárez-Villota, E.Y., C.A. Quercia, J.J. Nuñez, M.H. Gallardo, C.M. Hines, and G.J. Kenagy. 2018. Monotypic status of the South American relictual marsupial *Dromiciops gliroides* (Microbiotheria). Journal of Mammalogy 99: 803–812.

Sunquist, M.E., and J.F. Eisenberg. 1993. Reproductive strategies of female *Didelphis*. Bulletin of the Florida Museum of Natural History (Biol. Sci.) 36: 109–140.

Sunquist, M.E., S.N. Austad, and F. Sunquist. 1987. Movement patterns and home range in the common opossum (*Didelphis marsupialis*). Journal of Mammalogy 68: 173–176.

Sussman, R.W., and P.H. Raven. 1978. Pollination by lemurs and marsupials: an archaic coevolutionary system. Science 200: 731–736.

Szeplaki, E., J. Ochoa, and J. Clavijo A. 1988. Stomach contents of the greater long-nosed armadillo (*Dasypus kappleri*) in Venezuela. Mammalia 52: 422–425.

Taber, A.B., A.J. Novaro, N. Neris, and F.H. Colman. 1997. The food habits of sympatric jaguar and puma in the Paraguayan Chaco. Biotropica 29: 204–213.

Tantaleán, M., M. Díaz, N. Sánchez, and H. Portocarrero. 2010. Endoparásitos de micromamíferos del noroeste de Perú. 1. Helmintos de marsupiales. Revista Peruana de Biología 17: 207–213.

Tarifa, T., and S. Anderson. 1997. Two additional records of *Glironia venusta* Thomas, 1912 (Marsupialia, Didelphidae) for Bolivia. Mammalia 61: 111–113.

Tarquini, J., N. Toledo, L.H. Soibelzon, and C.C. Morgan. 2018. Body mass estimation for †*Cyonasua* (Procyonidae, Carnivora) and related taxa based on postcranial skeleton. Historical Biology 30: 496–506.

Tarquini, J., C.C. Morgan, N. Toledo, and L.H. Soibelzon. 2019. Comparative osteology and functional morphology of the forelimb of *Cyonasua* (Mammalia, Procyonidae), the first South American carnivoran. Journal of Morphology 280: 446–470.

Tate, G.H.H. 1933. A systematic revision of the marsupial genus *Marmosa* with a discussion of the adaptive radiation of the murine opossums (*Marmosa*). Bulletin of the American Museum of Natural History 66: 1–250 + 26 pls.

Terborgh, J. 1971. Distribution on environmental gradients: theory and a preliminary interpretation of distributional patterns in the avifauna of the Cordillera Vilcabamba, Peru. Ecology 52: 23–40.

Terborgh, J. 1983. Five New World primates, a study in comparative ecology. Princeton, NJ: Princeton University Press.

Terborgh, J. 1986. Community aspects of frugivory in tropical forests. Pages 371–384 *in* A. Estrada and T.H. Fleming, eds. Frugivores and seed dispersal. Dordrecht: W. Junk.

Terborgh, J., S.K. Robinson, T.A. Parker, III, C.A. Munn, and N. Pierpont. 1990. Structure and function of an Amazonian forest bird community. Ecological Monographs 60: 213–238.

Teta, P., and U.F.J. Pardiñas. 2007. Mammalia, Didelphimorphia, Didelphidae, *Chacodelphys formosa* (Shamel, 1930): range extension. Check List 3: 333–335.

Teta, P., U.F.J. Pardiñas, and G. D'Elía. 2006. Rediscovery of *Chacodelphys formosa*: a South American marsupial genus previously known from a single specimen. Mammalian Biology 71: 309–314.

Teta, P., E. Muschetto, S. Maidana, C. Bellomo, and P. Padula. 2007. *Gracilinanus microtarsus* (Didelphimorphia, Didelphidae) en la Provincia de Misiones, Argentina. Mastozoología Neotropical 14: 113–115.

Teta, P., J.A. Pereira, E. Muschetto, and N. Fracassi. 2009. Mammalia, Didelphimorphia, Chiroptera, and Rodentia, Parque Nacional Chaco and Capitán Solari, Chaco Province, Argentina. Check List 5: 144–150.

Thatcher, V.E. 2006. Os endoparasitos de marsupiais brasileiros. Pages 53–68 *in* N.C. Cáceres and E.L.A. Monteiro- Filho, eds. Os marsupiais do Brasil. Campo Grande: Editora UFMS.

Thielen, D.R., A. Arends, S. Segnini, and M. Fariñas. 1997a. Population ecology of *Marmosa xerophila* Handley and Gordon 1979 (Marsupialia: Didelphidae) in a semi-arid ecosystem from northern Venezuela. Zoocriaderos 2(1): 1–19.

Thielen, D.R., A. Arends, S. Segnini, and M. Fariñas. 1997b. Food availability and population dynamics of *Marmosa xerophila* Handley and Gordon 1979 (Marsupialia: Didelphidae). Zoocriaderos 2(2): 1–15.

Thielen, D.R., D.R. Cabello, G. Bianchi-Pérez, and P. Ramoni-Perazzi. 2009. Rearing cycle and other reproductive parameters of the xerophytic mouse opossum *Marmosa xerophila* (Didelphimorphia: Didelphidae) in the Peninsula of Paraguaná, Venezuela. Interciencia 34: 195–198.

Thompson, S.D. 1988. Thermoregulation in the water opossum (*Chironectes minimus*): an exception that "proves" a rule. Physiological Zoology 61: 450–460.

Thorén, S., P. Lindenfors, and P.M. Kappeler. 2006. Phylogenetic analyses of dimorphism in primates: evidence for stronger selection on canine size than on body size. American Journal of Physical Anthropology 130: 50–59.

Timm, R.M., and R.K. LaVal. 2000. Observations on Monteverde's mammals. Pages 235–236 *in* N.M. Nadkarni and N.T. Wheelwright, eds. Monteverde, ecology and conservation of a tropical cloud forest. New York: Oxford University Press.

Timm, R.M., and R.D. Price. 1985. A review of *Cummingsia* Ferris (Mallophaga: Trimenoponidae), with a description of two new species. Proceedings of the Biological Society of Washington 98: 391–402.

Timm, R.M., D.E. Wilson, B.L. Clauson, R.K. LaVal, and C.S. Vaughan. 1989. Mammals of the La Selva-Braulio Carrillo complex, Costa Rica. North American Fauna 75: 1–162.

Tipton, V.J., and C.E. Machado-Allison. 1972. Fleas of Venezuela. Brigham Young University Science Bulletin (Biol. Ser.) 17(6): 1–115.

Tomassini, R.L., C.I. Montalvo, C.M. Deschamps, and T. Manera. 2013. Biostratigraphy and biochronology of the Monte Hermoso Formation (early Pliocene) at its type locality, Buenos Aires Province, Argentina. Journal of South American Earth Sciences 48: 31–42.

Torres, E.J.L., A. Maldonado, Jr., and R.M. Lanfredi. 2009. Spirurids from *Gracilinanus agilis* (Marsupialia: Didelphidae) in Brazilian Pantanal wetlands, with a new species of *Physaloptera* (Nematoda: Spiruridae). Veterinary Parasitology 163: 87–92.

Tortato, M.A. 2009. Predação de cuíca-d'água (*Chironectes minimus*) por gavião-carijó (*Rupornis magnirostris*: Aves, Accipitridae). Mastozoología Neotropical 16: 491–493.

Trajano, E., and M. de Vivo. 1991. *Desmodus draculae* Morgan, Linares, and Ray, 1988, reported for southeastern Brazil, with paleoecological comments (Phyllostomidae, Desmodontinae). Mammalia 55: 456–459, pl. III.

Traub, R. 1985. Coevolution of fleas and mammals. Pages 295–437 *in* K.C. Kim, ed. Coevolution of parasitic arthropods and mammals. New York: John Wiley & Sons.

Travaini, A., M. Delibes, and O. Ceballos. 1998. Summer foods of the Andean hog-nosed skunk (*Conepatus chinga*) in Patagonia. Journal of Zoology London 246: 457–460.

Troyer, E.M., S.E.C. Devitt, M.E. Sunquist, V.R. Goswami, and M.K. Oli. 2014. Density dependence or climatic variation? Factors influencing survival, recruitment, and population growth rate of Virginia opossums. Journal of Mammalogy 95: 421–430.

Trupin, G.L., and B.H. Fadem. 1982. Sexual behavior of the gray short-tailed opossum (*Monodelphis domestica*). Journal of Mammalogy 63: 409–414.

Tschapka, M., and O. von Helversen. 1999. Pollinators of syntopic *Marcgravia* species in Costa Rican lowland rain forest: bats and opossums. Plant Biology 1: 382–388.

Tuttle, M.D., L.K. Taft, and M.J. Ryan. 1981. Acoustical location of calling frogs by *Philander* opossums. Biotropica 13: 233–234.

Tyndale-Biscoe, C.H., and R.B. Mackenzie. 1976. Reproduction in *Didelphis marsupialis* and *D. albiventris* in Colombia. Journal of Mammalogy 57: 249–265.

Tyndale-Biscoe, H., and M. Renfree. 1987. Reproductive physiology of marsupials. Cambridge, UK: Cambridge University Press.

Tyson, E. 1698. Carigueya, seu marsupiale americanum. Or, the anatomy of an opossum, dissected at Gresham-College by Edw. Tyson, M.D.[,] fellow of the College of Physicians and of the Royal Society, and reader of anatomy at the Chyrurgeons-Hall, in London. Philosophical Transactions 20: 105–164.

Udrizar Sauthier, D.E., M. Carrera, and U.F.J. Pardiñas. 2007. Mammalia, Marsupialia, Didelphidae, *Lestodelphys halli*: new records, distribution extension and filling gaps. Check List 3: 137–140.

Ujvari, B., N.R. Casewell, K. Sunagar, K. Arbuckle, W. Wüster, N. Lo, D. O'Meally, C. Beckmann, G.F. King, E. Deplazes, and T. Madsen. 2015. Widespread convergence in toxin resistance by predictable molecular evolution. Proceedings of the National Academy of Sciences 112: 11911–11916.

Ungar, P.S. 2010. Mammal teeth: origin, evolution, and diversity. Baltimore, MD: Johns Hopkins University Press.

Valdujo, P.H., C.C. Nogueira, L. Baumgarten, F.H.G. Rodrigues, R.A. Brandão, A. Eterovic, M.B. Ramos-Neto, and O.A.V. Marques. 2009. Squamate reptiles from Parque Nacional das Emas and surroundings, Cerrado of central Brazil. Check List 5: 405–417.

Valiente-Moro C., C. Chauve, and L. Zenner. 2005. Vectorial role of some dermanyssoid mites (Acari, Mesostigmata, Dermanyssoidea). Parasite 12: 99–109.

Valtierra-Azotla, M., and A. García. 1998. Mating behavior of the Mexican mouse opossum (*Marmosa canescens*) in Cuixmala, Jalisco, Mexico. Revista Mexicana de Mastozoología 3: 146–147.

VanBuren, C.S., and D.C. Evans. 2017. Evolution and function of anterior cervical vertebral fusion in tetrapods. Biological Reviews 92: 608–626.

VandeBerg, J.L. 1989. The gray short-tailed opossum (*Monodelphis domestica*) as a model didelphid species for genetic research. Australian Journal of Zoology 37: 235–247.

VandeBerg, J.L., and E.S. Robinson. 1997. The laboratory opossum (*Monodelphis domestica*) in laboratory research. ILAR Journal 38: 4–12.

van Schaik, C.P., J.W. Terborgh, and S.J. Wright. 1993. The phenology of tropical forests: adaptive significance and consequences for primary consumers. Annual Review of Ecology and Systematics 24: 353–377.

Varela, O., A. Cormenzana-Méndez, L. Krapovickas, and E.H. Bucher. 2008. Seasonal diet of the pampas fox (*Lycalopex gymnocercus*) in the Chaco dry woodland, northwestern Argentina. Journal of Mammalogy 89: 1012–1019.

Veevers, J.J. 2004. Gondwanaland from 650–500 Ma assembly through 320 Ma merger in Pangea to 185–100 Ma breakup: supercontinental tectonics via stratigraphy and radiometric dating. Earth-Science Review 68: 1–132.

Veilleux, C.C., and E.C. Kirk. 2014. Visual acuity in mammals: effects of eye size and ecology. Brain, Behavior, and Evolution 83: 43–53.

Vellard, J. 1945. Resistencia de los "*Didelphis*" (zarigueya) a los venenos ofídicos (nota prévia). Revista Brasileira de Biologia 5: 463–467.

Vellard, J. 1949. Investigaciones sobre inmunidad natural contra los venenos de serpientes. Publicaciones del Museo de Historia Natural "Javier Prado" Serie A (Zoología) 1(2): 1–61, figs. 1–20.

Vicente, J.J., H. de O. Rodrigues, D.C. Gomes, and R.M. Pinto. 1997. Nematóides do Brasil. Parte V: Nematóides de mamíferos. Revista Brasileira de Zoologia 14(suppl. 1): 1–452.

Vieira, E.M., and L.C. Baumgarten. 1995. Daily activity patterns of small mammals in a Cerrado area from central Brazil. Journal of Tropical Ecology 11: 255–262.

Vieira, E.M., and P. Izar. 1999. Interactions between aroids and arboreal mammals in the Brazilian Atlantic rainforest. Plant Ecology 145: 75–82.

Vieira, E.M., and E.L.A. Monteiro-Filho. 2003. Vertical stratification of small mammals in the Atlantic rain forest of south-eastern Brazil. Journal of Tropical Ecology 19: 501–507.

Vieira, E.M., and G. Paise. 2011. Temporal niche overlap among insectivorous small mammals. Integrative Zoology 6: 375–386.

Vieira, E.M., and A.R.T. Palma. 1996. Natural history of *Thylamys velutinus* (Marsupialia, Didelphidae) in central Brazil. Mammalia 60: 481–484.

Vieira, M.F., and R.M. de Carvalho-Okano. 1996. Pollination biology of *Mabea fistulifera* (Euphorbiaceae) in southeastern Brazil. Biotropica 28: 61–68.

Vignieri, S.N., J.G. Larson, and H.E. Hoekstra. 2010. The selective advantage of crypsis in mice. Evolution 64: 2153–2158.

Villarin, J.J., P.J. Schaeffer, R.A. Markle, and S.L. Lindstedt. 2003. Chronic cold exposure increases liver oxidative capacity in the marsupial *Monodelphis domestica*. Comparative Biochemistry and Physiology A 136: 621–630.

Volchan, E., C.D. Vargas, J.G. da Franca, A. Pereira, Jr., and C.E. da Rocha-Miranda. 2004. Tooled for the task: vision in the opossum. BioScience 54: 189–194.

von May, R., E. Biggi, H. Cárdenas, M.I. Díaz, C. Alarcón, V. Herrera, R. Santa-Cruz, et al. 2019. Ecological interactions between arthropods and small vertebrates in a lowland Amazon rainforest. Amphibian & Reptile Conservation 13: 65–77.

Voss, R.S. 1988. Systematics and ecology of ichthyomyine rodents (Muroidea): patterns of morphological evolution in a small adaptive radiation. Bulletin of the American Museum of Natural History 188: 259–493.

Voss, R.S. 2013. Opossums (Mammalia: Didelphidae) in the diets of Neotropical pitvipers (Serpentes: Crotalinae): evidence for alternative evolutionary outcomes? Toxicon 66: 1–6.

Voss, R.S., and L.H. Emmons. 1996. Mammalian diversity in Neotropical lowland rainforests: a preliminary assessment. Bulletin of the American Museum of Natural History 230: 1–115.

Voss, R.S., and D.W. Fleck. 2017. Mammalian diversity and Matses ethnomammalogy in Amazonian Peru. Part 2. Xenarthra, Carnivora, Perissodactyla, Artiodactyla, and Sirenia. Bulletin of the American Museum of Natural History 417: 1–118.

Voss, R.S., and S.A. Jansa. 2003. Phylogenetic studies on didelphid marsupials II. Nonmolecular data and new IRBP sequences: separate and combined analyses of didelphine relationships with denser taxon sampling. Bulletin of the American Museum of Natural History 276: 1–82.

Voss, R.S., and S.A. Jansa. 2009. Phylogenetic relationships and classification of didelphid marsupials, an extant radiation of New World metatherian mammals. Bulletin of the American Museum of Natural History 322: 1–177.

Voss, R.S., and S.A. Jansa. 2012. Snake-venom resistance as a mammalian trophic adaptation: lessons from didelphid marsupials. Biological Reviews 87: 822–837.

Voss, R.S., D.P. Lunde, and N.B. Simmons. 2001. The mammals of Paracou, French Guiana: a Neotropical rainforest fauna. Part 2. Nonvolant species. Bulletin of the American Museum of Natural History 263: 1–236.

Voss, R.S., A.L. Gardner, and S.A. Jansa. 2004a. On the relationships of *"Marmosa" formosa* Shamel, 1930 (Marsupialia: Didelphidae), a phylogenetic puzzle from the Chaco of northern Argentina. American Museum Novitates 3442: 1–18.

Voss, R.S., E. Yensen, and T. Tarifa. 2004b. An introduction to *Marmosops* (Marsupialia: Didelphidae), with the description of a new species from Bolivia and notes on the taxonomy and distribution of other Bolivian congeners. American Museum Novitates 3466: 1–40.

Voss, R.S., D.P. Lunde, and S.A. Jansa. 2005. On the contents of *Gracilinanus* Gardner and Creighton, 1989, with the description of a previously unrecognized clade of small didelphid marsupials. American Museum Novitates 3482: 1–34.

Voss, R.S., D.W. Fleck, and S.A. Jansa. 2009a. On the diagnostic characters, ecogeographic distribution, and phylogenetic relationships of *Gracilinanus emiliae* (Didelphimorphia: Didelphidae: Thylamyini). Mastozoología Neotropical 16: 433–443.

Voss, R.S., P. Myers, F. Catzeflis, A.P. Carmignotto, and J. Barreiro. 2009b. The six opossums of Félix de Azara: identification, taxonomic history, neotype designations, and nomenclatural recommendations. Bulletin of the American Museum of Natural History 331: 406–433.

Voss, R.S., E.E. Gutiérrez, S. Solari, R.V. Rossi, and S.A. Jansa. 2014. Phylogenetic relationships of mouse opossums (Didelphidae: *Marmosa*) with a revised subgeneric classification and notes on sympatric diversity. American Museum Novitates 3891: 1–70.

Voss, R.S., J.F. Díaz-Nieto, and S.A. Jansa. 2018. A revision of *Philander* (Marsupialia: Didelphidae), part 1: *P. quica, P. canus*, and a new species from Amazonia. American Museum Novitates 3891: 1–70.

Voss, R.S., D.W. Fleck, and S.A. Jansa. 2019. Mammalian diversity and Matses ethnomammalogy in Amazonian Peru. Part 3. Marsupials. Bulletin of the American Museum of Natural History 432: 1–87.

Wackermannová, M., L. Pinc, and L. Jebavý. 2016. Olfactory sensitivity in mammalian species: a review. Physiological Research 65: 369–390.

Wallis, I.R., A.W. Claridge, and J.M. Trappe. 2012. Nitrogen content, amino-acid composition, and digestibility of fungi from a nutritional perspective in animal mycophagy. Fungal Biology 116: 590–602.

Walls, G.L. 1939. Notes on the retinae of two opossum genera. Journal of Morphology 64: 67–87.

Walls, G.L. 1942. The vertebrate eye and its adaptive radiation (Cranbrook Institute of Science Bulletin 19). Bloomfield Hills, MI: Cranbrook Institute of Science.

Webb, S.D. 1976. Mammalian faunal dynamics of the Great American Interchange. Paleobiology 2: 220–234.

Webb, S.D. 1985. Late Cenozoic mammal dispersals between the Americas. Pages 357–386 *in* F.G. Stehli and S.D. Webb, eds. The Great American Biotic Interchange. New York: Plenum.

Webster, G.L. 1995. The panorama of Neotropical cloud forests. Pages 53–77 *in* S.P. Churchill, H. Balslev, E. Forero, and J.L. Luteyn, eds. Biodiversity and conservation of Neotropical montane forests. New York: New York Botanical Garden.

Weckel, M., W. Giuliano, and S. Silver. 2006. Jaguar (*Panthera onca*) feeding ecology: distribution of predator and prey through time and space. Journal of Zoology 270: 25–30.

Weisbecker, V., and A. Goswami. 2010. Brain size, life history, and metabolism at the marsupial/placental dichotomy. Proceedings of the National Academy of Sciences USA 107: 16216–16221.

Weisbecker, V., and A. Goswami. 2014. Reassessing the relationship between brain size, life history, and metabolism at the marsupial/placental dichotomy. Zoological Science 31: 608–612.

Weisbecker, V., and M. Nilsson. 2008. Integration, heterochrony, and adaptation in pedal digits of syndactylous marsupials. BMC Evolutionary Biology 8(160): 1–14.

Weisbecker, V., and D.I. Warton. 2006. Evidence at hand: diversity, functional implications, and locomotor prediction in intrinsic hand proportions of diprotodontian marsupials. Journal of Morphology 267: 1469–1485.

Weisbecker, V., S. Blomberg, A.W. Goldizen, M. Brown, and D. Fisher. 2015. The evolution of relative brain size in marsupials is energetically constrained but not driven by behavioral complexity. Brain, Behavior, and Evolution 85: 125–135.

Werner, R.M., and J.A. Vick. 1977. Resistance of the opossum (*Didelphis virginiana*) to envenomation by snakes of the family Crotalidae. Toxicon 15: 29–33.

Wesselingh, F.P., M. Räsänen, G. Irion, H. Vonhof, R. Kaandorp, W. Renema, L.R. Pittman, and M. Gingras. 2002. Lake Pebas: a palaeoecological reconstruction of a Miocene long-lived lake complex in western Amazonia. Cainozoic Research 1: 35–81.

Whitaker, J.O., Jr., and N. Wilson. 1974. Host and distribution lists of mites (Acari), parasitic and phoretic, in the hair of wild mammals of North America, north of Mexico. American Midland Naturalist 91: 1–67.

Whitaker, J.O., Jr., B.L. Walters, L.K. Castor, C.M. Ritzi, and N. Wilson. 2007. Host and distribution lists of mites (Acari), parasitic and phoretic, in the hair or on the skin of North

American wild mammals north of Mexico: records since 1974. Faculty Publications from the Harold W. Manter Laboratory of Parasitology 1: 1–173.

White, T.D. 1990. Gait selection in the brush-tailed opossum (*Trichosurus vulpecula*), the northern quoll (*Dasyurus hallucatus*), and the Virginia opossum (*Didelphis virginiana*). Journal of Mammalogy 71: 79–84.

Wilcove, D.S. 1985. Nest predation in forest tracts and the decline of migratory songbirds. Ecology 66: 1211–1214.

Wilf, P., N.R. Cúneo, I.H. Escapa, D. Pol, and M.O. Woodburne. 2013. Splendid and seldom isolated: the paleobiogeography of Patagonia. Annual Reviews of Earth and Planetary Sciences 41: 561–603.

Wilkinson, G.S., and J.W. Boughman. 1998. Social calls coordinate foraging in greater spear-nosed bats. Animal Behavior 55: 337–350.

Williams, G.C. 1957. Pleiotropy, natural selection, and the evolution of senescence. Evolution 11: 398–411.

Williamson, T.E., S.L. Brusatte, T.D. Carr, A. Weil, and B.R. Standhardt. 2012. The phylogeny and evolution of Cretaceous–Palaeogene metatherians: cladistic analysis and description of new early Palaeocene specimens from the Nacimiento Formation, New Mexico. Journal of Systematic Palaeontology 10: 625–651.

Williamson, T.E., S.L. Brusatte, and G.P. Wilson. 2014. The origin and early evolution of metatherian mammals: the Cretaceous record. ZooKeys 465: 1–76.

Wilson, D.E. 1970. Opossum predation: *Didelphis* on *Philander*. Journal of Mammalogy 51: 386–387.

Wilson, D.E., and D.A. Reeder, eds. 2005. Mammal species of the world (3rd ed.), 2 vols. Baltimore, MD: Johns Hopkins University Press.

Wilson, G.P., E.G. Ekdale, J.W. Hoganson, J.J. Calede, and A.V. Linden. 2016. A large carnivorous mammal from the Late Cretaceous and the North American origin of marsupials. Nature Communications 7: 13734.

Withers, P.C., C.E. Cooper, and A.N. Larcombe. 2006. Environmental correlates of physiological variables in marsupials. Physiological and Biochemical Zoology 79: 437–453.

Withers, P.C., C.E. Cooper, S.K. Maloney, F. Bozinovic, and A.P. Cruz-Neto. 2016. Ecological and environmental physiology of mammals. Oxford: Oxford University Press.

Wittenberger, J.F. 1979. The evolution of mating systems in birds and mammals. Pages 271–349 in P. Marler and J.G. Vandenbergh, eds. Handbook of behavioral neurobiology, vol. 3. New York: Plenum.

Woinarski, J.C.Z., A.A. Burbidge, and P.L. Harrison. 2015. Ongoing unravelling of a continental fauna: decline and extinction of Australian mammals since European settlement. Proceedings of the National Academy of Sciences 112: 4531–4540.

Woinarski, J.C.Z., M.F. Braby, A.A. Burbridge, D. Coates, S.T. Garnett, R.J. Fensham, S.M. Legge, N.L. McKenzie, J.L. Silcock, and B.P. Murphy. 2019. Reading the black book: the number, timing, distribution, and causes of listed extinctions in Australia. Biological Conservation 239: 108261.

Wolff, J.O. 1993. Why are female small mammals territorial? Oikos 68: 364–370.

Wolff, J.O., and D.W. Macdonald. 2004. Promiscuous females protect their offspring. Trends in Ecology and Evolution 19: 127–134.

Wolff, J.O., and J.A. Peterson. 1998. An offspring-defense hypothesis for territoriality in female mammals. Ethology, Ecology & Evolution 10: 227–239.

Woodburne, M.O. 2010. The Great American Biotic Interchange: dispersals, tectonics, climate, sea level, and holding pens. Journal of Mammalian Evolution 17: 245–264.

Woodburne, M.O., and J.A. Case. 1996. Dispersal, vicariance, and the Late Cretaceous to early Tertiary land mammal biogeography from South America to Australia. Journal of Mammalian Evolution 3: 121–161.

Woodburne, M.O., and W.J. Zinsmeister. 1984. The first land mammal from Antarctica and its biogeographic implications. Journal of Paleontology 58: 913–948.

Woods, H.A., II, and E.C. Hellgren. 2003. Seasonal changes in the physiology of male Virginia opossums (*Didelphis virginiana*): signs of the dasyurid semelparity syndrome? Physiological and Biochemical Zoology 76: 406–417.

Wright, D.D., J.T. Ryser, and R.A. Kiltie. 1995. First-cohort advantage hypothesis: a new twist on facultative sex ratio adjustment. American Naturalist 145: 133–145.

Wright, J.D., M.S. Burt, and V.L. Jackson. 2012. Influences of an urban environment on home range and body mass of Virginia opossums (*Didelphis virginiana*). Northeastern Naturalist 19: 77–86.

Wright, P.C. 1996. The Neotropical primate adaptation to nocturnality. Pages 369–382 *in* M.A. Norconk, A.L. Rosenberger, and P.A. Garber, eds. Adaptive radiations of Neotropical primates. New York: Plenum.

Wroe, S., M. Crowther, J. Dortch, and J. Chong. 2004. The size of the largest marsupial and why it matters. Proceedings of the Royal Society of London B (suppl.) 271: S34–S36.

Wroot, A.J. 1985. A quantitative method for estimating the amount of earthworm (*Lumbricus terrestris*) in animal diets. Oikos 44: 239–242.

Wunderle, J.M., Jr., M.R. Willig, and L.M.P. Henriques. 2005. Avian distribution in treefall gaps and understorey of *terra firme* forest in the lowland Amazon. Ibis 147: 109–129.

Wüster, W., M.G. Salomão, J.A. Quijada-Mascareñas, R.S. Thorpe, and BBBSP. 2002. Origins and evolution of the South American pitviper fauna: evidence from mitochondrial DNA sequence analysis. Pages 111–128 *in* G.W. Schuett, M. Höggren, M.E. Douglas, and H.W. Greene, eds. Biology of the vipers. Eagle Mountain, UT: Eagle Mountain Publishing.

Yotsu-Yamashita, M., J. Gilhen, R.W. Russell, K.L. Krysko, C. Melaun, A. Kurz, S. Kauferstein, D. Kordis, and D. Mebs. 2012. Variability of tetrodotoxin and of its analogues in the red-spotted newt, *Notophthalmus viridescens* (Amphibia: Urodela: Salamandridae). Toxicon 59: 257–264.

Youlatos, D. 2008. Hallucal grasping behavior in *Caluromys* (Didelphimorphia: Didelphidae): implications for primate pedal grasping. Journal of Human Evolution 55: 1096–1101.

Youlatos, D. 2010. Use of zygodactylous grasp by *Caluromys philander* (Didelphimorphia: Didelphidae). Mammalian Biology 75: 475–481.

Zangrandi, P.L., A.F. Mendonça, A.P. Cruz-Neto, R. Boonstra, and E.M. Vieira. 2019. The impact of botfly parasitism on the health of the gracile mouse opossum (*Gracilinanus agilis*). Parasitology 146: 1031–1021.

Zapata, S.C., D. Procopio, A. Travaini, and A. Rodríguez. 2013. Summer food habits of the Patagonian opossum, *Lestodelphys halli* (Thomas, 1921), in southern arid Patagonian shrub-steppes. Gayana 77: 64–67.

Zapata, T.R., and M.T. Arroyo. 1978. Plant reproductive ecology of a secondary deciduous tropical forest in Venezuela. Biotropica 10: 221–230.

Zarza, H., G. Ceballos, and M.A. Steele. 2003. *Marmosa canescens*. Mammalian Species 725: 1–4.

Zetek, J. 1930. The water opossum—*Chironectes panamensis* Goldman. Journal of Mammalogy 11: 470–471.

Zimicz, N. 2014. Avoiding competition: the ecological history of late Cenozoic metatherian carnivores in South America. Journal of Mammalian Evolution 21: 383–393.

Zullinger, E.M., R.E. Ricklefs, K.H. Redford, and G.M. Mace. 1984. Fitting sigmoidal equations to mammalian growth curves. Journal of Mammalogy 65: 607–636.

# Index

The letter "f" following a page number indicates a figure; the letter "t" indicates a table.

308    *Index*